AF897361

TRENDS IN VITAL FOOD AND CONTROL ENGINEERING

Edited by **Ayman Hafiz Amer Eissa**

Trends in Vital Food and Control Engineering
Edited by Ayman Hafiz Amer Eissa

Published by InTech
Janeza Trdine 9, 51000 Rijeka, Croatia

Copyright © 2012 InTech
All chapters are distributed under the Creative Commons Attribution 3.0 license, which allows users to download, copy and build upon published articles even for commercial purposes, as long as the author and publisher are properly credited, which ensures maximum dissemination and a wider impact of our publications. After this work has been published by InTech, authors have the right to republish it, in whole or part, in any publication of which they are the author, and to make other personal use of the work. Any republication, referencing or personal use of the work must explicitly identify the original source.

As for readers, this license allows users to download, copy and build upon published chapters even for commercial purposes, as long as the author and publisher are properly credited, which ensures maximum dissemination and a wider impact of our publications.

Notice
Statements and opinions expressed in the chapters are these of the individual contributors and not necessarily those of the editors or publisher. No responsibility is accepted for the accuracy of information contained in the published chapters. The publisher assumes no responsibility for any damage or injury to persons or property arising out of the use of any materials, instructions, methods or ideas contained in the book.

Publishing Process Manager Petra Nenadic

Technical Editor Teodora Smiljanic

Cover Designer InTech Design Team

Additional hard copies can be obtained from orders@intechopen.com

Trends in Vital Food and Control Engineering, Edited by Ayman Hafiz Amer Eissa
p. cm.
ISBN 978-953-51-0449-0

We are IntechOpen, the world's leading publisher of Open Access books

Built by scientists, for scientists

4,000+
Open access books available

116,000+
International authors and editors

120M+
Downloads

151
Countries delivered to

Our authors are among the
Top 1%
most cited scientists

12.2%
Contributors from top 500 universities

WEB OF SCIENCE™

Selection of our books indexed in the Book Citation Index
in Web of Science™ Core Collection (BKCI)

Interested in publishing with us?
Contact book.department@intechopen.com

Numbers displayed above are based on latest data collected.
For more information visit www.intechopen.com

For EU product safety concerns:
IN TECH d.o.o., Prolaz Marije Krucifikse Kozulić 3, 51000 Rijeka, Croatia,
info@intechopen.com or visit our website at intechopen.com.

Meet the editor

Professor Ayman Amer Eissa received his BSc in 1986, and MSc in 1992, both from Minoufiya University, Egypt. He received his PhD from Martin Luther University, Germany and Minoufiya University, in 1999, and completed fellowships in biosystems engineering and quality control in Germany, in 2007. Professor Amer Eissa previously served as the Director of the Center of Marketing Service, Minoufiya University, and is currently teaching food process engineering at the same University, as well as food process engineering at the Department of Agricultural Systems Engineering, College of Agriculture and Food Sciences, King Faisal University, Saudi Arabia. His research is directed at machine vision processing and the development of different package systems and transportation for food products. Professor Amer Eissa has authored and co-authored of over 40 Journal articles, and more than four books. He has served as a member of different honorary societies in food engineering ,and is a technical reviewer for most journals in the field. He also supervises more than 25 postgraduate, MSc and PhD students.

Contents

Preface

This book is an example of a successful addition to the literature of bioengineering and processing control within the scientific world. The book is divided into twelve chapters covering: selected topics in food engineering, advances in food process engineering, food irradiation, food safety and quality, machine vision, control systems and economics processing. All chapters have been written by renowned professionals working in food engineering and related disciplines.

The first chapter of the book deals with shelf life of food and processing, based on scientific evidence to the satisfaction of health authorities, countries that import it, and different stakeholders. A systematic approach for establishing by-product category shelf lives should be adopted for this. Research and technological centers, universities, food industries and sectorial associations should agree to the standardization of protocols in order to face this challenge. This could be applied to food production systems where a complex set of variables requires monitoring combined with complex reasoning to assure safety and quality.

The remaining chapters in Part I deal with the commercial application of the food irradiation process is consumer acceptance. From our point of view, as consumer safety questions are discussed, food preservation through radiation contribution to food safety will receive the same recognition as sterilization of medical products does, in terms of preventing the spread of infectious disease. Therefore, scientists have a responsibility to help the consumers understand the radiation process and its potential to improve our lives and protect our health.

Part II contains four comprehensive overviews. Many underutilized legumes grow well in tropical and sub-tropical countries in Africa, but the impending need is the capacity to carry out intensive co-ordinated research studies in these countries in order to improve their utilization as food for the improvement of nutritional status. The use value of sweetcorn and the process of cutting off kernel from cob cores is a major problem for the processing industry. Design steps of a simple fuzzy logic control for food processing is introduced, together with a full description of a food processing control that included most of the possible components required. Among other topics,machine vision to determine fresh fruit bunches (FFB) oil palm fruit maturity in harvestings process has been reviewed.

Part III includes a chapter that envisions the possibility of producing food for a healthier population and improving its health and wellbeing through nanoscale technology; because of this, the "nano-food" market is expected to grow very positively within the next few years. It also includes a chapter that deals with the anatomic parts, the structure of the cells and the chemical compounds of the aromatic medicinal plants, which can play a very important role in the growth and the development of the different microbial populations that are hosted in the plants. Food and pharmaceutical products can also be derived from them, as native or added microflora.

Finally, Part IV covers, among other topics, Coenzyme Q10 which is a key component in the inner mitochondrial membrane where it plays an important role in oxidative phosphorylation, designing new functional food and feed additives, Aflatoxins, health, new stainless steels for application in food production facilities and the economics of processing are also discussed.

We hope that this book, with its visionary approach, will be a valuable addition to the food engineering literature and will promote interest in food engineering research, development, and implementation. I consider that each of the authors has provided their extraordinary competence and leadership in their specific fields and that the publisher, with its enterprise and expertise, has enabled this project which includes various nations and continents. Thanks to them, I have the honor of being the editor of this book.

Prof. Dr Ayman Hafiz Amer Eissa
Professor of Food Process Engineering ,
Department of Agriculture Systems Engineering,
King Faisal University, Saudi Arabia and Minoufiya University,
Egypt

Part 1

Improving Food Safety and Quality for the Shelf Life in Foods

1

Principles and Methodologies for the Determination of Shelf–Life in Foods

Antonio Valero[1], Elena Carrasco[1,2] and Rosa Mª García-Gimeno[1]
[1]*University of Cordoba*
[2]*Centro Tecnológico del Cárnico (TEICA)*
Spain

1. Introduction

The establishment of validated methodologies for the determination of food shelf-life is currently demanded by both food industries and Health Authorities at national and international scale. It is well known that most foods are perishable, since they are subjected to modifications in their structure, composition and properties during storage before consumption. These changes are of physico-chemical origin attributed to food composition together with the action of intrinsic and extrinsic environmental factors, and also microbiological, where spoilage flora play an important role. These modifications are "translated into" sensorial deterioration at a specific time point. In this respect, food-borne bacteria, despite representing a threat for consumers´ health, do not affect sensorial changes.

Product "shelf-life" is defined, according to the American Heritage Dictionary of the English Language (Mifflin, 2006) as "the term or period during which a stored commodity remains effective, useful, or suitable for consumption". But, which is understood by "unsuitable"? Mifflin (2006) defines "unsuitability" as "the quality of having the wrong properties for a specific purpose". To know which is wrong and which is fine is not a straightforward question to answer, and often is subjected to individual perceptions. This issue is discussed later. Anyway, whichever the method to detect the unsuitability of a food product, once it is detected and established, its cause should be sought. Among the different elements which constitute and characterize a food product, generally only one is the responsible for the unsuitability of a product, namely, "the specific cause of unsuitability". With this, going back to the definition of shelf-life, we could redefine "product shelf-life" in the food field as "the term or period a product may be stored before a specific element of the product makes it unsuitable for use or consumption". This element could be of biological or physico-chemical nature.

In the last years, different procedures have been reported for the establishment of shelf-life, mainly based on the detection of microbial alteration, as well as physico-chemical and sensorial changes. The traditional approach consists of setting a cut-off point along the storage period at the time when any of the measured attributes exceeds a pre-established limit. Experimental work usually includes the storage of food product at different temperatures, performance of microbial analysis and the assesssment of spoilage by sensorial testing. In the case of foods whose shelf-lives might be conditioned by the presence and proliferation of pathogenic microorganisms, experiments also involve challenge testing with the target organism prior to storage. The cut-off point has been traditionally referred as quality limit (if deterioration of food is known to be produced by physico-chemical changes), or safety limit (if deteriorarion of food is due to the presence of nocive chemical substances and/or pathogenic microorganisms, parasites or virus at levels of concern). This method is usually labour-intensive and expensive.

Regarding microbiological proliferation of spoilage and/or pathogenic microorganisms, predictive microbiology is recognized as a reliable tool for providing an estimation of the course of the bacteria in the foods, and indirectly, provide an estimation of shelf-life of the product in the cases when the cause of food spoilage or unacceptability is known to be microbiological. Indeed, mathematical modelling is a science-based discipline which aims to explain a reality with a few variables, and whose applications have been extended beyond research as a real added-value industrial application (Brul et al., 2007; McMeekin et al., 2002; Peleg, 2006; McMeekin, 2007).

The main concept behind the application of predictive microbiology for the determination of shelf-life based on spoilage is the specific spoilage organisms (SSO), which are associated to sensorial changes and spoilage. As such, the end of shelf-life can be defined as the time needed for SSO to multiply from an initial contamination level to a spoilage level, or the time invested by SSO to produce a certain metabolite causing sensorial rejection (Koutsoumanis & Nychas, 2000). In the case of pathogens, challenge test protocols are available for the determination of kinetic parameters (named maximum growth rate [μ_{max}] and lag time [lag]) of *Listeria monocytogenes* in ready-to-eat foods (SANCO, 2008). In addition, European Regulation No. 2073/2005, recommends the performance of predictive microbiology studies in order to investigate compliance with the criteria throughout the shelf-life. In particular, this applies to ready-to-eat foods that are able to support the growth of *L. monocytogenes* and that may pose a *L. monocytogenes* risk for public health. In general, proliferation of pathogens for which absence is required although might be present in foods, growth/no growth models are generally accepted as useful tools for the determination of the probability of growth, whereas those pathogens for which hazardous levels (i.e. *Bacillus cereus*) or toxin-producing organisms (i.e. *Staphylococcus aureus*) have been set, growth kinetic models are more appropriate to estimate the time until reaching such levels.

Quality, in a very broad sense, means satisfaction of consumers´ expectations; in other words, quality experience delivered by a food should match quality expectations of a consumer (van Boekel, 2008). Quality aspects of foods, such as colour, nutrient content, chemical composition etc., are governed by biochemical reactions (oxidation, Maillard reactions, enzyme activity) together with physical changes (aggregation of proteins, coalescence, sedimentation etc.). As for microbials, there are kinetic models for quality attributes. However, these models only provide a representation of single biochemical reactions within a well-diluted ideal system; thus, they are not easily extrapolated to other more complex matrices like foods.

Sensory testing is designed to validate the length of time that a product will remain with the same "acceptable quality" level or presents "no changes in desired sensory characteristics" over the entire life of a product (IFTS, 1993; Kilcast & Subramaniam, 2000). Some product properties are difficult to measure objectively. Moreover, instrumental measurement alone cannot indicate consumer acceptability or rejection. It is very important to ensure no changes in sensory properties of foods during their shelf-lives, since consumers pay for a unconsciously established set of desired sensory characteristics.

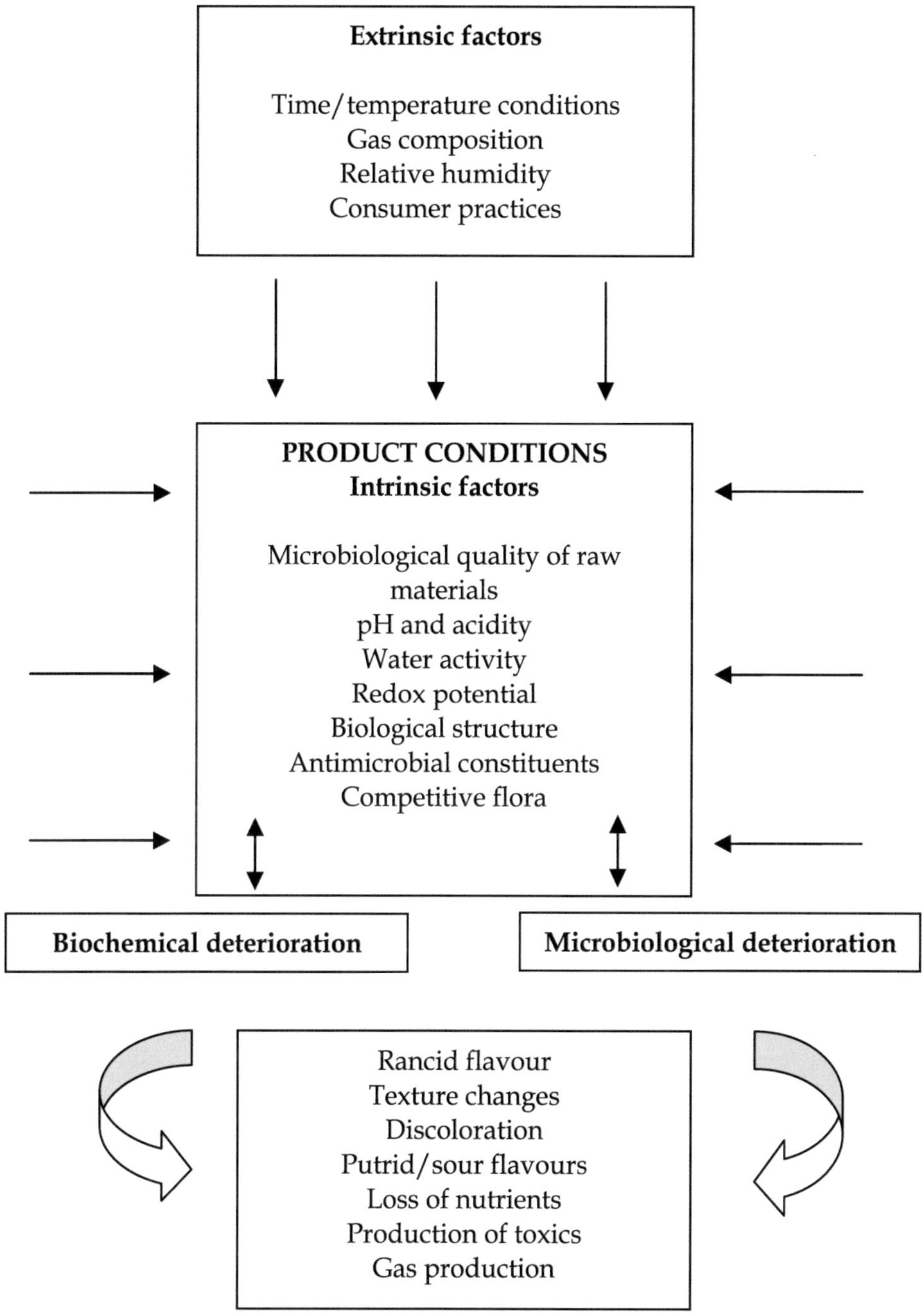

Fig. 1. Deterioration processes during food storage (adapted from Huis in`t Veld, 1996)

A unifying description of the interaction between microflora behaviour and physico-chemical changes undertaken in a food, represents a special challenge for food technologists. Such an integrated understanding of the interactions occurring during food storage should motivate the development of alternative preservation systems. A schematic representation of these phenomena is presented in Figure 1.

Throughout this chapter, the principles and methodologies for the establishment of shelf-life of foodstuffs are discussed. Also microbial, sensorial and physico-chemical parameters influencing shelf-life are analyzed. Subsequently, foods testing for data generation as well as procedures for assessment of shelf-life are reviewed. Finally, we will present some comments on future prospects.

2. Environmental factors affecting microbial shelf-life

Food spoilage is greatly influenced by environmental conditions concerning the food matrix, microbial characteristics, temperature, pH, water activity (a_w), processing time, etc. The main objective of studying the influence of environmental factors in food preservation is to inhibit spoilage due to microbial survival and growth and/or occurrence of chemical reactions. For this, a number of factors must be evaluated for different foods which could play an important role in the Hazard Analysis and of Critical Control Points (HACCP) system.

Regarding microbial growth, environmental conditions can affect largely the microbial load along the food chain. They can be classified as:

- Physical factors, such as temperature, food matrix.
- Chemical factors, such as pH, preservatives, etc.
- Biological factors, such as competitive flora, production of metabolites or inhibiting compounds etc.
- Processing conditions affecting foods (slicing, mixing, removing, washing, shredding etc.) as well as influencing transfer of microorganisms (cross-contamination events).

The application of one these factors alone can produce the desired effect on the food in terms of quality and safety, but this is not usual, especially in processed, ready-to-eat or perishable foods. Generally, the establishment of an adequate combination of more than one factor at moderate levels can offer the same result, but an improvement of the sensorial characteristics is achieved. This is the basic idea underlying the hurdle technology concept stated by Leistner (1995). Increasing the severity of one factor alone could produce a negative effect on food quality (e.g. chilling injury), while a moderate combination of several factors can lead to a shelf-life increase maintaining the original sensorial properties. For instance, in meat products, barriers such as the addition of salt, nitrite, modified atmosphere packaging, etc., can reduce the survival of pathogens (if present) and the proliferation of lactic acid bacteria during shelf-life.

2.1 Intrinsic factors

2.1.1 Microbiological quality of raw materials

Raw material entering the food industry represents a potential source of microbial contamination. The potential growth of pathogens and spoilage flora will be affected by the initial level of contamination and the efficacy of processing steps in eliminating bacteria in the food. Figure 2 represents the effect of various initial contamination levels on shelf-life of foods. At high contamination levels, less time would be needed by SSOs to reach the minimum spoilage level, thus, shelf-life would have to be reduced.

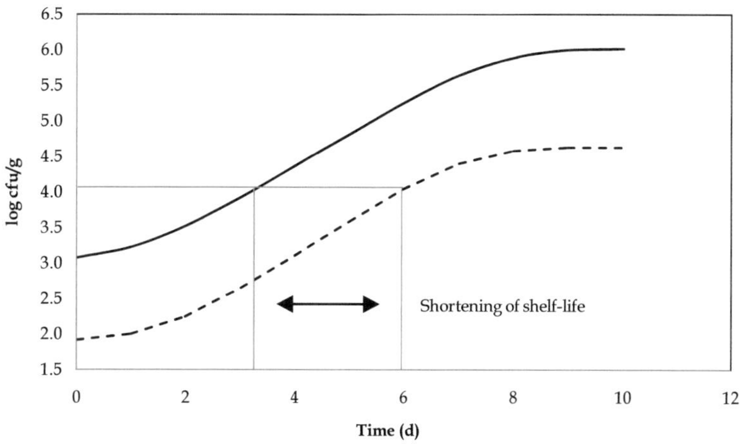

Fig. 2. Effect of the microbial initial contamination on shelf-life in a food.

A better raw material quality in terms of microbial contamination can be achieved by setting out a more strict suppliers control at primary production level, and optimizing sampling schemes. With an initial good quality of raw materials, processing operations, formulation of foods and storage conditions, shelf-life of foods could be extended.

2.1.2 pH and acidity

A widely used preservation method consists of increasing the acidity of foods either through fermentation processes or the addition of weak acids. The pH is a measure of the product acidity and is a function of the hydrogen ion concentration in the food product. It is well known that groups of microorganisms have pH optimum, minimum and maximum for growth in foods. Bacteria normally grow faster between pH ranges of 6.0 - 8.0, yeasts between 4.5 - 6.0 and moulds between 3.5 - 4.0.

An important characteristic of a food is its buffering capacity, i.e. its ability to resist changes in pH. Foods with a low buffering capacity will change pH quickly in response to acidic or alkaline compounds produced by microorganisms, whereas foods with high buffering capacity are more resistant to such changes. In any case, if low pH is a factor included in the preservation system of a food, control of pH and the application of a margin of safety are required for these foods.

2.1.3 Water activity (a_w)

The requirements for moisture by microorganisms are expressed in terms of water activity (a_w). The a_w is therefore one of the most important properties of food governing microbial growth and is defined as the free or available water in a food product (Fennema, 1996). Thus, the a_w of a food describes the fraction of water "not bounded" to the components of a food, i.e. the fraction of "available" water to participate in chemical/biochemical reactions and promote microbial survival and growth.

As for pH, control of a_w and the application of a margin of safety are required for these food products whose preservation system includes low a_w. Microorganisms respond differently to a_w depending on a number of factors. These factors can modify the minimum and maximum a_w values to grow. Generally, Gram (-) bacteria are more sensitive to changes in a_w than Gram (+) bacteria. The growth of most foodborne pathogens is inhibited at a_w values below 0.86.

2.1.4 Redox potential (E_h)

The oxidation-reduction potential (E_h) of a food is the ease by which it gains or loses electrons (Jay, 1992). The E_h value at which microorganisms will grow determines whether they require oxygen (i.e. aerobic/microaerophilic environment) for growth or not (i.e. anaerobic environment). According to their E_h value, microorganisms can be classified into the three following groups:

Aerobes	+500 to +300 mV;
Anaerobes	+100 to -250 mV; and
Facultative anaerobes	+300 to -100 mV

Although E_h as inhibitory factor is especially important in meat products (Leistner, 2000), product safety control should not solely focus on this factor since its measurement is subjected to limitations. Indeed, the E_h values are highly variable depending on the pH of the food, extent of microbial growth, packaging conditions, oxygen partial pressure in the storage environment, and ingredients. Measurement requirements for E_h in foods are reported by Morris (2000).

2.1.5 Biological structure

The natural surfaces of foods usually provide high protection against the entry and subsequent damage by spoilage organisms. External layers of seeds, the outer covering of fruits, and the eggshell are examples of biological protective structures. Several factors can influence the penetration of organisms through these barriers: (i) the maturity of plant foods enhances the effectiveness of the protective barriers; (ii) physical damage due to handling during harvest, transport, or storage, as well as invasion of insects can allow the penetration of microorganisms (Mossel et al., 1995); (iii) during the preparation of foods, processes such as slicing, chopping, grinding, and shucking destroy the physical barriers, thus favouring contamination inside the food.

Eggs can be seen as a good example of an effective biological structure that, when intact, will prevent external microbial contamination of the perishable yolk. Contamination by *Salmonella*, one of the most prevalent contaminant, is possible through transovarian infection before this shell structure is established. Additional factor is the egg white and its antimicrobial components. When there are cracks through the inner membrane of the egg, microorganisms may further penetrate into the egg. Factors such as storage temperature, relative humidity, age of eggs, and level of surface contamination will influence the internalization of microorganisms.

2.1.6 Antimicrobial constituents

Food products may contain substances (i.e. antimicrobial constituents) which have antimicrobial properties against the growth of specific microorganisms. There is a wide variety of substances with recognized antimicrobial activity in a wide variety of food products. Other antimicrobial constituents in food products are added as preservatives. Some forms of food processing will also result in the formation of antimicrobial substances in food products including:

- Smoking, e.g. fish and meat products;
- Fermentation, e.g. meat and dairy products;
- Condensation reactions between sugars and amino acids (i.e. Maillard reaction) during heating of certain foods.

2.1.7 Competitive flora (biopreservation)

Biopreservation techniques are the result of the development of a microorganism which may have an antagonistic effect on the microbial activity of other undesired microorganisms present in the food product (Mossel et al., 1995). There has been an increasing interest in biopreservation treatments in minimally processed foods in order to guarantee food safety whilst maintaining organoleptical properties.

Antagonistic processes include competition for essential nutrients, changes in pH value or E_h or the formation of antimicrobial substances like bacteriocins, which may negatively affect the survival or growth of other microorganisms (Huis in't Veld et al., 1996). Bacteriocins have been widely recognized as natural food biopreservatives, and latest advances on bacteriocin research have opened new fields to explore their role (García et al., 2010). The use of competitive microorganisms such as lactic acid bacteria against *L. monocytogenes* has been proven as a useful preservation technique in several food products (Brillet et al., 2005).

2.2 Extrinsic factors

2.2.1 Time/temperature conditions

All microorganisms have a defined temperature range within which they can grow. This range is limited by a minimum temperature needed for growth and a maximum temperature which does not support the growth. While the growth speed increases with increasing temperature, it tends to decline rapidly after the optimum temperature has been reached. An understanding of the interplay between time, temperature, and other intrinsic and extrinsic factors is crucial to select appropriate storage conditions for a food product. Temperature has a dramatic impact on the growth of microorganisms. Time plays an important role as a factor facilitating (long time) or avoiding (short time) the growth of microorganisms during food storage.

At low temperatures, two phenomena avoid microbial growth:

- Intracellular reactions, which become much slower; and
- Fluidity of the cytoplasmic membrane, interfering with transport mechanisms (Mossel et al., 1995).

At temperatures above the maximum temperature, structural cell components become denatured and heat-sensitive enzymes are inactivated.

2.2.2 Gas composition

Many scientific studies have demonstrated the antimicrobial activity of gases at ambient and sub-ambient pressures in foods (Loss & Hotchkiss, 2002). The oxygen availability is related to the E_h and plays a role in product packaging, together with CO_2 concentration. A variety of common technologies are used to inhibit the growth of microorganisms, and a majority of these methods have shown high inhibitory effect when applied with low temperatures. Technologies include Modified Atmosphere Packaging (MAP), Controlled Atmosphere Packaging (CAP), Controlled Atmosphere Storage (CAS) or Direct Addition of Carbon Dioxide (DAC) (Loss & Hotchkiss, 2002).

An adequate packaging formulation in food products should extend shelf-life without affecting their sensorial properties. In general, the inhibitory effect of CO_2 increases at low temperatures due to the fact that solubility of CO_2 becomes higher (Jay, 2000). Also, a synergistic effect has been observed when CO_2 has been applied together with low pH. The major concern when extending the shelf-life of a product by MAP is that the inhibition of spoilage bacteria could allow the proliferation of foodborne pathogens as they do not compete for nutrients or physical space anymore, and consequently foods, in spite of presenting an acceptable sensorial quality, may be unsafe.

There are several additional intrinsic and extrinsic factors affecting the efficacy of antimicrobial atmospheres, such as product temperature, product-to-headspace gas volume ratio, initial microbial load, package barrier properties, biochemical composition of the food, etc. By combining antimicrobial atmospheres with other preservation techniques, further enhancement of food quality and safety can be achieved (Leistner, 2000).

2.2.3 Relative humidity (RH)

Relative humidity (RH) is the quantity of moisture in the atmosphere surrounding a food product whether packaged or not. It is calculated as a percentage of the humidity required to completely saturate the atmosphere (i.e. saturation humidity). Typically, there will be an exchange of moisture between a food product and the surrounding atmosphere which continues until the food reaches equilibrium.

The RH is closely related with a_w and may actually alter the a_w of a food. It is important to assure that the product is stored with an environment where the RH prevents a_w changes. For example, if the a_w of a food is set at 0.60 to assure its stability, it is important that this a_w value remains during storage by establishing adequate RH conditions so that the food does not pick up moisture from the ambient air. When foods with low RH values are placed in environments of high RH, they pick up moisture until equilibrium is established. Likewise, foods with high a_w lose moisture when placed in an environment of low RH.

Determining the appropriate storage/packaging conditions for a food product is therefore essential for food safety and quality assurance. It is important to note that RH is related to temperature during storage (Esse & Humidipack, 2004).

2.2.4 Consumer practices

Consumer practices during dispatch, storage and use of food products in the home are obviously outside the control of food industries. However, food industries should take account of poor consumer handling in the establishment of product shelf-life, at least, those known to be popular. For example, many domestic refrigerators do not operate at refrigeration temperatures between 0-5ºC, being above this range; this can greatly affect the safety of food products and shorten their shelf-life (Carrasco et al., 2007). Food industries should wisely establish food shelf-lives according to real temperature records of domestic refrigerators, and not to theoretical refrigeration temperatures.

3. Role of spoilage microorganisms in foods

Foods are dynamic systems which experience changes in pH, atmosphere, nutrient composition and microflora over time. Each food product has its own unique flora, determined by the raw materials used, food processing parameters and subsequent storage conditions. The growth of SSOs in perishable foods is largely affected by several environmental factors already explained in the Section 2.

The main purpose of the application of one or a combination of intrinsic and extrinsic factors on a food product is to destroy and/or inhibit SSOs and pathogenic microorganisms, thus providing a more extended shelf-life, or to improve the sensorial characteristics of the food.

The microorganisms dominating a product can be predicted by understanding how the major preservation parameters affect microbial selection. The identification of SSOs are very important for the calculation of the remaining shelf-life of a given food product (Gram et al., 2002). Some examples are *Lactobacillus sakei* in cooked meat products (Devlieghere et al., 2001), *Photobacterium phosphoreum*, *Shewanella* spp and *Pseudomonas* spp in fishery products (Dalgaard et al., 1997; Gram and Dalgaard, 2002), *Brochothrix thermosphacta* in precooked chicken (Patsias et al., 2006), or gram negative bacteria in fresh-cut (Rico et al., 2007).

It is crucial to introduce quantitative considerations (Gram, 1989) since the spoilage activity of an organism is its quantitative ability to produce spoilage metabolites (Dalgaard, 1995; Dalgaard et al., 1993). In other words, it is necessary to evaluate if the levels of a particular organism reached in naturally spoiled foods are capable of producing the amount of metabolites associated with spoilage. In general, it requires a careful combination of microbiology, sensory analyses and chemistry to determine which microorganism(s) are the SSOs of a particular food product.

It is generally known that almost all microbial groups include species that could cause spoilage in a given food under specific conditions. Initially, different populations are present in a contaminated food, and some groups become more predominant after a certain storage period. This selection might depend on several physico-chemical factors (as discussed previously) and processing and storage conditions. For instance, in meat products stored at aerobic conditions, *Pseudomonas* spp. can produce metabolites and slime formation, but in an anaerobic atmosphere, the predominant SSOs are shift to lactic-acid bacteria, producing lactic acid and decreasing pH.

Vegetables are a special case due to the nutrient composition. The high pH (usually close to neutrality) will allow a range of Gram-negative bacteria to grow, but spoilage is specifically caused by organisms capable of degrading the vegetable polymer, pectin (Liao et al., 1997). These organisms, typically *Erwinia* spp. and *Pseudomonas* spp. are the SSO of several ready-to-eat vegetable products (Lund, 1992); also pectin degrading fungi can play a role in vegetable products (Pitt & Hocking, 1997). However, when decreasing pH due to the addition of organic acids in minimally processed fruits and vegetables, growth of fungi, yeasts and lactic-acid bacteria become the SSOs (Edwards et al., 1998).

After identification of the SSOs and the range of environmental conditions under which a particular SSO is responsible for spoilage, the next step in microbial shelf-life establishment is the decision about the microbial level of SSO above which spoilage occurs and shelf-life ends (Dalgaard, 1995; Koutsoumanis & Nychas, 2000). This step often requires a good understanding of the microbial evolution as a function of time. Initially, SSO is present in low quantities and constitutes only a minor part of the natural microflora. During storage, SSO generally produce the metabolites responsible for off-odours, off-flavours slime and finally cause sensory rejection. The cell concentration of SSO at rejection may be called the Minimal Spoilage Level (MSL) and the concentration of metabolites corresponding to spoilage is named Chemical Spoilage Index (CSI) (Dalgaard, 1993).

An example of these concepts is presented in Figure 3. In this case, the evolution of SSO together with histamine content occurs during storage of a seafood product. If MSL is set at 4 log cfu/g (if this limit is proven to cause food spoilage), the established shelf-life would be 4.2 days. However, this limit is largely different if the results are based on the histamine content, because if CSI is set at 200 ppm, the shelf-life would be 8.8 days. Generally, shelf-life may be determined based on the parameter that firstly produces alteration, in this example, microbial growth.

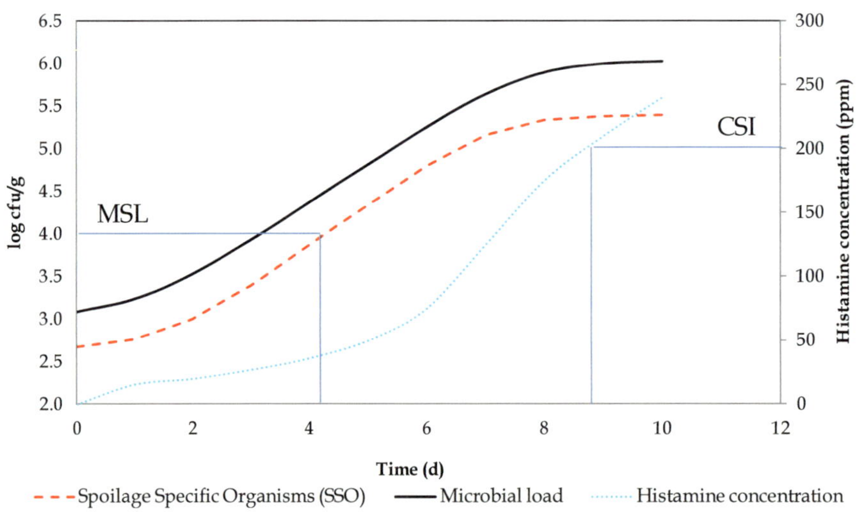

Fig. 3. Theoretical growth of SSO together with evolution of histamine content in a seafood product. MSL (Microbial Spoilage Level), CSI (Chemical Spoilage Index).

The limit of microbial growth that determines shelf-life differs according to the food type and storage conditions. SSO counts from 10^5 to 10^8 cfu/g are commonly considered as convenient quality limits. For microorganisms which produce toxins, 10^5 cfu/g has been established as a limit for risk management in processed foods (Rho & Schaffner, 2007). The time to reach hazardous levels by pathogens should determine the shelf-life limit. In these cases, shelf-life is greatly influenced by the initial contamination level. Thus, hygienic control measures must be taken in order to prevent their presence in raw material and processed food products.

Several studies relate limits for microbial groups defining the end of shelf-life with specific food products. In this sense, 10^7 cfu/g has been chosen for aerobic bacteria in minced chicken meat (Grandinson & Jennings, 1993), freshwater crayfish (Wang & Brown, 1983), or fresh-cut lettuce (Koseki & Ito, 2002). For other microbial groups such as coliforms in cottage cheese, yeasts in shredded chicory endive, or mesophilic bacteria in fresh-cut spinach, limit levels of 10^2, 10^5 or 10^8 cfu/g, respectively, have been proposed (Babic & Watada, 1996; Jacxsens et al., 2001; Mannheim & Soffer, 1996). To set these limits, a thorough knowledge about the particular microbial ecology and deterioration causes of foods during storage is required.

In relation with the microbiological limits previously mentioned, Regulation (EC) No. 2073/2005 regarding microbiological criteria in foodstuffs sets out two groups of criteria:

- Food safety criteria which define the acceptability of a product or a batch. They are applicable to foodstuffs placed on the market and throughout the shelf-life of the food.
- Process hygiene criteria which define the acceptability of the process. These apply only during the manufacturing process.

Other microbiological criteria for the end of shelf-life have been suggested by International Guidelines in Ready-To-Eat (RTE) foods. Some guidance documents are mainly focused on the study of growth of *L. monocytogenes* (SANCO, 2008) while other more general reports assess the microbiological quality of RTE foods placed in the market (Health Protection Agency, 2009; Refrigerated Food Association, 2009) or recommend specific principles to food industries (Food Safety Authority of Ireland, 2005; New Zealand Food Safety Authority, 2005). All of them establish microbiological criteria by specifying numerical limits, or admissible log increases during shelf-life.

4. Mathematical modelling of bacterial growth

It is a goal of food microbiologists to know in advance the behavior of microorganisms in foods under foreseeable conditions. If it was possible to predict the growth/survival/ inactivation of a microorganisms at given fixed conditions, it would be possible to systematically establish a shelf-life based on bacterial behaviour. The development of mathematical models describing the behavior of microorganisms under established conditions satisfactorily fulfill the challenge proposed. Already in 1921, Bigelow (1921) described the logarithm nature of thermal death, but it was not until the 1980s when mathematical models for predicting bacterial behavior experienced a great development. A new field emerged in the food microbiology area: "predictive microbiology", or as redefined recently, "modeling of microbial responses in foods". McKellar & Lu (2004a) presented a detailed review of predictive models published so far. Since primary to tertiary models, mathematical models have been extensively applied, in part due to the predictive friendly-software developed, where most models are implemented.

The advent of computer technology and associated advances in computational power have made possible to perform complex mathematical calculations that otherwise would be too time-consuming for useful applications in predictive microbiology (Tamplin et al., 2004). Computer software programs provide an interface between the underlying mathematics and the user, allowing model inputs to be entered and estimates to be observed through simplified graphical outputs. Behind predictive software programs are the raw data upon which the models are built. Some of the most popular software are: GlnaFiT (Geeraerd, et al., 2005), where different inactivation models are available (http://cit.kuleuven.be/ biotec/ downloads.php); DMFit (Baranyi & Roberts, 1994), with implementation of a dynamic growth primary model (http://www.combase.cc/index.php/en/downloads/category/11-dmfit); Pathogen Modeling Program (Buchanan, 1993), which incorporates a variety of models of different pathogens in broth culture and foods (http://pmp.arserrc.gov/ PMPOnline.aspx); Seafood Spoilage and Safety Predictor (SSSP) (Dalgaard et al., 2002), offering models for specific spoilage microorganisms and also for *L. monocytogenes* in seafood (http://sssp.dtuaqua.dk/); ComBase (Baranyi & Tamplin, 2004), a predictive tool for important foodborne pathogenic and spoilage microorganisms (http:// www.combase.cc / index.php/en/predictive-models); Sym´Previus (Leporq et al., 2005), a tool with a collection of models and data to be applied in the food industry context, e.g. strengthening HACCP plans, developing new products, quantifying microbial behavior, determining shelf-lives and improving safety (http://www.symprevius.net/); Microbial Responses Viewer (Koseki, 2009), a new database consisting of microbial growth/no growth data derived from ComBase, and also modelling of the specific growth rate of microorganisms as a function of temperature, pH and a_w (http:// mrv. nfri. affrc.go.jp/Default.aspx#/About); *Escherichia coli* fermented meat model (Ross & Shadbolt, 2004), describing the rate of inactivation of *Escherichia coli*, due to low a_w or pH or both, in fermented meats (http://www.foodsafetycentre.com.au/fermenter.php).

In a further step, a number of software called expert systems have been developed to provide more complex decision support based on a set of rules and algorithms for inference based on the relationships underlying these rules. In these systems, the core knowledge is stored as a series of IF-THEN rules that connect diverse evidence such as user input, data from databases, and the formalized opinions of experts into a web of knowledge. There are various examples of decision-support systems which encapsulate knowledge that is based in predictive microbiology, such as those applied in predicting food safety and shelf-life (Witjes et al., 1998; Zwietering et al., 1992); a step-wise system structured as a standard risk assessment process to assist in decisions regarding microbiological food safety (van Gerwen et al., 2000); or other systems for microbial processes described by Voyer & McKellar (1993) and Schellekens et al. (1994).

Whiting & Buchanan (1994) called the above integrated software models-based "tertiary models". They defined tertiary-level models as personal computer software packages that use the pertinent information from primary- and secondary-level models to generate desired graphs, predictions and comparisons. "Primary-level models" describe the change in microbial numbers over time, and "secondary-level models" indicate how the features of primary models change with respect to one or more environmental factors, such as pH, temperature and a_w. Below is a description of the most relevant primary and secondary models, as well as their uses and scope.

4.1 Primary models

Primary models aim to describe the kinetics of the process (growth or survival or inactivation) with as few parameters as possible, while still being able to accurately define the distinct stages of the process.

In the context of foods shelf-life, the process observed depends on the group of microorganisms studied (e.g. total aerobic flora, lactic acid bacteria, *Enterobacteriaceae*, etc.) and the characteristics and storage conditions of foods (e.g. presence of antimicrobials like nitrites, a_w, pH, temperature of storage, atmosphere inside the food package, etc.). Often, different groups of microorganisms behave differently in the same food product. For example, in fermented meat products, it is usual to observe growth of lactic acid bacteria against the survival or inactivation of other bacteria groups (González & Díez, 2002; Moretti et al., 2004; Rubio et al., 2007).

The application of primary models to a set of microbiological data proceeds as follows. In a first step, a mathematical model is assumed to explain the data, that is, how microbial counts change over time. In a second step, such model is fitted to microbiological data by means of regression (linear or non-linear). As a consequence of the fitting process, estimates for a number of kinetics parameters embedded in the model is provided, for example, for the rate of growth/inactivation or lag time (in growth processes) or "shoulder" (in inactivation processes). The dataset used to fit the model is assumed to be obtained under specific intrinsic and extrinsic factors. For this reason, the kinetic parameters provided after fitting solely apply for the specific intrinsic and extrinsic factors characterizing the dataset.

Several primary models have been published. Below is described the most important models used for growth and survival/inactivation.

4.1.1 Growth models

The first book devoted exclusively to the field of predictive microbiology (McMeekin et al., 1993) provide an excellent review and discussion of the classical sigmoid growth functions, especially the modified logistic and Gompertz equations. As they point out, these are empirical applications of the original logistic and Gompertz functions. Over the last years, a new generation of bacterial growth curve models have been developed that are purported to have a mechanistic basis: for example, the Baranyi model (Baranyi et al., 1993; Baranyi et al., 1994), the Hills model (Hills & Mackey, 1995; Hills & Wright, 1996), the Buchanan model (Buchanan et al., 1997), and the heterogeneous population model (McKellar, 1997).

a. Sigmoidal functions such as the modified logistic (Equation 1) and the modified Gompertz (Equation 2) introduced by Gibson et al. (1987), have been the most popular ones used to fit microbial growth data since these functions consist of four phases, similar to the microbial growth curve.

$$\mathrm{Log}x(t) = A + \left[C / \left(1 + e^{(-B(t-M))}\right)\right] \tag{1}$$

$$\text{Log}x(t) = A + C\exp\{-\exp[-B(t-M)]\} \tag{2}$$

where $x(t)$ is the number of cells at time t, A is the lower asymptotic value as t decreases to zero, C is the difference between the upper and lower asymptote, M is the time at which the absolute growth rate is maximum, and B is the relative growth rate at M.

The parameters of the modified Gompertz equation (A, C, B and M) can be used to characterize bacterial growth as follows:

$$\text{lag} = \text{M} - 1/\text{B} + [(\text{LogN}(0) - \text{A})/(\text{BC}/\text{e})] \tag{3}$$

$$\text{specific growth rate} = \text{BC}/\text{e} \text{ »BC}/2.18 \tag{4}$$

In order to simplify the fitting process, reparameterized versions of the Gompertz equation have been proposed (Zwietering et al., 1990; Willox et al., 1993):

$$\text{Log}_{10}x = A + C\exp\{-\exp[2.71(R_g/C)(\lambda - t) + 1]\} \tag{5}$$

where $A = \log_{10} x_0$ ($\log_{10}$ cfu x ml^{-1}), x_0 is the initial cell number, C the asymptotic increase in population density ($\log_{10}$ cfu x ml^{-1}), R_g the growth rate ($\log_{10}$ cfu h^{-1}), and λ is the lag-phase duration (h).

Although used extensively, some authors (Whiting & Cygnarowicz-Provost, 1992; Baranyi, 1992; Dalgaard et al., 1994; Membré et al., 1999) reported that the Gompertz equation systematically overestimated growth rate compared with the usual definition of the maximum growth rate. Note that there is also no correspondence between the lower asymptote and the inoculum size x_0. Baty et al. (2004), while comparing the ability of different primary models to estimate lag time, found that the Gompertz model is perhaps the least consistent. Nevertheless, this model is one of the most widely used to fit bacterial curves, notably by the Pathogen Modeling Program (Buchanan, 1993).

b. Baranyi et al. (1993) and Baranyi & Roberts (1994, 1995) introduced a mechanistic model for bacterial growth (Equation 6). In this model, it is assumed that during lag phase, bacteria need to synthesize an unknown substrate q critical for growth. Once cells have adjusted to the environment, they grow exponentially until limited by restrictions dictated by the growth medium.

$$\frac{dx}{dt} = \frac{q(t)}{q(t)+1} \cdot \mu_{\max} \cdot \left(1 - \left(\frac{x(t)}{x_{\max}}\right)^m\right) x(t) \tag{6}$$

where x is the number of cells at time t, x_{max} the maximum cell density, and $q(t)$ is the concentration of limiting substrate, which changes with time (Equation 7). The parameter m characterizes the curvature before the stationary phase. When $m = 1$ the function reduces to a logistic curve, a simplification of the model that is often assumed.

$$\frac{dq}{dt} = \mu_{\max} \cdot q(t) \tag{7}$$

The initial value of q (q_0) is a measure of the initial physiological state of the cells. A more stable transformation of q_0 may be defined as:

$$h_0 = \ln\left(1 + \frac{1}{q_0}\right) = \mu_{\max}\lambda \tag{8}$$

Thus, the final model has four parameters: x_0, the initial cell number; h_0; $x_{\max}$; and $\mu_{\max}$. The parameter h_0 describes an interpretation of the lag first formalized by Robinson *et al.* (1998). Using the terminology of Robinson et al (1998), h_0 may be regarded as the "work to be done" by the bacterial cells to adapt to their new environment before commencing exponential growth at the rate, μ_{max}, characteristic of the organism and the environment. The duration of the lag, however, also depends on the rate at which this work is done which is often assumed to be μ_{max}.

c. Hills & Mackey (1995) and Hills & Wright (1996) developed a theory of spatially dependent bacterial growth in heterogeneous systems, in which the transport of nutrients was described by the combination a structured-cell kinetic model with reaction-diffusion equations. The total biomass in the culture M at a time t is dependent on time and a rate constant A (Equation 9), while the total number of cells in the culture N depends on time and the rate constants A and K_n (Equation 10).

$$M(t) = M(0)\exp(At) \tag{9}$$

$$N(t) = N(0)\left[K_n \exp(At) + A\exp(-K_n t)\right] / (A + K_n) \tag{10}$$

The rate constants A and K_n depend on all the environment factors. The lag time and the doubling time have the following relationships:

$$t_{LAG} = A^{-1} - \log\left[1 + (A/K_n)\right] \tag{11}$$

$$t_{LAG} / t_D = (\ln 2)^{-1} \log\left[1 + (A/K_n)\right] \tag{12}$$

This shows that if the rate constants A and K_n have similar activation energies, the ratio of lag to doubling time (t_D) should be nearly independent of temperature. This model takes no account of possible lag behavior in the total biomass (M).

Hills model can also be generalized to spatially inhomogeneous systems such as food surfaces (Hills & Wright, 1996). If more detailed kinetic information on cell composition is available, more complex multicompartment kinetic schemes can be incorporated.

d. Buchanan et al. (1997) proposed a three-phase linear model. It can be described by three phases: lag phase, exponential growth phase and stationary phase, described as follows:

Lag phase:

$$\text{For } t \le t_{LAG}, N_t = N_0 \tag{13}$$

Exponential growth phase:

$$\text{For } t_{LAG} < t < t_{MAX}, N_t = N_0 + \mu(t - t_{LAG}) \tag{14}$$

Stationary phase:

$$\text{For } t > t_{MAX}, N_t = N_{MAX} \tag{15}$$

where N_t is the log of the population density at time t (log cfu ml^{-1}); N_0 the log of the initial population density (log cfu ml^{-1}); N_{MAX} the log of the maximum population density supported by the environment (log cfu ml^{-1}); t the elapsed time; t_{LAG} the time when the lag phase ends (h); and μ is the maximum growth rate (log cfu ml^{-1} h^{-1}).

In this model, the growth rate was always at maximum between the end of the lag phase and the start of the stationary phase. The μ was set to zero during both the lag and stationary phases. The lag was divided into two periods: a period for adaptation to the new environment (t_a) and the time for generation of energy to produce biological components needed for cell replication (t_m).

e. McKellar model assumes that bacterial population exists in two "compartments" or states: growing or nongrowing. All growth was assumed to originate from a small fraction of the total population of cells that are present in the growing compartment at $t = 0$. Subsequent growth is based on the following logistic equation:

$$\frac{dG}{dt} = G \cdot \mu \cdot \left(1 - \frac{G}{N_{MAX}}\right) \tag{16}$$

where G is the number of growing cells in the growing compartment. The majority of cells were considered not to contribute to growth, and remained in the nongrowing compartment, but were included in the total population. While this is an empirical model, it does account for the observation that growth in liquid culture is dominated by the first cells to begin growth, and that any cells that subsequently adapt to growth are of minimal importance (McKellar, 1997). The model derives from the theory that microbial populations are heterogeneous rather than homogeneous, existing two populations of cells that behave differently; the sum of the two populations effectively describes the transition from lag to exponential phase, and defines a new parameter G_0, the initial population capable of growing. Reparameterization of the model led to the finding that a relationship existed between μ_{max} and λ as described in Baranyi model. In fact, Baranyi & Pin (2001) stated that the initial physiological state of the whole population could reside in a small subpopulation. Thus, the McKellar model constitutes a simplified version of the Baranyi model, and has the same parameters.

The concept of heterogeneity in cell populations was extended further to the development of a combined discrete-continuous simulation model for microbial growth (McKellar & Knight, 2000). At the start of a growth simulation, all of the cells were assigned to the nongrowing compartment. A distribution of individual cell lag times was used to generate a series of discrete events in which each cell was transferred from nongrowing to the growing compartment at a time corresponding to the lag time for that cell. Once in the growing compartment, cells start growing immediately according to Equation 16. The combination of the discrete step with the continuous growth function accurately described the transition from lag to exponential phase.

At the present time it is not possible to select one growth model as the most appropriate representation of bacterial growth (McKellar & Lu, 2004b). If simple is better, then the three-phase model is probably sufficient to represent fundamental growth parameters accurately (Garthright, 1997). The development of more complex models (and subsequently more mechanistic models) will depend on an improved understanding of cell behavior at the physiological level.

4.1.2 Survival/inactivation models

Models to describe microbial death due to heating have been used since the 1920s, and constitute one of the earliest forms of predictive microbiology. Much of the early work centered on the need to achieve the destruction of *Clostridium botulinum* spores. This type of models has undertaken great advances from the classical linear inactivation model to others more sophisticated nonlinear models.

a. The classical linear model assumes that inactivation is explained by a simple first-order reaction kinetics under isothermal conditions (Equation 17):

$$\frac{dS_t}{dt} = -K'S_t \tag{17}$$

where S_t is the survival ratio (N_t/N_0) and k' is the rate constant. Thus the number of surviving cells decreases exponentially:

$$S_t = e^{-k't} \tag{18}$$

and when expressed as $\log_{10}$, gives:

$$\log S_t = -kt \tag{19}$$

where k = k′/ln 10. The well-known *D-value* (time required for a 1-log reduction) is thus equal to 1/k, where k is the slope. The *D-value* can also be expressed as:

$$D - value = \frac{t}{\log N_0 - \log N_t} \tag{20}$$

When log *D-values* are plotted against the corresponding temperatures, the reciprocal of the slope is equal to the *z-value*, which is the increase in temperature required for a 1-log decrease in *D-value*. The rate constant can also be related to the temperature by the Arrhenius equation:

$$k = N_0 e^{\left(-\frac{E_a}{RT}\right)} \tag{21}$$

where E_a is the activation energy, *R* the universal gas constant, and *T* is the temperature in Kelvin degrees.

Extensive work has been performed to calculate *D-values* and the *z-value* for most pathogens in broth culture and different food matrices (Mazzota, 2001; Murphy et al., 2002; van Asselt & Zwietering, 2006).

b. Although the concept of logarithmic death is still applied, nonlinear curves have been reported for many years, i.e. curves exhibiting a "shoulder" region at the beginning of the inactivation curve and/or a "tail" region at the end of the inactivation curve. Stringer et al. (2000) summarized the possible explanations for this behavior. Some of them are the variability in heating procedure, the use of mixed cultures, clumping, protective effect of dead cells, multiple hit mechanisms, natural distribution of heat sensitivity or heat adaptation.

We now know that cells do not exist simply as alive or dead, but may also experience various degrees of injury or sublethal damage, which may give rise to apparent nonlinear survival curves (Stringer et al., 2000). Survival modeling should also include a more complete understanding of the molecular events underpinning microbial resistance to the environment.

As heterogeneity is the most plausible reason for observing nonlinearity, the use of distributions to account for this nonlinearity such as gamma (Takumi et al., 2006) or Weibull (Coroller et al., 2006; van Boekel, 2002) have been suggested; from these, Weibull is the favored approach at the moment.

Also, "mirror images" of the primary growth models described above have been used to explain the inactivation process, such as the "mirror image" of the logistic function (Cole et al., 1993; Pruitt et al., 1993; Whiting, 1993), the "mirror image" of the Gompertz function (Linton et al., 1995) and the "mirror image" of the Baranyi model (Koutsoumanis et al., 1999).

4.2 Secondary models

These models predict the changes in the parameters of primary models such as bacterial growth rate and lag time as a function of the intrinsic and extrinsic factors. A description of the most popular secondary models is provided below. Two different approaches can be distinguished: i) the effects of the environmental factors are described simultaneously through a polynomial function; this type of model has probably been the most extensively used within predictive microbiology; and ii) the environmental factors are individually modeled, and a general model describes the combined effects of the factors; this approach is notably applied in the development of the increasingly popular square root and cardinal parameter-type models.

4.2.1 Polynomial models

Polynomial models describe the growth responses (μ_{max}, *lag*) or their transformations ($\sqrt{(\mu_{max})}$, $\ln(\mu_{max})$, …) through a polynomial function. An example of a polynomial model is that proposed by McClure et al. (1993) for *Brochothrix thermosphacta*:

$$\begin{aligned}\ln(\mu_{\max}) = a_0 + a_1T + a_2pH + a_3(\%NaCl) + a_4T \cdot pH + a_5T(\%NaCl) + \\ + a_6pH(\%NaCl) + a_7T^2 + a_7T^2 + a_8pH^2 + a_9(\%NaCl)^2\end{aligned} \tag{22}$$

Polynomial models allow in almost every case the development of a model describing the effect of any environmental factor including interactions between factors. Moreover, the fit of polynomial models does not require the use of advanced techniques such as the nonlinear regression. The disadvantages of polynomial models lie in the high number of parameters and their lack of biological significance.

4.2.2 Square root-type

Temperature was the first environmental factor taken into account in square root-type models. Based on the observation of a linear relationship between temperature and the square root of the bacterial growth rate in the suboptimal temperature range, Ratkowsky et al. (1982) proposed the following function:

$$\sqrt{\mu_{\max}} = b(T - T_{\min}) \tag{23}$$

where b is a constant, T is the temperature and T_{min} the theoretical minimum temperature for growth. This equation was extended to describe the effect of temperature in the entire temperature range allowing bacterial growth, the so-called 'biokinetic range' (Ratkowsky et al., 1983):

$$\sqrt{\mu_{\max}} = b(T - T_{\min})\left(1 - \exp\left(c\left(T - T_{\max}\right)\right)\right) \tag{24}$$

where T_{max} is the theoretical maximum temperature for growth and c a model parameter without biological meaning. These models were extended to take into account other environmental factors such as pH and water activity. McMeekin et al. (1987) proposed the following equation to describe the combined effects of temperature (suboptimal range) and water activity on the growth rate of *Staphylococcus xylosus*:

$$\sqrt{\mu_{\max}} = b(T - T_{\min})\sqrt{\left(a_w - a_{w\min}\right)} \tag{25}$$

where a_{wmin} is the theoretical minimum water activity for growth. More recently, square root models have also been expanded to include, for example, the effects of CO_2, (Devlieghere et al., 1998) or pH and lactic acid concentration (Presser et al., 1997; Ross et al., 2003).

4.2.3 The gamma concept and the cardinal parameter model

Zwietering et al. (1992) proposed a model called "Gamma model", describing the growth rate relative to its maximum value at optimal conditions for growth:

$$\mu_{\max} = \mu_{opt} \quad \gamma(T)\,\gamma(pH)\,\gamma(aw) \tag{26}$$

where μ_{opt} is the growth rate at optimum conditions, and $\gamma(T)$, $\gamma(pH)$, $\gamma(a_w)$ are the relative effects of temperature, pH and a_w, respectively. The concept underlying this model ("Gamma concept") is based on the following assumptions: i) the effect of any factor on the growth rate can be described, as a fraction of μ_{opt}, using a function (γ) normalized between 0 (no growth) and 1 (optimum condition for growth); and ii) the environmental factors act independently on the bacterial growth rate. Consequently, the combined effects of the environmental factors can be obtained by multiplying the separate effects of each factor (see Equation 26). Ross & Dalgaard (2004) considered that while this apparently holds true for growth rate under conditions where growth is possible, environmental factors do interact synergistically to govern the biokinetic ranges for each environmental factor.

At optimal conditions for growth, all γ terms are equal to 1 and therefore μ_{max} is equal to μ_{opt}. The γ terms proposed by Zwietering et al. (1992) for the normalized effects of temperature, pH and a_w are given in Equations 16 to 18.

$$\gamma(T) = \left(\frac{T - T_{\min}}{T_{opt} - T_{\min}} \right)^2 \tag{27}$$

$$\gamma(pH) = \frac{pH - pH_{\min}}{pH_{opt} - pH_{\min}} \tag{28}$$

$$\gamma(aw) = \frac{a_w - a_{w\min}}{1 - a_{w\min}} \tag{29}$$

γ-type terms for pH and lactic acid effects on growth rate were also included in square root-type models by Presser et al. (1997) and Ross et al. (2003).

Introduced by Rosso et al. (1993, 1995), the cardinal parameter models (CPMs) were also developed according to the Gamma concept. The relative effects of temperature, pH and a_w on the bacterial growth rate are described by a general model called CPM_n:

$$CM_n(X) = \begin{cases} 0 & , \quad X \le X_{\min} \\ \dfrac{(X - X_{\max})(X - X_{\min})}{(X_{opt} - X_{\min})^{n-1}\left[(X_{opt} - X_{\min})(X - X_{opt}) - (X_{opt} - X_{\max})((n-1)X_{opt} + X_{\min} - nX)\right]} & , \quad X_{\min} < X < X_{\max} \\ 0 & , \quad X \ge X_{\max} \end{cases} \tag{30}$$

where X is temperature, pH or a_w. X_{min} and X_{max} are respectively the values of X below and above which no growth occur. X_{opt} is the value of X at which bacterial growth is optimum. n is a shape parameter. As for the "Gamma model" of Zwietering et al. (1992), $CM_n(X_{opt})$ is equal to 1, $CM_n(X_{min})$ and $CM_n(X_{max})$ are equal to 0.

For the effects of temperature and pH, n is set to 2 and 1 respectively (Augustin et al., 2000a, 2000b; Le Marc et al., 2002; Pouillot et al., 2003 Rosso et al., 1993, 1995). For the effects of a_w, n is set to 2 (Augustin et al., 2000a, 2000b; Rosso & Robinson, 2001). The combined effects of the environmental factors are also obtained by multiplying the relative effects of each factor. Thus, the Cardinal Parameter Model for the effects of temperature, pH and a_w on μ_{max} can be written as:

$$\mu_{\max} = \mu_{opt}\, CM_2(T) CM_1(pH)\, CM_2(a_w) \tag{31}$$

or alternatively:

$$\mu_{\max} = \mu_{opt}\, CM_2(T) CM_1(pH)\, CM_1(a_w) \tag{32}$$

In many ways, CPMs resemble the square root model. Responses predicted by the two types of models can be almost identical (Oscar, 2002; Ross & Dalgaard, 2004; Rosso et al., 1993,

1995). The advantages of the CPMs lie in the lack of structural correlation between parameters and the biological significance of all parameters (Rosso et al., 1995).

Several attempts have been made to include in CPMs the effects of organic acids (Augustin et al., 2000a, 200b; Coroller et al., 2003; Le Marc et al., 2002) or other inhibitory substances (Augustin et al., 2000a, 200b).

4.2.4 Artificial neural networks

"Black boxes" such as artificial neural networks are an alternative to the models above and have been used to develop secondary growth (or lag time) models (Garcia-Gimeno et al, 2002, 2003, Geeraerd et al., 1998). For further details on the principles of this method in the context of predictive microbiology, consult Hajmeer et al. (1997), Hajmeer & Basher (2002, 2003).

4.2.5 Probabilistic models

Probabilistic growth/no growth models can be seen as a logic continuation of the kinetic growth models. If environmental conditions are getting more stressful, growth rate is decreasing until zero and lag phase is increasing until eternity or at least longer than the period of experimental data generation. At these conditions, the boundary between growth and no growth has been reached and the modeling needs to shift from a kinetic model to a probabilistic model. The output of a probabilistic model will be the chance that a certain event will take place within a certain given time under conditions specified by the user. The period of time is an important criterion, as evolving from 'no growth' to 'growth' is, in many cases, possible as long as the period of analysis is sufficiently extended.

In the case of probabilistic models, responses will be coded as either 0 (response not observed) or 1 (response observed), under a specific set of conditions. If replicate observations are performed at a specific condition, a probability between 0 and 100% will be obtained. To relate this probability to the predictor variables, logistic regression is used (McMeekin et al., 2000; Ratkowsky & Ross, 1995). The logit function is defined as:

$$\text{Logit}P = \log\left(P/\left(1 - P\right)\right) \tag{33}$$

where P is the probability of the outcome of interest.

Logit P is commonly described as some function Y of the explanatory variables, i.e.:

$$\text{Logit}P = Y \tag{34}$$

Equation 34 can be rearranged to:

$$1/\left(1 + e^{-Y}\right) = P$$

or

$$e^{Y}/\left(1 + e^{Y}\right) = P$$

where Y is the function describing the effects of the independent variables.

Several works have reported the application of probability models to different pathogens and spoilage microorganisms in foods (Lanciotti et al., 2001; Membré et al., 2001; Stewart et al., 2001).

5. Chemical spoilage

Chemical spoilage is mainly characterized by flavour and colour changes due to oxidation, irradiation, lipolysis (rancid) and heat. These changes may be induced by light, metal ions or excessive heat during processing or storage. Chemical processes may also bring physical changes such as increased viscosity, gelation, sedimentation or color change.

Biochemical reactions during food storage are relevant for the appearance of sensorial defects. Some of them are summarized as follows (Van Boekel, 2008):

- Lipid oxidation can lead to a loss of essential fatty acids and presence of rancidity, mainly due to the presence of free fatty acids. This phenomenon is associated to enzymatic oxidation, lipolysis, and discoloration.
- Non enzymatic and enzymatic browning can affect color, taste and aroma, nutritive value, and can contribute to the formation of toxicologically suspect compounds (acrylamide).
- Hydrolysis, lipolysis and proteolysis can cause changes in flavor, texture, vitamin content and formation of bitter taste.
- Separation and gelation are physical phenomena associated to sedimentation, creaming, gel formation and texture changes.

The first two processes normally characterize the majority of chemical spoilage phenomena in foods, and are explained below.

5.1 Lipid oxidation

Lipid oxidation is one of the most common causes of deterioration of food quality. Unsaturated fats are oxidized by free radical autoxidation, a chain reaction process catalyzed by the products of the reaction. The susceptibility and the rate of oxidation increase as the number of double bonds in the fatty acid increases.

Foods containing fat, e.g. milk, can undergo a number of subtle chemical and physical changes caused by lipolysis. Lipolysis can be defined as the enzymatic hydrolysis of fats by lipases. The accumulation of the reaction products, especially free fatty acids is responsible for the common off-flavour, frequently referred to as rancidity.

Fat stabilization is based on the use of physical methods (controlling temperature and light conditions during storage) and addition of antioxidants. In this sense, many foods or food ingredients contain components which with antioxidant properties (King et al., 1993). Examples of well known natural antioxidants are ascorbic acid, vitamin E (tocopherols), carotenoids and flavonoids. Antioxidants inhibit lipid oxidation by acting as hydrogen or electron donors, and interfere with the radical chain reaction by forming non radical compounds.

5.2 Non enzymatic and enzymatic browning

Non enzymatic browning resides in various chemical reactions occurring during food processing and storage. It is related to the formation of brown compounds and volatile substances that influence the sensorial quality of foods. Non enzymatic browning is in part associated to Maillard reactions. These include a number of complex reactions that are influenced by a large variety of factors (substrates nature, time/temperature combinations, pH, a_w, or presence of inhibitor compounds). Control and prevention measures of non enzymatic browning are based on the elimination of substrates, control of physico-chemical factors (shortening time and temperature, lowering pH) or addition of inhibitors such as sulfites, which delay the presence of pigments by means of fixing intermediate reaction products. Likewise, sulfites act against enzymatic browning and growth of microorganisms. Thus, they are used as preservatives in grape juices, wine, dehydrated fruits and concentrated juices.

Non enzymatic browning is a quality parameter for many dehydrated foods. For other products such as baked goods, coffee, molasses, and some breakfast foods, browning is desirable. However, in other foods browning is undesirable, decreasing the value or quality of the product. The browning include: decreased nutritional value from protein loss, off-flavor development, undesirable color, decreased solubility, textural changes, destruction of vitamins and increased acidity (Franzen et al., 1990). Other effects are destruction of essential amino acids such as lysine, formation of mutagenic compounds and modifications in the antioxidant capacity of foods.

Enzymatic browning is defined as the enzymatic transformation, in presence of oxygen at first stage, from phenolic compounds to colored polymers. Dark pigments formed as result of these reactions are known as melanines. Enzymatic browning is produced in fruits vegetables, which are rich in phenolic compounds, normally after harvesting and during processing conditions (washing, peeling, chopping, etc.). Polyphenoloxidases (PPO) and peroxidases found in fruit tissues can catalyze the oxidation of certain endogenous phenolic compounds to quinones that polymerize to form intense brown pigments (Lewis & Shibamoto, 1986).

Since enzymatic browning can cause important losses in agriculture production, control and prevention of the principal mechanisms of these reactions are crucial. These methods are based on the activity against enzymes, substrates and/or product reactions. Their use might depend on the specific food to be considered, as well as other nutritional, technical, economical, sensorial or ethical considerations.

6. Sensorial evaluation for determining shelf-life

As stated previously, sensorial evaluation is an essential task for establishing shelf-life of food products. Once a food product is designed (composition, packaging, foreseeable conditions of storage and use, target population), next step consists of estimating the period of time during which the food maintains its original sensorial properties. An immediate question arises from the last assertion: "how could I assess that sensorial properties do not change along time?" In other words, "how can these differences be detected?". There are different sensory tests to provide answer to this question; they are named sensory difference tests. However, at the same time, another question may come up to our minds: "Who should detect these difference (in case they are detectable)?". This is not a straightforward question to answer, as it greatly depends on the food quality management of the company. One could think that trained panellists are the most appropriate to detect possible differences in sensorial characteristics of foods during storage. After all, they are "sensorial instruments". However, actually, they do not represent the target population, and probably, target population in general would not detect differences in such a perfect manner as trained panellists would do. So, it is very important to design an adequate test fitted to the needs, potential and politics of the company.

A sensory difference test allows for the possible distinction between various foods. In the case of the application of these tests to shelf-life estimation, a distinction should be made between the food under storage (at the foreseeable conditions) and the food just produced. The most common sensory difference tests are the paired-comparison test, the duo-trio test, and the triangle test (O´Mahony, 1985).

The paired-comparison difference test presents two samples to a judge who has to determine which one has more of a given attribute (sweeter, spicier, et.). Should he consistently pick the right sample over replicate tests, it would indicate that he can distinguish between the two. Should he pick the right sample more often than the wrong one, a binomial test will determine the probability of picking the right sample this often by chance, so as to be able to make a decision about whether the judge really would tell the difference. The disadvantage of the paired-comparison test is that the experimenter has to specify which attribute the judge has to look for, and it is not easy to explain, for example, what is meant by "having more off-flavor". The duo-trio test does not require the judge to determine which of two samples has more of a given attribute; it is used to determine which of two samples are the same as a standard sample presented before the test. This saves the difficulty of having to describe the attribute under consideration. The test is analyzed statistically in the same way as the paired-comparison test, to see whether a judge demonstrated that he could distinguish the two samples, by consistently picking the sample that matched the standard.

An alternative to the duo-trio test, that also avoids the difficulty of having to define the attribute that is varying between the samples, is the triangle test. Rather than picking one of a pair that is the same as a standard, the judge has to pick out the odd sample from a group of three (two the same, one different). Again, a binomial test is the appropriate statistical analysis.

A summary of the sensory difference tests is presented in Table 1. By setting an adequate sensory analysis plan, it is possible to fix the day from which differences are detected at a pre-established confidence level.

Carpenter et al. (2000) provided a case-study of the application of the sensory analysis in the establishment of shelf-life of chocolate filled and covered tabs, where triangular test was applied. In order to provide with control samples (samples just produced) to be able to make comparisons at every analysis day, a number of samples just produced was freezed at the beginning of the study, and at every analysis day, the necessary control samples were left to defrost. Previously, it was demonstrated that freezing and defrosting of samples did not influence the sensory characteristics of the product.

Paired-comparison	A *vs* B	
Which is greater in a given attribute?	A or B?	One-tailed binomial test, $p = q = 1/2$ or two tails, $p = q = 1/2$.
Is there a difference between the two?	Same of different?	Two-tailed binomial test, $p = q = 1/2$.
Duo-trio	A: A *vs* B	
Which is the same as the standard?	A or B?	One-tailed binomial test, $p = q = 1/2$
Triangle	A B B	
Which is the odd sample?	A or B?	One-tailed binomial test, $p = 1/3$, $q = 2/2$

Table 1. Sensory difference tests and their analyses.

When sensory difference testing aims at detecting differences by consumers, the result of these tests will tell us the day from which differences are detected; however, these tests do not provide any information about the changes in sensorial attributes of a food. For a through knowledge of the food product, it is necessary to carry out sensory studies able to provide sensorial information about the food. These studies are called sensory descriptive tests. Murray et al. (2001) reviewed the state-of-the-art of descriptive sensory analysis. One of the most relevant descriptive tests is the Quantitative Descriptive Analysis (QDA). In the QDA, colour, odour, flavour, texture and other attributes are, in a preliminary stage, examined by an expert panel, which generates descriptive terms for the attributes. Then, the leader of the sensory analysis gathers and organized the terms, and prepares a draft list which is subsequently presented to the expert panel for discussion and fine-tuning. A definition and a scale for each attribute are established. The aim of this preliminary stage is to agree on a lot of standard attributes for a specific food, so that panellists can use it systematically. A definitive questionnaire is made up and, in a second stage, different sessions are organized according to the sensory analysis plan where every panellist fills in the questionnaire individually. QDA data obtained from every session can be statistically processed in different ways. The most common statistical tests are the *t* test (comparison between two products A and B), and the ANOVA test (comparison between more than two products), although other more sophisticated test designed to summarize a high number of sensory data can also be applied such as principal components analysis, correspondence analysis or discriminant analysis. Some guidelines like that published by Donelly & Mitchell, (2009) propose a scale from 1 to 4 to different attributes (e.g. for odour, 1 "normal odour; 2 = slight off-odour; 3 = moderate off-odour, and 4 = strong off odour), stating that a mean of 2.5 or below indicates an acceptable product, and a mean score of 2.5 marks the end of the product shelf-life.

In the case-study published by Carpenter et al. (2000), a QDA was performed in parallel to the triangle test. In this way, it was possible to identify the nature and magnitude of the changes in sensorial characteristics during storage. Apart from the statistical tests to analyze and interpret QDA data, a graphical representation of scores of the different attributes along time provides a "picture" of the evolution of attributes during storage.

Some authors have modelled the relationship between the time when defects appear or when differences between control and studied samples are detected, and temperature of storage (Labuza & Fu, 1993; Mataragas et al., 2006; Valero et al., 2006), applying the following model:

$$t_s = t_0 \exp\left(-bxT\right) \tag{35}$$

where t_s is the time (d) when defects appear at any temperature T within the examined range; t_0, the time (d) when defects appear at 0°C; and b, the slope of the regression line of plot ln t_s = f(T). Labuza & Fu (1993) stated that when the temperature range of concern is relatively narrow (0 to 12°C) then the shelf-life of a product can be determined by using the lot shelf-life model (Equation 35).

For sensory evaluation, it is very important to establish the objective of the study, and depending on the company quality management, resources and duration of the study, an analysis plan should be properly designed.

7. Food testing for determination of shelf-life

When designing a shelf-life study, several questions arise, such as proper food samples storage, testing time, quantity of samples or tests which should be applied. The key steps for designing a shelf-life study are focused on the duration time of the study, sampling frequency and controls which will be carried out until the food presents a significant spoilage. Usually, controls regarding microbiology, sensory and physico-chemical analysis are established. To design a shelf-life study, factors such as temperature, relative humidity, illumination conditions, etc., should be strictly controlled, preferably in a certified laboratory. Additionally, the design must be simple, easy to follow and interpretable by food operators.

Preliminary information is very useful for refining the experimental design. This information can be obtained from multiple sources like historical data (i.e. processing and storage conditions, formulation of the food, historical microbiological data, etc.), or published studied performed in similar products.

Current International legislation does not provide the application of standard protocols aiming at establishing the shelf-life of food products. However, various National and International Guidelines for this purpose are published (Food Safety Authority of Ireland, 2005; Health Protection Agency, 2009; New Zealand Food Safety Authority, 2005; Refrigerated Food Association, 2009). These Guidelines suggest different methods and analyses protocols for shelf-life establishment, being mainly focused on RTE foods, as mentioned in Section 3.

7.1 Steps to be followed for the design of a shelf-life study

There are common steps to follow when designing a shelf-life study. The study can differ according to the previous knowledge one has with regards to the food product.

7.1.1 Identification of the main causes of food spoilage

Each food product includes a series of factors that limit its shelf-life. Some of them were described in Sections 2 and 5 and they can affect food quality and safety during storage and distribution if they are not adequately controlled. Throughout this first step, identification of all related factors that can potentially influence food shelf-life from production to consumption is crucial. Consequently, the entire food chain must be examined.

7.1.2 Planning the shelf-life study

Once factors are identified, the shelf-life studies must be planned in order to give answer to the following questions:

a. Which type of analyses must be carried out? A number suitable tests should be performed during a predetermined storage period. These comprise sensorial, microbiological and physico-chemical analyses. Regarding sensorial aspects, different attributes have to be evaluated such as odour, appearance, flavour, texture, as well as other organoleptical characteristics by a trained panel. In section 6, several sensory tests are proposed. Microbial analyses must be performed along the storage time, aiming at monitoring the changes in microbial concentration. The most common microbial groups analyzed are aerobic mesophilic bacteria, coliforms, yeasts, lactic-acid bacteria, and some indicators such as *E. coli*, *Staphylococcus aureus* or sporulated bacteria in thermal treated foods (*Bacillus* spp., *Clostridium* spp.). Physical analyses include texture changes, package type, appearance, etc., and they are usually considered as sensorial attributes. Regarding chemical analyses, changes in concentration of metabolites or chemical parameters (pH, nitrogen content, salt, organic acids, etc.) are monitored during the storage period. These tests are recommended to be performed at three different storage conditions as minimum: one established by the food operator (commercial conditions), one intermediate condition and one abuse condition (in case of breaking of the cold chain). For temperature control during storage, dataloggers should be used in refrigeration cameras.
b. How many analyses must be carried out? This mainly depends on the pre-established shelf-life or prior knowledge, food type and storage conditions. In principle, for those foods with a long expected shelf-life (i.e. more than 3 months) analyses can be more sporadic. On the contrary, perishable and RTE foods have shorter shelf-lives (between one week and one month); in this case, microbial growth is normally the main reason of food spoilage, so it is necessary to increase the frequency of analyses in order to provide an accurate estimation of shelf-life. In general, for both long and short shelf-lives, it is desirable to set around 7-10 analytical points regularly distributed, although this number will change according to the specific case to study. Sampling points must be separated between 15-20% of the total shelf-life period. Besides, one final point has to be performed once shelf-life is expired (after 15-20% of the shelf-life time). In this way, the total analytical period will comprise 115-120% of the shelf-life period. For instance, if shelf-life is calculated in 30 days, analyses are recommended to be performed in days 0, 6, 12, 18, 24, 30 and 36. Obviously, frequency of analyses is influenced by the storage temperature.
c. How many samples need to be withdrawn per analytical point? It is recommended to analyze, at least, two samples at each analytical point and storage condition. For instance, 7 analytical points x 2 temperature conditions x 2 samples = 28 samples. Normally, the samples destined to microbial or physico-chemical analyses, if not completely destructed, can be used for sensorial analyses.

d. Which period is the most appropriate to perform the study? In any case, shelf-lives studies have to be repeated to control the variability of foods, but the most suitable period corresponds to summer months.

7.1.3 Shelf-life establishment

Shelf-life ends when the food loses its original quality and/or safety attributes. Based on the information collected in previous steps, the maximum storage time of the food at given conditions assuring the maintenance of appropriate characteristics, has to be decided. Estimations will be different if decisions are based on microbial or sensorial or physico-chemical changes. In any case, the most conservative approach should be chosen as the best one, unless there are other factors needing to be considered.

7.1.4 Shelf-life monitoring

It is advisable to validate the established shelf-life of a food as well as to repeat the analyses in several occasions to include shelf-life variability due to effects of different environmental variables.

7.2 Shelf-life studies based on *Listeria monocytogenes* growth in RTE foods

L. monocytogenes is a pathogen which may cause disease in humans and it is typically transmitted as a food-borne pathogen. *L. monocytogenes* is frequently present in the environment, soil, vegetation and faeces of animals. The organism can be found in raw foods such as fresh meat, raw milk and fish. The ubiquitous occurrence and the increased ability to grow or survive in a chilled environment compared to other microorganisms, makes *L. monocytogenes* a significant challenge in food production. This concern is of particular relevance in Ready-To-Eat (RTE) foods, as they do not received further treatment after production and their intrinsic characteristics support the growth of *L. monocytogenes*.

It is crucial that producers of RTE foods take actions to control the contamination by *L. monocytogenes* as well as its growth throughout the shelf-life. Knowledge on the growth potential of *L. monocytogenes* in RTE foods is needed, and this must be taken into account when setting their shelf-lives.

A guidance document was launched by the European Commission in 2008 regarding shelf-life studies for *L. monocytogenes* in RTE foods (SANCO, 2008). This document addressed the question stated in Annex II of the Regulation (EC) No 2073/2005, where food operators must demonstrate that *L. monocytogenes* cannot exceed 100 cfu/g throughout the shelf-life of RTE foods able to support the growth of *L. monocytogenes* and that may pose a risk for public health. In this document, microbiological procedures for determining the growth of *L. monocytogenes* through challenge tests were described, as well as durability studies in the frame of the application of the Regulation (EC) No. 2073/2005. The microbiological procedures include:

a. Challenge tests
- assessing a growth potential (δ)
- assessing the maximum growth rate (μ_{max})

b. Durability studies

8. Assessment of shelf-life scientific-based

Food testing renders a data set of different nature (microbiological, physico-chemical, sensorial) which should be wisely managed to produce an accurate estimation of shelf-life of foods. It is a decision of the quality department of food companies to define the criteria upon which product shelf-life should be established.

At the beginning of the chapter, "product shelf-life" was defined as "the term or period a product may be stored before a specific element of the product makes it unsuitable for use or consumption". Unsuitability can be evaluated by sensory testing, while the cause for unsuitability could be of biological or physico-chemical nature. Whichever characteristic is analyzed (microbiological, physico-chemical or sensorial) when a batch of product is analyzed at different time intervals during the period of storage, a "survival curve" can be constructed (Figure 4). A "survival curve" is a representation of the percentage of suitable food along time. In other words, a batch of product, even though belonging to the same lot, fails as storage time increases.

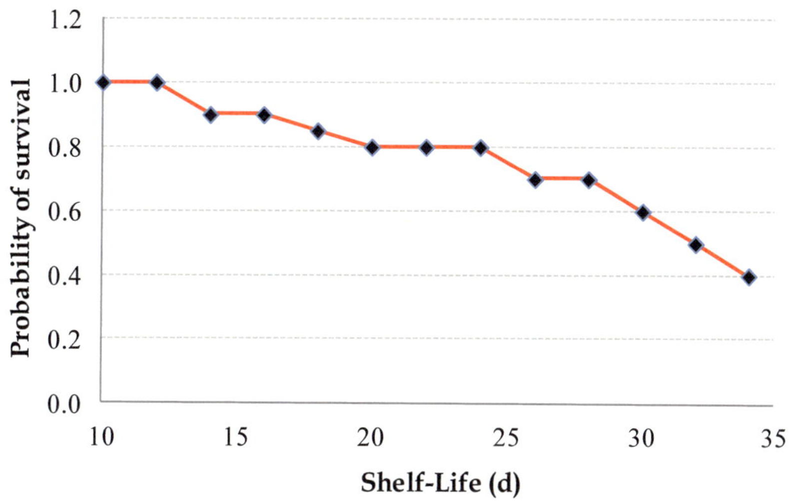

Fig. 4. Survival curve for a Food product.

A number of probability distributions have been proposed to model failure time of a product (Hough et al., 2003). Typically, such a distribution is defined by a small number of parameters and a mathematical equation. Some commonly used statistical distributions are the exponential, Weibull or log-normal distribution. The goodness of fit of these distributions to data can be assessed through the Anderson-Darling statistics; the smaller the value of this test statistics, the better the fit. In the shelf-life context, probability distributions represent storage time *vs* probability. Figure 5 represents the Weibull distribution fitted to a data set, where probability of rejection (1 - probability of survival) is plotted against shelf-life. To use these curves to predict shelf-life, the experimenter needs to define the largest acceptable proportion of defects that can be tolerated. A horizontal line can then be drawn on the plot to determine the corresponding shelf-life. Mataragas et al. (2007) used this approach to estimate the shelf-life of a sliced, cooked, cured meat product.

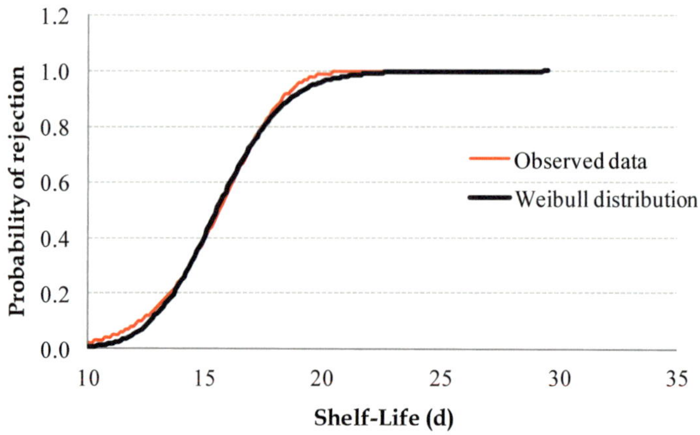

Fig. 5. Weibull distribution to predict shelf-life of a food product.

Statistical tools are of great value for the food industry and to the satisfaction of the different stakeholders, to demonstrate a science-based shelf-life, assuring the safety while maintaining the original sensorial characteristics of the food.

9. Prospects

Shelf-life of foods are being reformulating based on scientific evidence to the satisfaction of Health Authorities, import countries and different stakeholders. For this, a systematic approach for establishing by-product category shelf-lives should be adopted. Research and technological centres, Universities, food industries and sectorial associations should agree on standardized protocols to face this challenge.

Food industries could incorporate informatic tools to systematically produce foods meeting the quality level set up for them. The more homogeneous the food lots, the less deviations from established shelf-lives. A number of possibilities exist for the development of decision-support systems based on rule-based expert systems. Examples (Paoli, 2001, as cited in Tamplin et al., 2004) include:

- Virtual inspection: Simulating interaction with an inspector to allow establishments to self-assess their facility or a food processing operation. This may be of interest to large companies and regulators who find their quality control or inspection resources inadequate for the number of establishments and that they are required to assess.
- Process deviation assessment: Tools which incorporate expert knowledge regarding the best actions to take in case of process deviations, in terms of assessing the seriousness of the deviation and in recommending corrective actions for the implicated product and/or process.
- In-line real-time expert systems: Examples exist (though not in predictive microbiology) of expert systems that received real-time data and provide continuous assessments of the status of systems, based on the combination of these data and embedded knowledge-based rules that interpret the data for display to operators. This could be applied to food production systems where a complex set of variables requires monitoring combined with complex reasoning to assure safety and quality.

10. Acknowledgments

CTS-3620 Project of Excellence from the Andalusia Government, AGL 2008-03298/ALI project from the Spanish government, FP7-KBBE-2007-2A n° 222738 project from the VII Framework Programme and European ERDF funding are greatly acknowledged for providing material and specially human resources, making possible the continuation of Risk Assessment and Management activities at national and European level by our research group AGR170 "HIBRO".

11. References

Augustin, J.C. & Carlier, V. (2000a). Modelling the growth rate of *Listeria monocytogenes* with a multiplicative type model including interactions between environmental factors. *International Journal of Food Microbiology*, Vol.56, pp. 53-70, ISSN 0168-1605

Augustin, J.C. & Carlier, V. (2000b). Mathematical modelling of the growth rate and lag time for *Listeria monocytogenes*. *International Journal of Food Microbiology*, Vol.56, pp. 29-51, ISSN 0168-1605

Babic, I. & Watada, A.E. (1996). Microbial populations of fresh-cut spinach leaves affected by controlled atmospheres. *Postharvest Biology and Technology*, Vol. 9, pp, 187-193. ISSN 0925-5214

Baranyi, J. & Roberts, T.A. (1994). A dynamic approach to predicting bacterial growth in food. *International Journal of Food Microbiology*, Vol.23, pp. 277-294, ISSN 0168-1605

Baranyi, J. & Tamplin, M.L. (2004). ComBase: A Common Database on Microbial Responses to Food Environments. *Journal of Food Protection*, Vol.67, pp. 1967-1971, ISSN 0362-028X

Baranyi, J. (1992). Letters to the Editor: a note on reparameterization of bacterial growth curves. *Food Microbiology*, Vol.9, pp. 169-171, ISSN 0740-0020

Baranyi, J. Roberts, T.A. & McClure, P. (1993). A non-autonomous differential equation to model bacterial growth. *Food Microbiology*, Vol.10, pp. 43-59, ISSN 0740-0020

Baranyi, J. Robinson, T.P. & Mackey, B.M. (1995). Predicting growth of *Brochothrix thermosphacta* at changing temperature. *International Journal of Food Microbiology*, Vol.27, pp. 61-75, ISSN 0168-1605

Baty, F. & Delignette-Müller, M.L. (2004). Estimating the bacterial lag time: which model, which precision? *International Journal of Food Microbiology*, Vol.91, pp. 261-277, ISSN 0168-1605

Bigelow, W.D. (1921). The logarithmic nature of thermal death time curves. *The Journal of Infectious Diseases*, Vol.29, pp. 528-536, ISSN 0022-1899

Brillet, A. Pilet, M.F. Prévost, H. Cardinal, M. & Leroi, F. (2005) Effect of inoculation of *Carnobacterium divergens* V41, a biopreservative strain against *Listeria monocytogenes* risk, on the microbiological, and sensory quality of cold-smoked salmon. *International Journal of Food Microbiology*, Vol. 104, pp, 309-324. ISSN 0168-1605

Brul, S. van Gerwen, S. & Zwietering, M. (2007). Modelling microorganisms in food. Cambridge, Woodhead Publishing Limited.

Buchanan, R.L. (1993). Developing and Distributing User-Friendly Application Software. Journal of Industrial Microbiology, Vol.12, pp. 251-255, ISSN 0169-4146

Carpenter, R.P. Lyon, D.H. & Hasdell, T.A. (2000). Guidelines for Sensory Analysis in Food Product Development and Quality Control (2nd Edition), ISBN 978-0-8342-1642-6, Kluwer Academic/Plenum Publishers, New York, USA

Carrasco, E. Pérez-Rodríguez, F. Valero, A. García-Gimeno, R.M. & Zurera, G. (2007). Survey of temperature and consumption patterns of fresh-cut leafy green salads: risk factors for listeriosis. *Journal of Food Protection*, Vol. 70, pp, 2407-2412. ISSN 0362-028X

Cole, M.B. Davies, K.W. Munro, G. Holyoak, C.D. & Kilsby, D.C. (1993). A vitalistic model to describe the thermal inactivation of Listeria monocytogenes. *Journal of Industrial Microbiology and Biotechnology*, Vol.12, 232-239, ISSN 1367-5435

Commission of the European Communities (2008). Technical Guidance Document on *Listeria monocytogenes* shelf-life studies for ready-to-eat foods, under Regulation (EC) No 2073/2005 of 15 November 2005 on microbiological criteria for foodstuffs. EU Community Reference Laboratory for *Listeria monocytogenes*. SANCO/1628/2008 ver. 9.3 (26112008)

Commission Regulation (EC) No 2073/2005 of 15 November 2005 on microbiological criteria for foodstuffs. Official Journal of the European Union, L338/1-L338/26. Available from:
http://eur-lex.europa.eu /LexUriServ /LexUriServ.do?uri= OJ:L:2005: 338:0001: 0026:EN:PDF

Commission Regulation (EC) No 852/2004 of the European Parliament and of the Council of 29 April 2004 on the Hygiene of Foodstuffs. Official Journal of the European Union, L339/1. Available from:
http://eur-lex.europa.eu /LexUriServ /LexUriServ.do? uri=OJ:L: 2004: 139: 0001: 0054:en:PDF

Coroller, L. Guerrot, V. Huchet, V. Le Marc, Y. Sohier, D. & Thuault, D. (2003). Growth kinetics modelling of pathogen bacteria as function of different acids. In: Predictive Modelling in foods-Conference proceedings, J.F.M. Van Impe, A.H. Geerarerd, I. Leguerinel & P. Mafart (Eds.), pp 120-122, Kathoelieke Universiteit Leuven/BioTec, Belgium

Coroller, L. Leguerinel, I. Mettler, E. Savy, N. & Mafart, P. (2006). General model, based on two mixed Weibull distributions of bacterial resistance, for describing various shapes of inactivation curves. *Applied and Environmental Microbiology*, Vol.72, pp. 6494-6502, ISSN 0099-2240

Dalgaard, P. (1995). Qualitative and quantitative characterization of spoilage bacteria from packed fish. *International Journal of Food Microbiology*, Vol. 26, pp, 319–333. ISSN 0168-1605

Dalgaard, P. Mejlholm, O. & Huss, H.H. (1997). Application of an iterative approach for development of a microbial model predicting the shelf-life of packed fish. *International Journal of Food Microbiology*, Vol. 38, pp, 169-179. ISSN 0168-1605

Dalgaard, P.; Buch, P. & Silberg, S. (2002). Seafood Spoilage Predictor-development and distribution of a product specific application software. *International Journal of Food Microbiology*, Vol.73, pp. 343-349, ISSN 0168-1605

Dalgaard, P. Ross, T. Kamperman, L. Neumeyer, K. & McMeekin, T.A. (1994). Estimation of bacterial growth rates from turbidimetric and viable count data. *International Journal of Food Microbiology*, Vol.23, pp. 391-404, ISSN 0168-1605

Dalgaard. P. (1993) Evaluation and prediction of microbial fish spoilage. PhD thesis, Royal Veterinary and Agricultural University. Copenhagen.

Devlieghere, F. Van Belle, B. & Debevere, J. (2001). Shelf life of modified atmosphere packed cooked meat products: a predictive model. *International Journal of Food Microbiology*, Vol. 46, pp, 57–70. ISSN 0168-1605

Devlieghere, F. Debevere, J. & Van Impe, J.F.M. (1998). Concentration of carbon dioxide in the water-phase as a parameter to model the effect of a modified atmosphere on microorganisms. *International Journal of Food Microbiology*, Vol.43, pp. 105-113, ISSN 0168-1605

Donelly, C.W. & Mitchell, M.M. (2009). Refrigerated Foods Association (RFA) Standardized Protocol for Determining the Shelf-Life of Refrigerated Ready-To-Eat (RTE) Foods, In: *Refrigerated Foods Association*, 10.10.2011, Available from: http://www.refrigeratedfoods.org/food-safety-resources

Edwards, C.G. Haag, K.M. Collins, M.D. Hutson, R.A. & Huang, Y.C. (1998). *Lactobacillus kunkeei* sp. nov.: a spoilage organism associated with grape juice fermentations. *Journal of Applied Microbiology*, Vol. 84, pp, 698–702. ISSN 1364-5072

Esse, R. & Humidipak, A. (2004). Shelf-life and moisture management. In: Understanding and Measuring the Shelf-Life of Food, (ed. Steele, R.). Published by Woodhead Publishing Ltd., Cambridge, England, pp, 24-41.

Fennema, O.R. (1996). Water and ice. In: Fennema, O.R. (ed.) Food Chemistry (3rd Edition), Marcel Dekker Inc., New York. pp. 17-94.

Food Safety Authority of Ireland (2005). Guidance Document on the determination of product shelf-life. Available from: http://www.fsai.ie/faq/shelf_life.html.

Franzen, K. Singh, R.K. & Okos, M.R. (1990). Kinetics of nonenzymatic browning in dried skim milk. *Journal of Food Engineering*, Vol. 11, pp, 225-239. ISSN 0260-8774

García, P. Rodríguez, L. Rodríguez, A. and Martínez, B. (2010). Food biopreservation: promising strategies using bacteriocins, bacteriophages and endolysins. *Trends in Food Science and Technology*, Vol. 21, pp, 373-382. ISSN 0924-2244

Garcia-Gimeno, R.M. Hervas-Martinez, C. & de Simoniz, M.I. (2002). Improving artificial neural networks with a pruning methodology and genetic algorithms for their application in microbial growth predictions in food. *International Journal of Food Microbiology*, Vol.72, pp. 19-30, ISSN 0168-1605

Garcia-Gimeno, R.M. Hervas-Martinez, C. Barco-Alcaia, E. Zurena-Cosano, G. & Sanz-Tapia, E. (2003). An artificial neural network approach to *Escherichia coli* O157:H7 growth estimation. *Journal of Food Science*, Vol.68, pp. 639-645, ISSN 1750-3841

Garthright, W.E. (1997). The three-phase linear model of bacterial growth: a response. *Food Microbiology*, Vol.14, 395-397, ISSN 0740-0020

Geeraerd, A.H. Herremans, C.H. Cenens, C. & Van Impe, J.F.M. (1998). Application of artificial neural networks as a non linear modular modeling technique to describe bacterial growth in chilled food products. *International Journal of Food Microbiology*, Vol.44, pp. 49-68, ISSN 0168-1605

Geeraerd, A.H. Valdramidis, V.P. & Van Impe, J.F. (2005.) GInaFiT, a freeware tool to assess non-log-linear microbial survivor curves. *International Journal of Food Microbiology*, Vol.102, pp. 95-105, ISSN 0168-1605

Gibson, A.M. Bratchell, N. & Roberts, T.A. (1987). The effect of sodium chloride and temperature on the rate and extent of growth of *Clostridium botulinum* tupe A in pasterurized pork slurry. *Journal of Applied Bacteriology*, Vol.62, pp. 479-490, ISSN 1364-5072

González, B. & Díez, V. (2002). The effect of nitrite and starter culture on microbiological quality of "chorizo" - a spanish dry cured sasusage. *Meat Science*, Vol.60, pp. 295-298, ISSN 0309-1740

Gram, L. (1989). Identification, characterization and inhibition of bacteria isolated from tropical fish. Ph.D. Thesis. Danish Institute for Fisheries Research, Lyngby and The Royal Veterinary and Agricultural University, Copenhagen, Denmark.

Gram, L., & Dalgaard, P. (2002). Fish spoilage bacteria - problems and solutions. *Current Opinion in Biotechnology*, Vol. 13, pp, 262–266. ISSN 0958-1669

Gram, L. Ravn, L. Rasch, M. Bruhn, J.B. Christensen, A.B. & Givskov, M. (2002). Food spoilage- interactions between food spoilage bacteria. *International Journal of Food Microbiology*, Vol. 78, pp, 79– 97. ISSN 0168-1605

Grandinson, A.S. & Jennings, A. (1993). Extension of the shelf-life of fresh minced chicken meat by electron bean irradiation combined with modified atmosphere packaging, *Food Control*, Vol. 4, pp. 83-88. ISSN 0956-7135

Hajmeer, M.N. & Basher, I. (2003a). A probabilistic neural network approach for probabilistic modelling of bacterial growth/ no growth data. *Journal of Microbiological Methods*, Vol.51, pp. 217-226, ISSN 0167-7012

Hajmeer, M.N. & Basher, I.A. (2003b). A hybrid Bayesian-neural network approach for probabilistic modelling of bacterial growth/ no growth interface. *International Journal of Food Microbiology*, Vol.82, pp. 233-243, ISSN 0168-1605

Hajmeer, M.N. Basher, I.A. & Najjar, Y.M. (1997). Computational neural networks for predictive microbiology. 2. Application to microbial growth. *International Journal of Food Microbiology*, 34, pp. 51-66, ISSN 0168-1605

Health Protection Agency (2009). Guidelines for Assessing the Microbiological Safety of Ready-to-Eat Foods Placed on the market. Available from:

Hough, G. Langohr, K. Gomez, G. & Curia, A. (2003). Survival analysis applied to sensory shelf-life of foods. *Journal of Food Science*, Vol.68, pp. 359-362, ISSN 0022-1147 http://www.hpa.org.uk/webc/HPAwebFile/HPAweb_C/1259151921557.

Huis In`t Veld, J. (1996). Microbial and biochemical spoilage of foods: an overview. *International Journal of Food Microbiology*, Vol. 33, pp, 1-18. ISSN 0168-1605

Institute of Food Science and Technology (1993). Shelf life of Foods-Guideline for its Determination and Prediction. IFST, London.

Jacxsens, L. Devlieghere, F. Van der Steen, C. & Debevere, J. (2001). Effect of high oxygen modified atmosphere packaging on microbial growth and sensorial qualities of fresh-cut produce. *International Journal of Food Microbiology*, Vol. 71, pp, 197-210. ISSN 0168-1605

Jay, J.M. (1992). Intrinsic and extrinsic parameters of foods that affect microbial growth. In: Jay, J.M. (ed.), Modern Food Microbiology 4th Edition, Chapman & Hall, New York. pp, 38-62.

Jay, J.M. (2000). Modern food microbiology. 6th ed. Gaithersburg (MD): Aspen, pp, 679.

Kilcast, D. & Subramaniam, P. (2000). Stability and Shelf-life of Foods. pp, 1-22, 79-105. New York: CRC Press.

King, D.L. Hahm, T.S. & Min, D.B. (1993) Chemistry of antioxidants in relation to shelf life of foods. In: G. Charalambous (editor), Shelf-life Studies in Foods and Beverages. Elsevier, Amsterdam, pp, 629-707.

Koseki, S. (2009). Microbial Responses Viewer (MRV): a new ComBase-derived database of microbial responses to food environments. *International Journal of Food Microbiology*, Vol.134, pp. 75-82, ISSN 0168-1605

Koseki, S. & Itoh, K. (2002). Effect of nitrogen gas packaging on the quality and microbial growth of fresh-cut vegetables under low temperatures. *Journal of Food Protection*, Vol. 65, pp, 326-332. ISSN 0362-028X

Koutsoumanis, K. & Nychas, G.J.E. (2000). Application of a systematic experimental procedure to develop a microbial model for rapid fish shelf-life prediction. *International Journal of Food Microbiology*, Vol. 60, pp, 171-184. ISSN 0168-1605

Koutsoumanis, K. Lambropoulou, K. & Nychas, G.J.E. (1999). A predictive model for the non-thermal inactivation of *Salmonella enteritidis* in a food model system supplemented with a natural antimicrobial. *International Journal of Food Microbiology*, Vol.49, pp. 63-74, ISSN 0168-1605

Labuza, T.P. & Fu, B. (1993). Growth kinetics for shelf-life predictions: theory and practice. *Journal of Industrial Microbiology and Biotechnology*, Vol.12, pp. 309-323, ISSN 1367-5435

Lanciotti, R. Sinigaglia, M. Guardini, F. Vannini, L. & Guerzoni, M.E. (2001). Growth/no growth interface of Bacillus cereus, Staphylococcus aureus and Salmonella enteritidis in model systems based on water activity, pH, temperature and ethanol concentration. *Food Microbiology*, Vol.18, pp. 659-668, ISSN 0740-0020

Le Marc, Y. Huchet, V. Bourgeois, C.M. Guyonnet, J.P. Mafart P. & Thuault, D. (2002). Modelling the growth kinetics of *Listeria* as a function of temperature, pH and organic acid concentration. *International Journal of Food Microbiology*, Vol.73, pp. 219-237, ISSN 0168-1605

Leistner, L. (1995). Principles and applications of hurdle technology. In: Gould GW, editor. New methods of food preservation. London: Blackie Academic & Professional, pp, 1-21.

Leistner, L. (2000). Minimally processed, ready-to-eat, and ambient-stable meat. In Shelf-Life Evaluation of Foods (ed. Man, C.M. & Jones, A.A.), pp, 242-262. Aspen Publishers, Gaithersburg, USA.

Leporq, B. Membré, J.M. Dervin, C. Buche, P. & Guyonnet, J.P. (2005). The "Sym´Previus" software, a tool to support decisions to the foodstuff safety. *International Journal of Food Microbiology*, Vol.100, pp. 231-237, ISSN 0168-1605

Lewis, D.C. & Shibamoto, T. (1986). Shelf-life of fruits, In: G. Charamlabous (editor), Handbook of food and beverage stability. Academic Press, Orlando, pp, 353-391.

Liao, C.H. Sullivan, J. Grady, J., Wong, & L.J.C. (1997). Biochemical characterization of pectate lyases produced by fluorescent pseudomonads associated with spoilage of fresh fruits and vegetables. *Journal of Applied Microbiology*, Vol. 83, pp, 10– 16. ISSN 1365-2672

Lindqvist, R. & Lindblad, M. (2008). Inactivation of *Escherichia coli, Listeria monocytogenes* and *Yersinia enterocolitica* in fermented sausages during maturation/storage. *International Journal of Food Microbiology*, Vol.129, pp. 59-67, ISSN 0168-1605

Linton, R.H. Carter, W.H. Pierson, M.D. & Hackney, C.R. (1995). Use of a modified Gompertz equation to model nonlinear survival curves for *Listeria monocytogenes* Scott A. *Journal of Food Protection*, Vol.58, pp. 946-954, ISSN 0362-028X

Loss, C.R. & Hotchkiss, J.H. (2002). Inhibition of microbial growth by low-pressure and ambient pressure gasses. In: Juneja VK, Sofos JN, editors. Control of food-borne microorganisms. New York: Marcel Dekker, pp, 245-79. Forthcoming.

Lund, B.M. (1992). Ecosystems in vegetable foods. *Journal of Applied Bacteriology*, Vol. 73, pp, 115S- 126S. ISSN 1364-5072

Mannhein, C.H. & Soffer, T. (1996). Shelf-life extension of cottage cheese by modified atmosphere packaging. *LWT-Food Science and Technology*, Vol. 29, pp, 767-771. ISSN 1365-2672

Mataragas, M. & Drosinos, E.H. (2007). Shelf life establishment of a sliced, cooked, cured meat product base don quality and safety determinants. *Journal of Food Protection*, Vol.70, pp. 1881-1889, ISSN 0362-028X

Mataragas, M. Drosinos, E.H. Vaidanis, A. & Metaxopoulos, I. (2006). Development of a predictive model for spoilage of cooked cured meat products and its validation under constant and dynamic temperature storage conditions. *Journal of Food Science*, Vol.71, pp. M157-M167, ISSN 0022-1147

Mazzota, A.S. (2001). Thermal inactivation of stationary-phase and acid-adaptaded *Escherichia coli* O157-H7, *Salmonella*, and *Listeria monocytogenes* in fruit juices. *Journal of Food Protection*, Vol.64, pp. 315-320, ISSN 0362-028X

McClure, P.J. Baranyi, J. Boogard, E. Kelly, T.M. & Roberts, T.A. (1993). A predictive model for the combined effect of pH, sodium chloride and storage temperature on the growth of *Brochothrix thermosphacta*. *International Journal of Food Microbiology*, Vol.19, pp. 161-178, ISSN 0168-1605

McKellar, R.B. & Lu, X. (Eds.). (2004a). *Modeling microbial responses in food*. CRC Press, ISBN 0-8493-1237-X, Boca Raton, Florida

McKellar, R.B. & Lu, X. (2004b). Primary models, In: *Modelling Microbial Responses in Foods*, R.C. McKellar & X. Lu (Eds.), pp. 21-62, CRC Press, ISBN 0-8493-1237-X, Boca Raton, Florida

McKellar, R.C. & Knight, K.P. (2000). A combined discrete-continuous model describing the lag phase of *Listeria monocytogenes*. *International Journal of Food Microbiology*, Vol.54, pp. 171-180, ISSN 0168-1605

McKellar, R.C. (1997). A heterogeneous population model for the analysis of bacterial growth kinetics. *International Journal of Food Microbiology*, Vol.36, pp. 179-186, ISSN 0168-1605

McMeekin, T.A. (2007). Predictive microbiology: quantitative science delivering quantifiable to the meat industry and other food industries. *Meat Science*, Vol. 77, pp, 17-27. ISSN 0309-1740

McMeekin, T.A. Olley, J. Ratkowsky, D.A. & Ross, T. (2002). Predictive microbiology: towards the interface and beyond. *International Journal of Food Microbiology*, Vol. 73, pp, 395-407. ISSN 0168-1605

McMeekin, T.A. Chandler, R.E. Doe., P.E. Gardland, C.D. Olley, J. Putro, S. & Ratkowsky, D.A. (1987). Model for the combined effect of temperature and salt/water activity on growth rate of *Staphylococcus xylosus*. *Journal of Applied Bacteriology*, Vol.62, pp. 543-550, ISSN 1364-5072

McMeekin, T.A. Olley, J.N. Ross, T. & Ratkowsky, D.A. (1993). *Predictive Microbiology: Theory and Application*, Research Studies Press, ISBN 10: 0863801323, Taunton, UK

McMeekin, T.A. Presser, K. Ratkowsky, D.A. Ross, T. Salter, M. & Tienungoon, S. (2000). Quantifying the furdle concept by modelling the bacterial growth/no growth interface. *International Journal of Food Microbiology*, Vol.55, pp. 93-98, ISSN 0168-1605

Membré, J.M. Kubaczka, M. & Chene, C. (2001). Growth rate and growth-no-growth interface of *Penicillium brevicompactum* as functions of pH and preservative acids. *Food Microbiology*, Vol.18, pp. 531-538, ISSN 0740-0020

Membré, J.M. Ross, T. & McMeekin, T.A. (1999). Behaviour of *Listeria monocytogenes* under combined chilling processes. *Letters in Applied Microbiology*, Vol.28, pp. 216-220, ISSN 1472-765X

Mifflin, H. (Ed.). (2006). *The American Heritage Dictionary of the English Language* (fourth edition), Houghton Mifflin Company, ISBN-10: 0618701729, ISBN-13/EAN: 9780618701728, Florida

Moretti, V.M. Madonia, G. Diaferia, C. Mentasti, T. Paleary, M.A. Panseri, S. Pirone, G. & Gndini, G. (2004). Chemical and microbiological parameters and sensory attributes of a typical Sicilian salami ripened in different conditions. *Meat Science*, Vol.66, pp. 845-854, ISSN 0309-1740

Morris, J.G. (2000). The effect of redox potential. In: Lund BL, Baird-Parker TC, Gould GW, editors. The microbiological safety and quality of food. Volume 1. Gaithersburg (MD): Aspen. pp, 235-250.

Mossel, D.A.A. Corry, J.E.L. Struijk, C.B. & Baird, R.M. (1995). Essentials of the microbiology of foods: a textbook for advanced studies. Chichester (England): John Wiley and Sons. 699 pp.

Murphy, R.Y. Duncan, L.K. Johnson, E.R. Davis, M.D. & Smith, J.N. (2002). Thermal inactivation *D*- and *z*-values of *Salmonella* serotypes and *Listeria innocua* in chicken patties, chicken tenders, franks, beef patties, and blended beef and turkey patties. *Journal of Food Protection*, Vol.65, pp. 53-60, ISSN 0362-028X

Murray, J.M. Delahunty, C.M. & Baxter, I.A. (2001). Descriptive sensory analysis: past, present and future. *Food Research International*, Vol.34, pp. 461-471, ISSN 0963-9969

New Zealand Food Safety Authority (2005). A guide to calculating shelf-life of foods. Information booklet for the food industry. Available from http://www.nzfsa.govt.nz.

O'Mahony, M. (1985). *Sensory evaluation of food. Statistical Methods and Procedures.* Marcel Dekker, Inc., ISBN 0-8247-7337-3, New York

Oscar, T.P. Development and validation of a tertiary simulation model for predicting the potential growth of *Salmonella typhimurium* on cooked chicken. (2002). *International Journal of Food Microbiology*, Vol.76, pp. 177-190, ISSN 0168-1605

Paoli, G. (2001). Prospects for expert system support in the application of predictive microbiology, *Proceedings of Decission-Supot Tools for Microbial Risk Assessment*, Seattle, WA

Patsias, A. Chouliara, I. Badeka, A. Savvaidis, I.N. & Kontominas, M.G. (2006). Shelf-life of a chilled precooked chicken product stored in air and under modified atmospheres: microbiological, chemical, sensory attributes. *Food Microbiology*, Vol. 23, pp, 423-429. ISSN 0740-0020

Peleg, M. (2006). Advanced quantitative microbiology for foods and biosystems: models for predicting growth and inactivation. Boca Ratón, FL, CRC, Taylor & Francis Group.

Pitt, J.I. & Hocking, A.D. (1997). Fungi and Food Spoilage, 2nd ed. Blackie Academic and Professional, 596 pp

Pouillot, R. Albert, I. Cornu, M. & Denis, J.B. (2003). Estimation of uncertainty and variability in bacterial growth using Bayesian inference. Application to *Listeria monocytogenes*. *International Journal of Food Microbiology*, Vol.81, pp. 87-104, ISSN 0168-1605

Presser, K.A. Ratkowsky, D.A. & Ross, T. (1997). Modelling the growth rate of *Escherichia coli* as a function of pH and lactic acid concentration. *Applied and Environmental Microbiology*, Vol.63, pp. 2355-2360. ISSN 0099-2240

Pruitt, K.M. & Kamau, D.N. (1993). Mathematical models of bacterial growth, inhibition and death under combined stress conditions. *Journal of Industrial Microbiology and Biotechnology*, Vol.12, pp. 221-231, ISSN 1367-5435

Ratkowsky, D.A. & Ross, T. (1995). Modelling the bacterial growth/no growth interface. *Letters in Applied Microbiology*, Vol.20, pp. 29-33, ISSN 1472-765X

Ratkowsky, D.A. Lowry, R.K. McMeekin, T.A. Stokes, A.N. & Chandler, R.E. (1983). Model for the bacterial culture growth rate throughout the entire biokinetic temperature range. *Journal of Bacteriology*, Vol.154, pp. 1222-1226, ISSN 0021-9193

Ratkowsky, D.A. Olley, J. McMeekin, T.A. & Ball, A. (1982). Relationship between temperature and growth rates of bacterial cultures. *Journal of Bacteriology*, Vol.149, pp. 1-5, ISSN 0021-9193

Refrigerated Foods Association (RFA) (2009). Standardized Protocol for Determining the Shelf Life of Refrigerated Ready-To-Eat (RTE) Foods (revised January 2009). Available from: https://email.rutgers.edu/pipermail/fen_members/attachments/20100614/7f670f6b/RFA_ShelfLifeProtocol-09-0001.pdf.

Rho, M.J. & Schaffner, D.W. (2007). Microbial risk assessment of staphylococcal food poisoning in Korean *kimbab*. *International Journal of Food Microbiology*, Vol. 116, pp, 332-338. ISSN 0168-1605

Rico, D. Martín-Diana, A.B. Barat, J.M. & Barry-Ryan, C. (2007). Extending and measuring the quality of fresh-cut fruit and vegetables: a review. *Trends in Food Science & Technology*, Vol. 18, pp, 373-386. ISSN 0924-2244

Robinson T.P. Ocio, M.J. Kaloti, A. & Mackey B.M. (1998). The effect of the growth environment on the lag phase of *Listeria monocytogenes*. *International Journal of Food Microbiology*, Vol. 44, pp. 83-92, ISSN 0168-1605

Ross T. & Dalgaard, P. (2004). Secondary models. In: *Modelling Microbial Responses in Foods*, R.C. McKellar & X. Lu (Eds.), pp. 63-150, CRC Press, ISBN 0-8493-1237-X, Boca Raton, Florida

Ross, T. & Shadbolt, C.T. (2004). Predicting *Escherichia coli* inactivation in uncooked comminuted fermented meat products, In: *Meat &Livestock Australia*, ISBN 095852548, 20.10.2011, Available from: http://www.foodsafetycentre.com.au/docs/Salami%20Final%20report.pdf

Ross, T. Ratkowsky, D.A. Mellefont, L.A. & McMeekin, T.A. (2003). Modelling the effects of temperature, water activity, pH and lactic acid concentration on the growth rate of

Escherichia coli. International Journal of Food Microbiology, Vol.82, pp. 33-43, ISSN 0168-1605

Rosso, L. & Robinson, T.P. (2001). A cardinal model to describe the effect of water activity on the growth of moulds. *International Journal of Food Microbiology*, Vol.63, pp. 265-273, ISSN 0168-1605

Rosso, L. Lobry, J.R. & Flandrois, J.P. (1993). An unexpected correlation between cardinal temperatures of microbial growth highlighted by a new model. *Journal of Theoretical Biology*, Vol.162, pp. 447-463, ISSN 0022-5193

Rosso, L. Lobry, J.R. Bajard, S. & Flandrois, J.P. (1995). A convenient model to describe the combined effects of temperature and pH on microbial growth. *Applied and Environmental Microbiology*, Vol.61, pp. 610-616, ISSN 0099-2240

Rubio, B. Martínez, B. Sánchez, M.J. García-Cachán, M.D. Rovira, J. & Jaime, I. (2007). Study of the shelf life of a dry fermented sausage "salchichon" made from raw material enriched in monounsaturated and polyunsaturated fatty acids and stored under modified atmospheres. *Meat Science*, Vol.76, pp. 128-137, ISSN 0309-1740

Schellekens, M. Martens, T. Roberts, T.A. Mackey, B.M. Nicolai, M.B. Van Impe, J.F. & De Baerdemaeker, J. (1994). Computer-aided microbial safety design of food processes. *International Journal of Food Microbiology*, Vol.24, pp. 1-9, ISSN 0168-1605

Stewart, C.M. Cole, M.B. Legan, J.D. Slade, L. Vandeven, M.H. & Schaffner, D.W. (2001). Modeling the growth boundary of Staphylococcus aureus for risk assessment purposes. *Journal of Food Protection*, Vol.64, 51-57, ISSN 0362-028X

Stringer, S.C. George, S.M. & Peck, M.W. (2000). Thermal inactivation of *Escherichia coli* O157:H7. *Symposium Series (Society for Applied Microbiology)*, Vol.88. 79S-89S, ISSN 1467- 4734

Takumi, K. de Jonge, R. & Havelaar, A. (2000). Modelling inactivation of *Escherichia coli* by low pH: application to passage through the stomach of young and elderly people. *Journal of Applied Microbiology*, Vol.89, pp. 935-943, ISSN 1364-5072

Tamplin, M. Baranyi, J. & Paoli, G. (2004). Software programs to increase the utility of predictive microbiology information, In: *Modelling Microbial responses in Foods*, R.C. McKellar & X. Lu (Eds.), pp. 233-242, CRC Press, ISBN 0-8493-1237-X, Boca Raton, Florida

US Food and Drug Administration (2009). Safe Practices for Food Processes. Chapter 6. Microbiological Challenge Testing: Evaluation and Definition of Potentially Hazardous Foods. Available from: http://www.fda.gov/Food/ScienceResearch/ResearchAreas/SafePracticesforFoodProcesses/ucm094154.htm. Accessed 06/18/2009.

Valero, A. Carrasco, E. Pérez-Rodríguez, F. García-Gimeno, R.M. Blanco, C. & Zurera, G. (2006). Monitoring the sensorial and microbiological quality of pasteurized white asparagus at different storage temperatures. *Journal of the Science of Food and Agriculture*, Vol.86, pp. 1281-1288, ISSN 0022-5142

Van Asselt, E.D. & Zwietering, M.H. (2006). A systematic approach to determine global thermal inactivation parameters for various food pathogens. *International Journal of Food Microbiology*, Vol.107, pp. 73-82, ISSN 0168-1605

Van Boekel, M.A.J.S. (2008). Kinetic modeling of food quality: a critical review. *Comprehensive Reviews in Food Science and Food Safety*, Vol. 7, pp, 144-158. ISSN 1541-4337

Van Boekel, M.A.J.S. (2002). On the use of the Weibull model to describe thermal inactivation of vegetative cells. *International Journal of Food Microbiology*, Vol.74, pp. 139-159, ISSN 0168-1605

Van Gerwen. S.J.C. te Giffel, M.C. Van´t Riet, K. Beumer, R.R. & Zwietering, M.H. (2000). Stepwise quantitative risk aassessment as a tool fod characterization of microbiological safety. *Journal of Applied Microbiology*, Vol.88, pp. 938-951, ISSN 1364-5072

Vestergaard, E.M. (2001). Building product confidence with challenge studies. *Dairy, Food and Environmental Sanitation*, Vol. 21, pp, 206-209. ISSN 1043-3546

Voyer, R. & McKellar, R.C. (1993). MKES tool: a microbial kinetics expert system for developing and assessing food production systems. *Journal of Industrial Microbiology & Biotechnology*, Vol.12, pp. 256-262, ISSN 1367-5435

Wang, M.Y. & Brown, W.D. (1983). Effects of elevated CO_2 atmosphere on storage of freshwater crayfish (*Pacifastacus leniusculus*). *Journal of Food Science*, Vol. 48, pp, 158-162. ISSN 1750-3841

Whiting, R.C. & Buchanan, R.L. (1994). Microbial Modeling. *Food Technology*, Vol.48, pp. 113-120, ISSN 0015-6639

Whiting, R.C. & Cygnarowicz-Provost, M. (1992). A quantitative model for bacterial growth and decline. *Food Microbiology*, Vol.9, pp. 269-277, ISSN 0740-0020

Whiting, R.C. (1993). Modeling bacterial survival in unfavorable environments. *Journal of Industrial Microbiology and Biotechnology*, Vol.12, 240-246, ISSN 1367-5435

Willocx, F. Mercier, M. Hendrickx, M. & Tobback, P. (1993). Modelling the influence of temperature and carbon dioxide upon the growth of *Psedomonas fluorescens*. *Food Microbiology*, Vol.10, 159-173, ISSN 0740-0020

Witjzes, T. Van´t Riet, K. in´t Veld, J.H.J. & Zwietering, M.H. (1998). A decision support system for the prediction of microbial food dsfety and food quality. *International Journal of Food Microbiology*, Vol.42, pp. 79-90, ISSN 0168-1605

Zwietering, M.H. Jongenburger, I. Rombouts, F.M. & van´t Riet, D. (1990). Modelling of the bacterial growth curve. *Applied and Environmental Microbiology*, Vol.56, pp. 1876-1881, ISSN 0099-2240

Zwietering, M.H. Witjzes, T. Wit, J.C. de & Van´t Riet, K. (1992). A decision support system for prediction of the microbial spoilage in foods. *Journal of Food Protection*, Vol.55, pp. 973-979, ISSN 0362-028X

2

The Potential of Food Irradiation: Benefits and Limitations

Hossein Ahari Mostafavi[1*],
Seyed Mahyar Mirmajlessi[2] and Hadi Fathollahi[1]
[1]*Nuclear Science and Technology Research Institute, Agricultural, Medical and Industrial Research School, Karaj,*
[2]*Dept. Plant Pathology, College of Agriculture, Tarbiat Modares University, Tehran, Iran*

1. Introduction

Preservation of food has been a major anxiety of man over the centuries. Contamination with microorganisms and pests causes considerable losses of foods during storage, transportation and marketing (15% for cereals, 20% for fish and dairy products and up to 40% for fruits and vegetables). Particularly, pathogenic bacteria are an important cause of human suffering and one of the most significant public health problems all over the world. The World Health Organization (WHO) stated, the infectious and parasitic diseases represented the most frequent cause of death worldwide (35%), the majority of which happened in developing countries in 1992 (Loaharanu, 1994). Numerous processing techniques have been developed to control food spoilage and raise safety. The traditional methods have been supplemented with pasteurization (by heat), canning, freezing, refrigeration and chemical preservatives (Agrios, 2005). Another technology that can be added to this list is irradiation. Food irradiation is the process of exposing amount of energy in the form of speed particles or rays for improving food safety, eliminating and reducing organisms that destroy the food products. This is a very mild treatment, because a radiation dose of 1 kGy represents the absorption of just enough energy to increase the temperature of the product by 0.36°C. It means that, heating, drying and cooking cause higher nutritional losses. Moreover, heterocyclic ring compounds and carcinogenic aromatic produced during thermal processing of food at high temperatures were not identified in irradiated foods (Tomlins, 2008). More than one century of research has gone into the understanding of the effective use of irradiation as a safety method. It has been repeatedly considered and judged suitable on available evidence. The international bodies including the Food and Agriculture Organization (FAO), the International Atomic Energy Agency (IAEA), WHO and Codex Alimentarius Commission (CAC) investigate projects on food irradiation to verify the safety and quality of different irradiated products. It has shown that irradiation used on alone or in combination with other methods could improve the microbiological safety and extend shelf-

* Corresponding Author

life (IAEA, 2009). Furthermore, people are very confused to distinguish irradiated foods from radioactive foods. At no time during the irradiation process does the food come into contact with the radiation source and, it is not possible to induce radioactivity in the food by using gamma rays or electron beams up to 10 MeV (Farkas, 2004). The differences in chemical and physical structure of organisms, environmental factors, moisture content, temperature during irradiation, presence or absence of oxygen and in their ability to recover from the radiation injury cause the distinctions in their sensitivity to radiation. According to long-term animal feeding studies, radiation-pasteurized or sterilized foods are safe and nutritious also for humans (Thayer & Boyd, 1999). Irradiation is used for a variety of reasons, such as disinfesting food, reducing or eliminating food borne pathogens, shelf life extending and may serve as a quarantine treatment for many fruits, vegetables, nuts, cut flowers and animal origin products to facilitating international trade of foods (Marcotte, 2005). But, not all foods are appropriate for irradiation. For instance, some fruits (such as cucumbers, grapes, and some tomatoes) are sensitive to radiation. Nowadays, over 60 countries use irradiation for one or more food products. But, the misconceptions and irrational fear of nuclear technologies mostly caused the lack of acceptance of food irradiation. It should be noted that if irradiated products offer clear advantages, and the science-based information on the process is readily available, consumers would be ready to accept more irradiated products.

2. Effects of Ionizing radiation

Irradiation can be effect direct, caused by reactive oxygen-centred (•OH) radicals originating from the radiolysis of water or indirect on organisms and food products. An indirect effect (the damage to the nucleic acids) occurs when radiation ionizes a neighboring molecule, which in turn reacts with the genetic material. Since, water is a major component of most foods and microbes; it is often ends up producing a lethal product (Hallman, 2001). The optimum dose is a balance between that what is needed and that what can be tolerated by the products (European Food Safety Authority [EFSA], 2011).

2.1 Effects on bio-organisms (microorganisms and insects) and viruses

The biological effects caused by irradiation are primarily the result of disruption of the DNA or RNA in the nuclei of cells. Since, the DNA is much larger than the other molecular structures and very long ladder twisted into a double helix in a cell, if it becomes injured, by either primary ionizing or through secondary free radical attack, the induced biological and chemical changes can prevent replication and destroy cells (Scott Smith & Suresh, 2004; The Institute of Food Science and Technology [IFST], 2006). Therefore microorganisms, insect gametes, and plant meristems are prevented from their reproduction, which consequently results in various preservative effects as a function of the absorbed radiation dose (Scott Smith & Suresh, 2004).

Since, the size of the DNA molecule generally increases with the complexity of an organism, viruses are more radiation resistant than other organisms (Koopmans & Duizer, 2004). Viruses do not grow in food directly, but can contaminate host bacteria (Deeley, 2002). Poliomyelitis viruses and infectious hepatitis can be transmitted by means of infected raw milk and shellfish (DeWit et al., 2003; Frankhauser et al., 2002). More studies showed that the combination of irradiation with heating can be used successfully to inactivation of viruses (Koopmans & Duizer, 2004).

The spore-forming bacteria are resistant to irradiation and other treatments. Doses used below 10 kGy may only give a 2-3 $\log_{10}$ reduction in spore numbers, and this is not sufficient to foods shelf-life (Patterson et al., 2006). Yeasts are more radiation resistance than molds and vegetative bacteria. So, they are important in the spoilage of irradiated meat products (such as sausages) stored at refrigeration (Patterson et al., 2006; Scott Smith & Suresh, 2004).

Insects, mites and other such pests are higher level, multicellular organisms responsible for considerable loss of fresh produce and grains. Also, they can provide as vectors for carrying pathogenic bacteria and parasites (Ahari et al., 2010). The best control of insects in agricultural products can be achieved by using fumigants (such as ethylene bromide). But, the use of these pesticides has been forbidden or severely restricted in most countries (World Health Organization, 2005). Consequently, radiation has been suggested as an alternative to them. Irradiation as direct control is one aimed primarily at reducing insects population. For the first time in the early 1957s irradiation as a commercial insect control technique was applied to spices in the Federal Republic of Germany (Diehl, 1995). Basis of practical experience, the necessary radiation dose to pests control according to different growth stages is in the range of 100-1000 Gy. A dose level of 250 Gy can be used for quarantine treatment of fruit flies, while a dose of 500-1000 Gy can control all stages of other pests (Landgraf, et al., 2006; Marcotte, 2005). Indirect control, on the other hand is the sterile insect technique (SIT). The principle of the technique is to introduce sterility by rearing great numbers of the pest and release them into the target area. When the released males mate with wild females, the females generate no viable offspring. With a constant releasing rate of sterile males over several generations, this results a rapid decline in the overall population. The sterilization process (in this method) is important in determining the quality of the released pests and their ability to compete with the wild population. SIT has been used against a number of pest species like as Mediterranean fruit fly *Ceratitis capitata* (Wiedemann), melon fly *Bactrocera cucurbitae* (Coquilett), pink bollworm *Pectinophora gossypiella* (Saunders), codling moth *Cydia pomonella* (L.) and tsetse fly *Glossina austeni* Newstead (Hendrichs et al. 2005; Klassen & Curtis, 2005).

2.2 Mutagenic and toxin production effects

Mutation in micro-organisms is a well known event. So, selection of spontaneous mutants and production of mutant strains by various mutagenic agents are practiced for scientific and industrial purposes. Ionizing radiation is able to induce mutations, as a number of physical or chemical antimicrobial agents. Some evidences showed that the pathogenicity of infectious organisms is diminished by irradiation (EFSA, 2011). Studies on transcriptional changes of *S. Typhimurium* and *Vibrio* spp., showed that, the expression of the virulence genes in Salmonella irradiated mutants was reduced, and expression of toxin genes of *Vibrio* irradiated mutants did not increase, compared with non-irradiated counterparts (Lim et al., 2007). The wholesomeness (lack of teratogenicity, mutagenicity, and toxicity) of irradiated products has been studied expansively (IAEA, 2009). Multigenerational feeding studies did not show any confirmation of toxicological effects in mammals due to intake of irradiated foods. It means, multigenerational studies with animals have showed that the nutritive value remains essentially unchanged and that the ingestion of irradiated foods is completely safe (Thayer & Boyd, 1999; IAEA, 2009). Food irradiation has not provided any evidence of increased threat of mycotoxin formation in irradiated food. Therefore these do not show any specific hazard in relation to mycotoxin production (Kottapalli et al., 2006).

2.3 Radiation resistance

Radiation resistance in organisms is different. For instance, *Alternaria* sp. and *Fusarium* sp. are more resistant than the *penicillium* sp. and *Aspergillus* sp. Fusarium and Alternaria Spores are multicellular. If only one cell survives, the spore may still have the ability to germinate. So, these spores are more radiation resistant (Patterson et al., 2006). The effect of repeated doses of γ-radiation indicated a progressive increase of the radiation resistance of the organisms when exposed at sublethal dose. But, under industrial conditions, no situation is conceivable whereby a population of organisms would be repeatedly resuscitated after sublethal irradiation (Levanduski & Jaczynski, 2008). Some studies on food irradiation facilities and many other radiation-emitting sources showed no evidence for increased occurrence of resistant strains in the environment of these facilities (Scientific Committee on Food [SCF], 2003).

2.4 Selective effects on the microbial flora

Vegetative food pathogens and non pathogenic microorganisms are sensitive to radiation. The medium-dose irradiation processes reduce their populations by several log10 cycles. The possibility that decrease of the competitive microflora could assist growth of food pathogens after irradiation was studied. The study did not show any significant difference in the bacterial growth in irradiated food compared to non-irradiated one. Results indicated that, the absence of the spoilage microflora in food did not provide a competitive advantage to the bacterial growth (Prendergast et al., 2009).

3. Effects on food products

Ionizing radiation produces chemical changes by primary and secondary radiolysis effects. The effect of chemical reactions depends on the absorbed dose, dose rate and facility type, presence or absence of oxygen and temperature. Generally, most food micronutrients (mainly water-soluble and fat-soluble vitamins) and macronutrients (carbohydrates, proteins, and lipids) are not affected by 10 kGy-range -range ionizing dose with regard to their nutrient contents. But, with radiation doses above 10 kGy, the properties of fibrous carbohydrates can be degraded structurally and lipids can become somewhat rancid (Miller, 2005; Brewer, 2009). The physical status of food (frozen or fresh, solid, liquid or powder) and also its composition influence the reactions induced by the radiation (IAEA, 2009). Direct absorption of energy by the irradiated food can produce chemical changes via primary and secondary indirect radiolysis effects. Irradiation effects water molecule to free an electron producing HO+. This product reacts with other water molecules to create a quantity of compounds, including hydrogen and hydroxyl radicals (OH•), hydrogen peroxide (HO), molecular hydrogen and oxygen. Because the hydroxyl radical is an oxidizing agent, and the hydrated electron is a reducing agent, free radical attack can be expected to cause oxidizing and reducing reactions in food (Black & Jaczynski, 2008; Calucci

2003). Chemical reactions and the products generated from major food components like fat, proteins, carbohydrates and vitamins are described as fallow:

3.1 Carbohydrates

The use of gamma-irradiation (up to 6.2 kGy/h) to starches from maize, wheat, rice or potato, induced the aldehydes such as malonaldehyde, formaldehyde, and acetaldehyde, formic acid and hydrogen peroxide as main radiolytic products (Stefanova et al., 2010). Recent studies showed that, gamma irradiated solutions of fructose, glucose, sucrose and starch (at 3 kGy at 5ºC) produced malonaldehyde, But, the accumulation of aldehydes formed after irradiation of fruit juices decreased by reducing the presence of oxygen and using low temperature (Fan, 2003; Fan & Thayer, 2002). In general, irradiation modifies Mono and polysaccharides, but thermal treatment can produce more modifications (Fan, 2005).

3.2 Proteins

The study showed that, the irradiation of proteins could produce chemical reactions depend on the protein structure, state (native or denatured), physical status, amino acid composition, the presence of other substances and the radiation treatment. The most important changes include dissociation, aggregation, cross-linking and oxidation. For example gamma irradiation of hazelnuts at 10 kGy induced protein aggregation and denaturation (Dogan et al., 2007). Also, the reported decrease in pectinase activity (the most sensitive enzyme to irradiation) at 20 kGy was in the range 20% to 50%, (Duliu et al., 2004). Indeed, low and medium doses induce only a small breakdown of food proteins into lower molecular weight protein parts and amino acids. As a result, experiments indicated that such treatments cause less chemical reactions than steam heat sterilization (Fan & Sommers, 2006).

3.3 Lipids

Due to free radicals formed during irradiation, it has been proven to increase lipid oxidation (Stewart, 2009). The chemical effects are more relevant in foods with larger fat content, physical status (liquid or solid), presence of antioxidants, environmental conditions (light, heat, oxygen, moisture, pH), the irradiation treatment, type of storage (vacuum, modified atmosphere, *etc.*), storage conditions (time, temperature, light, *etc.*) and high unsaturated fatty acids content (EFSA, 2011). The use of low temperatures, the presence of oxygen, antioxidants and suitable packaging type showed a great capacity to minimize lipid oxidation (Stefanova et al., 2010). Phytosterols (include sterols and stanols), which are naturally present in cereals, nuts, seeds, fruits and vegetables have a structure similar to cholesterol. These compounds can be oxidized by standard heating treatments and also by irradiation and produce oxyphytosterols (Da Trindade et al., 2009; Johnsson & Dutta, 2006).

3.4 Vitamins

The primary effects of radiation on vitamins at low and medium doses are not considerable. But, the combination of free radicals produced through irradiation with antioxidant vitamins can lose some of their influence (Stefanova et al., 2010). In the case of water soluble vitamins, Thiamine is the most sensitive and significant losses can occur in irradiated meats.

Irradiation of chicken meals with 1 kGy resulted in a 16% decrease in thiamine compared with non-irradiated meals (Stewart, 2009). Mitchell et al. (1992) showed that, irradiation of capsicums (green and red), cucumbers, apples, lemons, lychees, mandarins, mangoes, nectarines, papayas, peaches, persimmons, and zucchinis at doses of 75 - 300 Gy can induce small changes in some parameters, such as vitamin C and dehydroascorbic acid after 3 to 4 weeks of storage at 1–7°C. However, storage effects were higher than irradiation effects. Dionísio et al. (2009) demonstrated that, Strawberries (Shasta variety, *Fragaria* sp.) submitted to 1.0-2.0 kGy doses did not show any significant decrease in vitamin C levels during two and 11 days of storage at 5°C. Folic acid levels decreased 20–30% following irradiation with a dose of 2 kGy, and no additional decrease was observed at the higher dose of 4 kGy (Galan et al., 2010). After gamma irradiation at a dose of 10 kGy, significant losses (10 to 34%) of total ascorbate and carotenoids (about 40-60%) have been reported for cinnamon, nutmeg, black pepper, oregano, parsley, rosemary, bird pepper, and sage. (Calucci et al., 2003). No transforms have been showed for Riboflavin, vitamin B6, vitamin B12 and niacin after gamma irradiation with up to 5 kGy in wheat, maize, mung beans and chick peas (Kilcast, 1994). The different sensitivity levels of food vitamins to processing are shown in Table 1. More studies showed that after high dose gamma-irradiation, E-beam irradiation or heat sterilization treatments, vitamin losses in the food are within similar ranges in all cases (Galan et al., 2010; EFSA, 2011).

Highly sensitive	Moderatly sentitive	Little sensitive
A (retinol)	b-carotene	Folic Acid
B1(thiamin)	K (in meat)	Pantothenic Acid
C (ascorbic acid + dehydro-ascorbic)		B2 (Riboflavin)
E (a-tocopherol)		B3 (Niacin)
		B6 (Pyridoxine)
		B10 (Biotin)
		B12 (Cobalamin)
		Choline
		D
		K (in vegetables)

Table 1. Different sensitivity levels of food vitamins to processing (Dionísio et al., 2009)

3.5 Inorganic salts

Primary radicals are moderately uncreative towards inorganic anions, excepting for nitrates which are reduced by solvated electrons to nitrites. In frozen muscle foods this is considered to be a rare event because of the competition for electrons by the other matrix ingredients (SCF, 2003).

In summary, carbohydrates, proteins, lipids (the macronutrients), vitamins and inorganic salts are not significantly affected by low and medium range doses according to their nutrient content and digestibility. Indeed, traditional food preservation methods (heating, drying and cooking) may cause upper nutritional losses.

4. Main objectives and limitations of food irradiation

The type of food being processed and the desired effects determine the radiation dose used in food processing. The examples of recommended dose ranges and the main purposes of food irradiation for different purposes are listed in Table 2.

Purpose and effects	Dose range (kGy)*
Inhibition of sprouting of stored tubers, roots and bulbs	0.05-0.15
Prevention of post-harvest losses by destruction of insects in stored cereals, fresh and dried fruits, nuts, oilseeds and pulses, or phytosanitary (quarantine) treatment for insect pests infesting fresh fruits and vegetables	0.15-1
Delay of ripening of fruits	0.2-1
Shelf-life extension of fruit and vegetables, meat, poultry, fish and ready meals by reduction of micro-organisms that cause spoilage	0.5-3
Inactivation/destruction of various food-borne parasites	0.3-6
Prevention of food-borne illness by destruction of non-sporeforming pathogenic bacteria (e.g. Salmonella, Campylobacter, Listeria) in fresh or frozen foods	3-7
Shorten drying and cooking times of vegetables and fruits	3-10
Reduction in viable counts of microorganisms in spices and other dry ingredients to minimize contamination of food to which the ingredients are added	5-10
Production of microbiologically shelf-stable, vacuum-packaged meat, poultry and ready-to-eat meals by heat-inactivating of their tissue-enzymes and sterilizing them by irradiation in deep-frozen state	up to 50

*The maximum doses reported are intended for good irradiation practice and not for consumer safety purposes.

Applications up to 1 kGy, between 1 and 10 kGy, and higher than 10 KGy are referred to as low-dose, medium-dose, and high-dose irradiation, respectively.

Table 2. Main purposes of food irradiation and examples of recommended dose ranges (EFSA, 2011).

4.1 Fruits and vegetables

Several food-borne outbreaks have reported in fruit products such as fresh fruit salads and non pasteurized fruit juices (Lynch et al., 2009; Vojdani et al., 2008). Applicable doses on fruit products are limited by their impact on quality (Arvanitoyannis et al., 2009). Irradiation should anticipate 2 to more than 5 log10 reduction of pathogenic (non spore forming bacteria), in some fresh and many processed fruits. But, less than 1.5 log10 reductions should be expected for the most radiation sensitive fruits. In many cases, surface decontamination of fruits, pre-washed and packaged vegetables and ready-to-eat fresh processed products with chemical agents should be less efficient (by at most 2 log10 reduction) than food irradiation (Arvanitoyannis et al., 2009). Most studies indicated that, the irradiation of fresh fruits led to a reduction in firmness. The maximum doses which can be applied on fruits and vegetables range between 1 and 2 kGy. However these maximum values depend on the type of products and might modify with new, resistant cultivars (Zhu, et al., 2009). For example, some salad vegetables could resist up to 4000 Gy without physical-chemical damages and considerable quality loss (Nunes et al., 2008) whereas, for some products, such as lettuce leaves and apple fruit (Gala and Fuji variety) radiation doses should not exceed 600 Gy (Dionísio et al., 2009; Niemira et al., 2002).

Some products have a very short shelf-life. For instance, within a day at 10°C the cap of mushroom opens, the stem elongates, and the gills darken. Research showed that mushroom irradiated at a dose of 1.5 to 2.0 kGy soon after harvest had a shelf-life at 5° C of 11 days, and at 10°C, of 7 days. Most investigations show that sensory quality of mushroom is unaffected by irradiation (IAEA, 2001). For powdered dehydrated vegetables higher treatments (3-5 kGy) were suggested (Dong-Ho et al., 2002). Treatments for instance calcium chloride could decrease the effect of irradiation on fruit firmness (Prakash et al., 2007). Also, D10 values for various Salmonella strains in orange juice were between 0.6 and 0.8 KGy (Niemira & Lonczynski, 2006). Since, viruses are more resistant to irradiation, the doses applicable to most fresh fruits and fruit products will be of limited efficacy (Koopmans & Duizer, 2004).

4.2 Cereal products

Cereals, grains and dry legume seeds are usually consumed after a wide range of processing. Operations and irradiation (tested at doses between 1 and 10 kGy) can modify the quality and technological properties of cereals and cereal products positively or negatively (Lynch et al., 2009). Irradiation (as a pest control method) has some advantages include the absence of undesirable residues in the foods treated, no resistance development by pest insects and few significant changes in the physicochemical properties or the nutritive value of the treated products (Ahmed, 2001). Several studies had been done on the use of irradiation (as an approved method) to control stored-product pests in wheat, flour and dry legume seeds in many countries (Azelmat et al., 2005; Boshra & Mikhaiel, 2006). Lower radiation doses can not cause immediate death of adults but can prevent an increase in pest populations through lethal effects on the immature stages of adults (Hosseinzadeh & Shayesteh, 2011). The major safety concern since 1986 for cereals has been mycotoxin producing fungi. Some recent studies have shown irradiation up to 10 kGy can reduce the colonization of cereals with mycotoxin producing fungi. For instance, 6 to 8 kGy reduces the risk of mycotoxin accumulation during barley germination for malt production. However, such doses have no impact on toxins already present in the cereals (Kottapalli et al., 2006).

4.3 Tubers

Starchy tubers are an important source of starch used in the food industry. In order to provide consumers a year-round supply of various sprouting foods, storage durations of up to several months are often necessary. Sprouting can be inhibited by refrigeration and the application of various chemicals. But, refrigeration is expensive and particularly in the tropical and subtropical regions of the world. Whereas, the chemical treatments are relatively inexpensive and efficient, they do leave residues and many countries have banned their usage for health reasons (Ahari et al., 2010). In such instances, irradiation can be recommended as a reasonable alternative. The first use of food irradiation for a fresh commodity was on potatoes in Canada in 1965 to inhibit sprouting (Diehl, 1995). Starchy tubers are also normally consumed after heat treatments and have rarely been the cause of food borne disease. Starch can be a significant source of spores of pathogenic bacteria such as *Bacillus cereus* in processed foods (Guinebretiere et al., 2003; Guinebretiere & Nguyen-The, 2003). Irradiation can be recommended for sprouting inhibition (in the range of 50 -200 Gy) and disinfestations purposes (at doses similar to those used for other dry foods) in sprouting foods such as potatoes, garlic, onions and yams (Marcotte, 2005).

4.4 Spices and condiments

Spices and condiments naturally, contain a great number of microorganisms. They often originate in developing countries where harvest and storage conditions are insufficiently controlled. Thus, Spices and condiments may have been contaminated by a high level of mesophylic, sporogenic,and asporogenic bacteria, hyphomycetes, and faecal coliforms. Also, the public health significance microorganisms such as Salmonella, Escherichia coli, Clostridium perfringens, Bacillus cereus, can be present. (Bendini et al., 1998). Since moist heat treatment is not generally suitable for such dry products, many commercial food processors fumigate spices and condiments with methyl bromide and ethylene oxide to eliminate insects or bacteria and moulds. Nowadays it has been appeared that, methyl bromide and ethylene oxide are high toxic compounds because of safety and environmental concerns. So, ethylene oxide has been banned in many countries and methyl bromide is being phased out globally (Marcotte, 2005). Based on the scientific evidence, irradiation of spices and condiments with a dose in the range of 5-10 kGy permits successful reduction of food-borne pathogens. The accurate dose applied would depend on the desired inactivation factor and the contamination (Farkas, 2006). Producers are now increasingly turning to ionizing radiation. Considering its antimicrobial activity and relatively minor effects on quality, irradiation is a feasible method for reducing the microbial load on spices and condiments (Sádecká et al., 2005). It is confirmed that the treatment with ionizing energy is more effective against bacteria than the thermal treatment and it is also less harmful to the spices than heat sterilization, which implicates the loss of thermo labile aromatic volatiles and/or causes additional thermally induced changes (e.g. thermal decomposition or production of thermally induced radicals). (Loaharanu, 1994; Olson, 1998; Alam Khan, 2010).

4.5 Sea foods and fresh meats

Several diseases originated by pathogenic Vibrios, *Listeria monocytogenes,* parasites and viruses are generally related to consumption of raw products (Venugopal, 2005). Irradiation of sea foods and fresh meats is proposed to extend shelf-life, inactivate parasites and decrease pathogen load (Norhana et al., 2010). Many studies indicated that irradiation at doses of 3 kGy, should yield 2 to 5 log10 reduction of pathogenic, non spore forming bacteria (Guinebretiere et al., 2003; Lim et al., 2007). The applicable dose to fresh meats and sea foods should be adapted to the pathogen reduction and the product (O'Bryan et al., 2008; Norhana et al., 2010). For example, Shrimp is separate from fish and shellfish given that certain pathogens (i.e *Listeria monocytogenes*) for several log10 reductions need doses in excess of 3 kGy. In frozen shrimp the dose required to reduce *Aeromonas hydrophila* and *Vibrio* by 10-4 per gram was about 3 kGy, while 3.5 kGy was needed for *L. monocytogenes* and 1.0 kGy for Salmonella (Hatha et al., 2003; Pinu et al., 2007). Irradiation of fresh meat can change colour, odour and taste. Some recent studies have shown, irradiation imparts undesirable organoleptic attributes to high-fat products. So, content of fat in the fresh meats and sea foods is a limiting factor (Norhana et al., 2010). These changes can be reduced by adapted atmosphere packaging and minimized fat levels and mainly prevented by irradiating in the frozen state. In frozen poultry products, the average dose of 7 kGy would be adequate to provide at least a 5-log10 reduction in the number of vegetative pathogens. There is no evidence that treatment by ionizing radiation made induce allergenicity. But, some studies showed that application of moderate radiation doses can make foods (which have some natural allergenic factors) less allergenic (EFSA, 2011). For instance, irradiation of eggs and products containing eggs has been used for the reduction of allergenicity and improving egg white foaming ability (Song et al., 2009).

4.6 Other food classes/commodities

Ready-to-eat foods are very various and their consumption has increased. These foods may cause a specific danger to consumers when they do not endure a process pathogen reduction (Sommers & Boyd, 2006). To improve microbiological safety of ready-to-eat foods by irradiation, extensive research has been devoted, specifically to eliminate non spore-forming pathogens from ready-to-eat meat products. Irradiation is an effective treatment that can be used in many of these products after packaging (Romero et al., 2005). Also, Sommers and Boyd (2006) demonstrated that doses of 2 to 4 kGy inactivate food-borne pathogens including *Salmonella* spp., *Listeria monocytogenes, Staphylococcus aureus, Escherichia coli* O157:H7 and *Yersinia enterocolitica* in a variety of ready-to-eat food products.

Cheeses are also included in ready-to-eat foods. The pathogenic and toxin-producing microorganisms (considered in studies on irradiation of cheeses) are *L. monocytogenes*, pathogenic *E. coli*, Salmonella, Clostridium, Staphylococcus (mycotoxins), Brucella and Mycobacterium (Tsiotsias et al., 2002). The treatment of Camembert cheeses with gamma irradiation at doses up to 2.5 kGy (from a health point of view) was acceptable (SCF, 2003). Tsiotsias et al. (2002) showed the feasibility of gamma irradiation at doses of 2-4 kGy for reducing *L. monocytogenes* inoculated into the freshly produced soft whey cheese. Also, Ju-Woon et al. (2005) studied irradiation of fried-frozen cheese balls and demonstrated that irradiation at dose of 3 kGy is a successful treatment to make sure microbiological safety.

4.7 Quarantine treatments

The amount of transportation in all types of food product reaches millions of tons, annually. While world economic markets shrunk substantially, the cost of all traded agricultural products increased 19% in 2008 (WTO, 2010). During transportation, some invasive species that may cause economic and environmental damage (especially in areas of the world where they do not currently exist) can be transmitted. However, plant pests are not only transported through agricultural goods; they can be found in any commodity. For instance, important tree-infesting beetles (Bostrichid beetles) have been transported in solid wood packing material and pallets arriving from other countries to U.S. states (Haack, 2006). A number of techniques have been investigated as quarantine treatments, and those most commonly used for commercial purposes involve extreme temperatures and fumigants (Heather & Hallman, 2008; IPPC, 2009). Until the late 1920s, phytosanitary treatments were based on fumigation or nonsynthetic pesticide applications. The chemicals used then were largely not safe enough to be used on food but were used on nursery stock and other nonfood items that could carry invasive species. In 1929, nonchemical treatments (heated air and cold) were used as quarantine treatments against the fruit flies. The first international use of Phytosanitary irradiation was in December, 2004, when Australia sent one-half ton of irradiated mangoes to New Zealand. For a phytosanitary treatment to be commercially possible, the treated commodities must tolerate it. Most food products can tolerate radiation at the doses used for quarantine treatments (150 to 400 Gy) and spice disinfection is the chief use of food irradiation worldwide today with over 185600 tons irradiated each year (Heather & Hallman, 2008; Kume et al., 2009). Phytosanitary irradiation differs from other commercial treatments. The end point of the treatment is not acute mortality but it prevents further biological development and reproduction (Table 3).

Treatment	End point	Commodity tolerance	Cost	Certified organic?	Speed	Logistics	Commonly treated commodity
Cold	Mortality	Moderate	Low	Yes	Very slow	Easy	Citrus, apple
Heated air	Mortality	Moderate	Moderate	Yes	Moderate	Moderate	Mango, papaya
Hot water immersion	Mortality	Moderate	Low	Yes	Fast	Moderate	Mango
Methyl bromide fumigation	Mortality	Moderate	Low	No	Fast	Easy	Citrus
Irradiation	Stop development	High	Moderate	No	Fast	Moderate	Mango, guava

Table 3. Subjective comparison of major phytosanitary treatments (Hallman, 2011).

5. Association with other preservation methods

Irradiation in combination with other treatments may suppress the growth of surviving microorganisms during storage (Fan et al., 2006; Caillet et al., 2006). For example, the effect of irradiation was increased by packaging vegetables in atmospheres enriched with carbon dioxide or containing essential oils (Thayer & Boyd, 1999). Sensory changes on high-fat products can be reduced by vacuum packaging associated to refrigeration and depend on the type of products (Zhu et al., 2009). Ahari et al. (2011) demonstrated that the combination of gamma irradiation and antagonist treatment (*Pseudomonas fluorescens*) was more effective in reducing *Penicillium expansum* growth, than either treatment alone and that integration of irradiation. Whereas, there is a dose rate limitation for application of gamma irradiation to control of postharvest disease on fruits and vegetables, the combination of irradiation and biocontrol agent increase applied range of irradiation for postharvest control by decreasing of dose rate (Ahari et al., 2011). Lee et al. (2005) demonstrated that the combined treatment of gamma irradiation and rosemary extract powder (genus Rosmarinus) improve the quality of a ready-to-eat hamburger steak by changing the storage condition from frozen (−20°C) to a chilled temperature (4°C). Fan et al. (2006) applied dose of 0.5 kGy in combination with mild heat treatment for Shelf-life extension of fresh-cut cantaloupe. Also, Niemira et al.

(2003) used freezing orange juice concentrates after irradiation, to reduction of Salmonella populations. The effect of irradiation on vegetables and sprouted seeds could be completed by washing with water (Rajkowski & Fan, 2008). Chiasson et al. (2005) studied the effect of the combination of irradiation and different atmosphere conditions (CO2, air, MAP with 10% N2, 30% CO2 and 60% O2, and vacuum), a mix of tetrasodium pyrophosphate (TSP) and carvacrol on the reduction of *E. coli* and *Salmonella typhi* in minced beef.

6. Packaging for irradiated foods

The radiation treatment can be applied after packaging, so re-contamination or re-infestation of the products is avoided. In 2001, the FDA illustrated gamma rays, E-beam and X-ray to be equal in conditions of levels and types of radiolysis products formed in the packaging materials (Deeley, 2002; Goulash et al., 2004). Polymers such as polyethylene, polypropylene, poly vinyl chloride, polystyrene, polyethylene terepthalate and polyamide are some of the most common plastic packaging materials presently available. They all contain additives that vary in nature and quantity for obtaining certain useful properties. Some studies have shown, Irradiation can change some physical and chemical properties of polymeric packaging materials and the changes depend on the type of polymer, irradiation conditions and processing exposure (Ozen & Floros, 2001; Hammad et al., 2006). For this reason any packaging materials must be confirmed by FDA before use in food irradiation (Crook & Boylston, 2004; Pentimalli et al., 2000). Nowadays, flexible packages have been developed which tend to be multi-layer films with different barrier properties because, no single flexible material has all the chemical, physical and protective characteristics needed for packaging radiation-processed food (Chytiri et al., 2005; Twaroski et al., 2006).

6.1 Physical effects

Physical effects include changes in crystallinity, permeability, surface structures and post irradiation aging effects. Radiation changes in the physical properties of a packaging material should not delay its function (Han et al., 2004). Since the Permeability is measure of the ease with which gases or vapors can penetrate through the polymer materials, it is a major consideration in the selection of a polymeric material for food packaging (Han, 2007). Kale (1994) showed that, physical properties such as permeability, crystallinity, mechanical strength and IR spectra of irradiated polymers were unaffected at high doses carried out on the effect of radiation on polymeric films both single such as polypropylene, low density polyethylene (LDPE), PET and laminates (BOPP/LDPE, PET/LDPE, PET/PET/LDPE, PET/metalized PE/LDPE, PET/HDPE-LDPE, Polyolefin-Tie layer-Nylon-Tielayer-LDPE and 5 layer nylon coextruded film/Metalized PET). Study on different packaging films, permeating gases and analytical techniques have confirmed that, there is no modification in the permeability, crystallinity and shrinkage in regularly used packaging materials such as low density polyethylene (LDPE), high density polyethylene (HDPE), polypropylene, polyethylene terephthalate (PET), poly vinyl chloride and poly vinylidene chloride in the dose range of 0-8 kGy (Goulas et al., 2002). Some researches have shown that, changes in mechanical properties of certain polymers e.g. polyethylene can be minimized (when subjected to higher doses of radiation) using suitable stabilizers (Buchalla et al., 1997; Han, 2007). Also deterioration with respect to dosage was reduced after laminating of some polymers (e.g. polypropylene) with LDPE.

6.2 Chemical effects

Most food packaging materials are originated from polymers. They may be susceptible to chemical changes (after ionizing radiation) that are the result of two reactions, cross-linking (polymerization) and chain scission (degradation). Both reactions are generally relative to dose, and depend on dose rate and the oxygen content of the atmosphere in which the polymer is irradiated (Ahn et al., 2003). Unique chemical markers present in irradiated packaging have not been recognized (Ahari et al., 2010). Chemical effects include evolution of radiolysis products, migration of radiolytic products of the polymers and degradation of antioxidants (Twaroski et al., 2006). Radiolytic degradation should neither be toxic nor affect the sensory qualities of the packed product. The vacuum condition or an inert atmosphere prevents radiation cross-linking of polymers. Also in the presence of oxygen (during irradiation of polymers) chain scission dominates. It is important because it has served as the basis for recent approvals under 21 CFR 170.39 (table 4) for packaging materials irradiated under non-oxygen atmospheres in contact with food (FDA, 2006a). During irradiation, ions and free radicals are produced. These highly reactive materials are responsible for the colour changes in irradiated polymers and could migrate into food and affect taste, odor and safety (Welle et al., 2002; Franz & Welle, 2004; Stoffers et al., 2004). Also post irradiation aging effects could be the result of trapped radicals in crystalline regions of polymers (Buchalla et al., 1993a, 1993b). It must be noted that the radiation-induced changes depend on the polymer composition (additives), processing history of the plastic and the irradiation conditions (presence of oxygen, temperature, dose and dose rate). For example, plastic films containing a phenyl group or an amide linkage (are known to stabilize the polymer) are the most radiation resistant, due to their increased resonance energy (Variyar et al., 2000). Therefore, radiation resistant of polystyrene, polyester and polyamide is due to the high energy requirement to form cross-links (Crook & Boylston, 2004).

21 CFR Citation	Packaging Materials	Max Dose [kGy]
	Nitrocellulose-coated cellophane	10
	Glassine paper	10
	Wax-coated paperboard	10
	Polyolefin film	10
Section 179.45(b)	Kraft paper	0.5
	Polyethylene terephthalate film (basic polymer)	10
	Polystyrene film	10
	Rubber hydrochloride film	10
	Vinylidene chloride-vinyl chloride copolymer film	10
	Nylon 11 [polyamide-11]	10
Section 179.45(c)	Ethylene-vinyl acetate copolymer	30
	Vegetable parchment	60
	Polyethylene film (basic polymer)	60
Section 179.45(d)	Polyethylene terephthalate film	60
	Nylon 6 [polyamide-6]	60
	Vinyl chloride-vinyl acetate copolymer film	60

Table 4. Packaging materials listed in 21 CFR 179.45 for use during irradiation of prepackaged foods (FDA, 2006a).

Any new packaging material not yet listed in 21 CFR 179.45 is subject to a pre-market safety evaluation by FDA prior to irradiation. FDA (2006b) suggests that information be generated in accordance with the available guidance documents (http://www.cfsan.fda.gov/~dms/guidance.html). Nowadays, most of the food packaging materials are normally resistant to doses used in food irradiation practice (EFSA, 2011).

7. Irradiation facilities

Ionizing radiation for food processing is limited to high energy photons (gamma rays) of radio nuclides ^{60}Co or ^{137}Cs, X-rays from machine sources with energies up to 5 MeV and accelerated electrons with energies up to 10 MeV generated by electron accelerating machines (Hvizdzak et al., 2010). These kinds of Ionizing radiation are preferred due to: i) the suitable food preservative effects; ii) do not generate radioactivity in foods or packaging materials; iii) available at costs as commercial use of the irradiation process (Farkas, 2004). The strength of the source and the length of time the food is exposed to the ionizing energy determine the irradiation dose, measured in grays (Gy) or kilo grays (1kGy = 1,000 Gy). One gray is equal with one joule of energy absorbed in a mass of one kilogram (EFSA, 2011).

7.1 Gamma irradiation facilities

Gamma rays with energies of 1.17 and 1.33 MeV are emitted by the cobalt-60 or energy of 0.66 MeV is emitted by caesium-137. The Co-60 is a radioactive metal that decays with a half-life of around 5.3 years. Although Cs-137 has a longer half-life of around 30.1 years, few commercial gamma facilities use Cs-137 as a gamma ray source. Because, Cs-137 emits gamma rays that are approximately half the energy of those emitted by Co-60 (Suresh et al., 2005). Gamma irradiation has higher penetration than Electron beams. Therefore, it is as suitable for treating large bulk packages of food. Gamma facilities are the majority of food irradiation facilities worldwide (Fig. 1). For instance, 27 of the 33 food irradiation facilities (23 in the EU) approved in 2010 are Co-60 gamma facilities and six are E-beam (EFSA, 2011). The gamma radiation can not be switched off and when not being used to treat food, must be stored in a water pool to absorb the radiation energy and protect workers from exposure if they must enter the irradiation room (Hvizdzak et al., 2010).

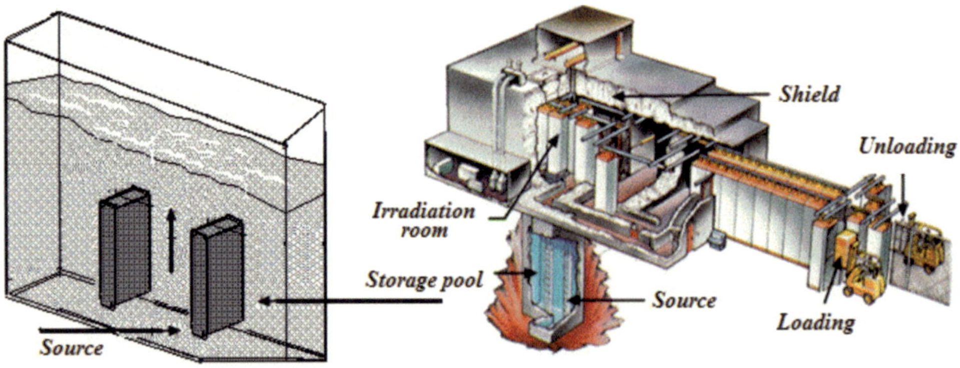

Fig. 1. Schematic diagram of a commercial gamma irradiator. The sealed sources are stored in water and raised into the air to irradiate a product that may be moved into the irradiation room on a conveyor system.

It should be noted that, all radioactive irradiators must be licensed and meet all applicable safety requirements under the Atomic Energy Act of 1954. A license must be obtained from the Nuclear Regulation Commission (NRC) or an Agreement State prior to beginning construction of a new irradiator (NRC, 2009).

7.2 E-beam facilities

E-beams are produced by machines and not by radioactive material. They are generated by accelerating a stream of electrons, which are focused into a narrow beam-spot (Fig. 2b). As food is travels perpendicular to the beam direction, this spot of incident electrons is scanned across food (Suresh et al., 2005). E-beam radiation offers three distinct advantages compared with gamma: first, the need to carry radioactive materials around the country is eliminated. Second, it can be turned off when not in use and the last; E-beam characterized by its low penetration and high dosage rates. It performs best when used on low-density, uniformly packaged products. Therefore, it can effectively inactivate foodborne pathogens on the surface of the slices, with the least negative effect (Jeong-Ok et al., 2008; Hvizdzak et al., 2010).

7.3 X-ray facilities

X-rays are generated by machines and can be switched off. In this case, electrons are accelerated at a metallic target (*e.g.*, tungsten or gold) to generate a stream of X-rays (Fig. 2a). In this process much of the E-beam energy is lost as heat; however the X-ray efficiency can be increased with atomic number of the target material and also with increasing E-beam energy (Kume et al., 2009). Nevertheless, X-ray facilities can process large bulk packages without the need for radioactive material. Nowadays, very few food products are irradiated by X-rays (Follett, 2004). As the technology advances, X-ray irradiation will become more extensive in future. But, it seems that, the commercial food irradiation facilities will continue to be gamma cells for a long period (Kume et al., 2009).

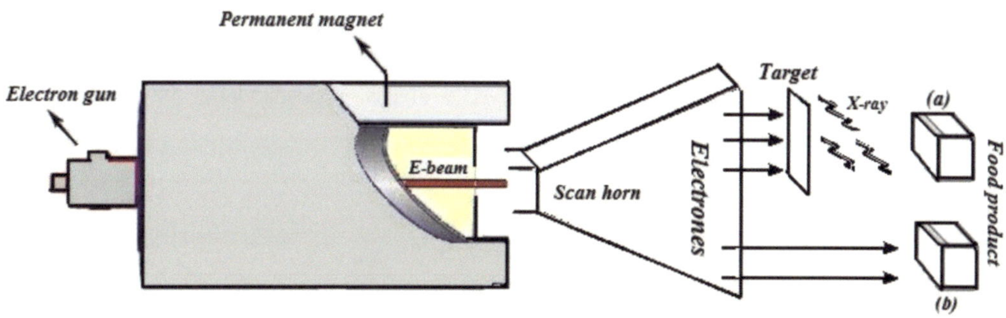

Fig. 2. Schematic diagram of a typical x-ray (a) and E-beam (b) irradiators

8. General aspects of dosimetry

Radiation dosimetry is the measurement of the absorbed dose in matter and products resulting from the exposure to radiation (Codex Alimentarius Commission [CAC], 2003). The estimated range of absorbed dose used in food processing is from 0.01 to 100 kGy. The irradiation dose absorbed by a product is not uniform due to the limits in penetration capacity of ionising radiations. This can reduce the overall efficacy of the food irradiation in practice (EFSA, 2011). It is necessary to determine the ability of an irradiation facility to deliver the absorbed dose, prior to the irradiation of the food product. Also, it is essential to monitor and document the absorbed dose during each production run (Moreno et al., 2008).

Dosimetry provides important functions in radiation processing, where large absorbed doses and dose rates have to be measured with practical accuracy. When dosimeters are exposed to radiation, they undergo a physical or chemical change in properties that can be recorded. This change is related to dose and dose rate, and it can be calibrated to a standard dose. For example, the basis of a common dosimetry system is plastic film containing a dye that changes color in proportion to dose. The color change can be calculated using a spectrophotometer and the degree color change can be calibrated to a dose standard. Verified dosimetry systems are extensively used to perform radiation measurements, quality control and validation of processes (Moreno et al., 2008).

The success of food irradiation depends on:

1. Calculations of the absorbed dose delivered to the food product;

In radiation processing, the absorbed dose is a key quantity in the processing of food. Therefore, dosimetry is a fundamental affair in radiation processing. Since the radiation absorbed dose is related to the desired effect in a particular food, the need for appropriate and precise dose measurement techniques must not be underestimated (IAEA, 2002). For instance, the dose received by every part of the product must be between Dmax (limit) and Dmin (Fig. 3).

2. Determination of the dose distribution patterns in the product package (process qualification);

Process load geometries are commonly limited to conventional shapes and sizes (e.g., rectangular cartons containing cylindrical, spherical or rectangular unit packages) of commercial food packaging. The importance of dosimetry in process qualification is to ensure that the absorbed dose requirements for a specific product can be satisfied. This process (dose mapping) verifies the scale and regions of the maximum and minimum doses (Fig. 3), and helps set up all the parameters required to achieve the absorbed dose (Mehta, 1992).

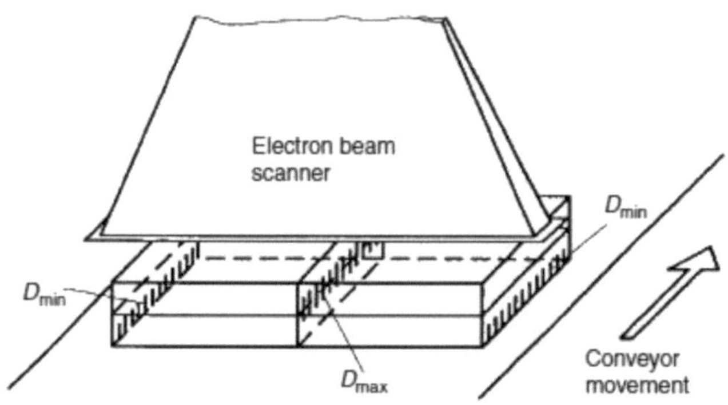

Fig. 3. Regions of Dmin and Dmax (indicated by hatching) for a rectangular process load for tow pass, tow sided electron beam irradiation (IAEA, 2002).

3. Control of the routine radiation process (control procedures);

In order to ensure that the irradiation process is under control and the correct irradiation dose is being received by a specific product, routine dosimeters are used in radiation processing facilities. The routine dosimeters, traceable to national or international standards, are important and necessary in process control of food irradiation (Miller et al., 1989). Theses should be made during operation and records kept of such measurements can be used as supporting evidence (IAEA, 2002).

9. Acceptance and trade

The use of the treatment as a commercial food process depends on its acceptance by consumers. Frequently, consumers are conservative and they are reluctant to accept products processed by new technologies like as food irradiation method. This is often related to the fear and confusion about radiation itself and the lack of understanding of the process. The main worries of consumer organizations included safety, nutrition, detection, and labeling of irradiated products (P. Junqueira-Gonc-alves et al., 2011). The concern about the food irradiation appears to centre on the safety of the process. Giving science-based information on food irradiation leads to positive consumer approaches (Fox, 2002). Many consumers are primarily hostile to irradiation. By other means, "People think the irradiated product is radioactive," but when the process is made clear to them they will become more in favor (Landgraf et al. 2006; Marcotte, 2005). Fox (2002) reported that, consumer awareness of food irradiation was 29%. Also, 80% of consumers were unsure about the safety of irradiated foods. Only 11% of the interviewers expressed that irradiated foods are safe. It should be noted that upon hearing a benefit statement of food irradiation, level of positive attitude increased significantly (62%) towards irradiated foods. On the other hand, unfavorable description of irradiation has a great negative effect on consumer acceptance of irradiated foods. Recent study showed a low level of awareness among consumers about the food irradiation processing. 76.5% of the interviewed people did not know that irradiation could be used as a method for food preservation. 46% of them expressed that irradiated food means the same as radioactive food. Nevertheless, 91% stated that if they knew that "irradiated" is not "radioactive" and that proper irradiation enhances food safety they would become consumers of irradiated food. 95.8% of the interviewers were not familiar with the "Radura" symbol (Fig. 4). However, 55.8% of people expressed that they would buy irradiated food because of the Radura symbol (Junqueira-Gonc-alves et al., 2011). It shows that the labeling of irradiated products (the Radura symbol coupled with either "treated with ionizing radiation" or "irradiated") is the key issue with the consumers. Similar result was observed by Stephen et al. (2009). Because the words "radiation" and "irradiation" may have negative connotations, the labeling requirement has been viewed as an obstacle to consumer acceptance. Some researchers believe that an alternative wording, e.g. "electronically pasteurized," would be helpful (Morehouse & Komolprasert, 2004). These results confirm the importance of training the public on the controversy, technology, and the benefits of irradiation.

Several countries have regulatory approvals in place for irradiation of one or more food products. But all these countries are not practically using the technology for a restricted number of food products (Johnson & Reynolds, 2004; Marcotte, 2005). China, USA and Ukraine make up about three quarters of the whole food irradiated in the world. China is the most important country in the use of food irradiators. More than 200 facilities are reported from China and has about 100 irradiators with designed capacity of >300 kCi and more than 40 irradiators with designed capacity >1000 kCi. During 2006-2010 China established 20 new irradiators (EFSA, 2011). The number of radiation processing facilities and commercial food irradiation is showing an increasing trend around the world.

Fig. 4. The ''Radura''symbol (in general it is presented in green color). The plant-like structure represents agricultural products in a closed package (the circle) broken in the upper half by penetrating ionizing rays or particles.

10. Conclusions

Food irradiation is approved for use in over 60 countries worldwide for different products, such as grains, herbs and spices, poultry, seafood, and ground beef (Kume et al. 2009). Irradiation does not cause any significant loss of macronutrients. Proteins, fats and carbohydrates undergo little modify in nutritional value through irradiation even with doses over 10 kGy, though there may be sensory changes. In the same way, the essential amino acids, essential fatty acids, minerals and trace elements are also unchanged. There can be a decrease in certain vitamins (mostly thiamin) but these are of the same order of magnitude as occurs in other manufacturing processes such as drying or canning (thermal sterilization) (Ahari et al., 2010). In the world today, 500000 tons of foods are irradiated every year among spices, meat, fresh fruits and vegetables. Despite the clear benefits of the food irradiation, this method remains under estimated in the food trade. It has not been commonly accepted and approved yet (Guy, 2011; Kume et al., 2009). The main factor in the way of commercial application of the food irradiation process is consumer acceptance. From our point of view, as consumer safety questions are discussed, food preservation throughout radiation contribution to food safety will reach the same recognition as the sterilization of medical products has in terms of preventing the spread infectious disease. For this reason, scientists have the responsibility to help the consumer understand the radiation process and it's potential to improve our lives and protect our health.

11. References

Ahari, M. H.; Fathollahi, H.; Motamedi, F. & Mirmajlessi, S. M. (2010). *Food irradiation: Applications, public acceptance and global trade*: a review. African J. Biotech., Vol. 9, no. 20, pp. 2826-2833.

Ahari, M. H.; Mirmajlessi, S. M.; Fathollahi, H.; Minasyan, V. & Mirjalil S. M. (2011). *Evaluation of gamma irradiation effect and Pseudomonas flourescens against Penicillium expansum*, African J. Biotech., Vol. 10, no. 54, pp. 11290-11293.

Ahmed, M. (2001). *Disinfestation of stored grain, pulses, dried fruits and nuts, and other dried foods*, In: Molins, R. (ed.) *Food Irradiation Principles and Applications*, Wiley, New York, pp. 77-112.

Ahn, H. J.; Jo, C.; Lee, J. W.; Kim, J. H.; Kim, K. I. & Byun, M. W. (2003). *Irradiation and modified atmosphere packaging effects on residual nitrite, ascorbic acid, nitro-somyoglobin, and color in sausage*, J. Agri. & Food Chem., Vol. 51, pp. 1249-1253.

Alam Khan, K. & Abrahem, M. (2010). *Effect of irradiation on quality of spices.* International Food Res. J., Vol. 17, pp. 825-836.

Arvanitoyannis, I. S.; Stratakos, A. C. & Tsarouhas P. (2009). *Irradiation applications in vegetables and fruits*: a review. Critical Reviews in Food Science and Nutrition, Vol. 49, no. 5, pp. 427- 462.

Azelmat, K.; Sayah, F.; Mouhib, M.; Ghailani, N. & Elgarrouj, D. (2005). *Effects of gamma irradiation on forth-instar Plodia interpunctella* (Hubner) (Lepidoptera: Pyralidae). J. Stored Prod. Res., Vol. 41, pp. 423-431.

Black, J.,L. & Jaczynski, J. (2008). *Effect of water activity on the inactivation kinetics of Escherichia coli O157: H7 by electron beam in ground beef, chicken breast meat, and trout fillets.* Inter. J. Food Sc. & Tech., Vol. 43, no. 4, pp. 579-586.

Boshra, S. A. & Mikhaiel, A. A. (2006). *Effect of gamma radiation on pupal stage of Ephestia calidella* (Guenee). J. Stored Prod. Res., Vol. 42, pp. 457-467.

Brewer, M. S. (2009). *Irradiation effects on meat flavor*: a review, Meat Science, 1-14.

Buchalla, R.; Schuttler, C. & Werner, B. (1993a). *Effects of ionizing radiation on plastic food packaging materials*, A review. Chemical and physical changes. J. Food Prot., Vol. 56, pp. 991.

Buchalla, R.; Schuttler, C. & Werner, B. (1993b). *Effects of ionizing radiation on plastic food packaging materials*, A review. Global migration, sensory changes and the fate of additives. J. Food Prot., Vol. 56, pp. 998.

Buchalla, R.; Boess, C. & Bogl, K. W. (1997). *Radiolysis products in gamma –irradiated plastics by thermal desorption GC-MS*, Part 1 Bundesinstitut fur Gesundheitlichen Verbraucherschutz und Veterinarmedizin, Berlin (BgVV-Hefte 04/1997).

Codex Alimentarius Commission (CAC), (2003). Report of the thirty fifth session of the codex committee on foot hygiene. Orlando, Florida, US., available at http://www.codexalimentarius.net/download/report/117/Al0313ae.pdf

Caillet, S.; Millette, M.; Salmieri, S. & Lacroix, M. (2006). *Combined effects of antimicrobial coating, modified atmosphere packaging, and gamma irradiation on Listeria innocua present in ready-touse carrots (Daucus carota)*, J. Food Prot., Vol. 69, pp. 80-85.

Calucci, L.; Pinzino, C.; Zandomeneghi, M.; Capocchi, A.; Ghiringhelli, S.; Saviozzi, F.; Tozzi, S. & Galleschi, L. (2003). *Effects of gamma-irradiation on the free radical and antioxidant contents in nine aromatic herbs and spices*, J. Agri. & Food Chem., Vol. 51, pp. 927-934.

Chiasson, F.; Borsa, J. & Lacroix, M. (2005). *Combined effect of carvacrol and packaging conditions on radiosensitivity of Escherichia coli and Salmonella typhi in ground beef*, J. Food Prot., Vol. 68, pp. 2567-2570.

Chytiri, S.; Goulash, A. E.; Badeka, A.; Riganakos, K. A. & Kontominas, M. G. (2005). *Volatile and non-volatile radiolysis products in irradiated multilayer coextruded food –packaging films containing a buried layer of recycled low-density polyethylene*, Food additives & contaminants, Vol. 22, pp. 1264-1273.

Crook, L. R. & Boylston, T. D. (2004). *Flavor characteristics of irradiated apple cider during storage: Effect of packaging materials and sorbate addition*, J. Food Sci., Vol. 69, pp. 557—563.

Da Trindade, R. A.; Mancini, J & Villavicencio, A. (2009). *Effects of natural antioxidants on the lipid profile of electron beam-irradiated beef burgers*, European J. Lipid Sci. and Tech., Vol. 111, pp. 1161-1168.

Deeley, C. (2002). *Food irradiation: setting new standards or a slippery slope?* Food Sci. Technol., Vol. 16, pp. 52-55.

De Wit, M. A.; Koopmans, M. P. & Van Duynhoven, Y. T. (2003). *Risk factors for novo virus, Sapporo-like virus and Group A rotavirus gastroenteritis*, Emerging Infect. Dis., Vol. 9, no. 12, pp. 163-170.

Diehl, J. F. (1995). *Safety of irradiated foods*, Marcel Dekker, Inc., New York, pp 1.

Dionísio, A. P.; Gomes, R. T. & Oetterer, M. (2009). *Ionizing radiation effects on food vitamins* – A Review. Brazilian Archives Biology Tech., Vol. 52, n. 5, pp. 1267-1278.

Dogan, A.; Siyakus, G. & Severcan, F. (2007). *FTIR spectroscopic characterization of irradiated hazelnut (Corylus avellana L.)*, Food Chem., Vol. 100, pp. 1106-1114.

Dong-Ho, K.; Hyun-Pa, S.; Hong-Sun, Y.; Young-Jin, C.; Yeung-Ji, K. & Myung-Woo, B. (2002). *Distribution of microflora in powdered raw grains and vegetables and improvement of hygienic quality by gamma irradiation*, J. Korean Society Food Sci. & Nutrition, Vol. 31, no. 4, pp. 589-593, Vol. 31, pp. 589-593.

Duliu, O. G.; Ferdes, M. & Ferdes, O. S. (2004). *EPR study of some irradiated food enzymes*, J. Radioanalytical & Nuclear Chem., Vol. 260, pp. 273-277.

EFSA (European Food Safety Authority), (2011). *Statement summarizing the conclusions and recommendations from the opinions on the safety of irradiation of food adopted by the BIOHAZ and CEF panels*, The EFSA J., Vol. 9, no. 4, pp. 2107.

Fan, X. (2003). *Ionizing radiation induces formation of malondialdehyde, formaldehyde, and acetaldehyde from carbohydrates and organic acid*, J. Agri. & Food Chem., Vol. 51, pp. 5946–5949.

Fan, X. T. (2005). *Formation of furan from carbohydrates and ascorbic acid following exposure to ionizing radiation and thermal processing*, J. Agri. and & Food Chem., Vol. 53, pp. 7826-7831.

Fan, X.; Annous, B. A.; Sokorai, K. J.; Burke, A. & Mattheis, J. P. (2006). *Combination of hot-water surface pasteurization of whole fruit and low-dose gamma irradiation of fresh-cut cantaloupe*, J. Food Prot., Vol. 69, pp. 912-919.

Fan, X. T. & Sommers, C. H. (2006). *Effect of gamma radiation on furan formation in ready-to-eat products and their ingredients*, J. Food Sci., Vol. 71, pp. C407-C412.

Fan, X. & Thayer, D. W. (2002). *Formation of malonaldehyde, formaldehyde, and acetaldehyde in apple juice induced by ionizing radiation*, J. Food Sci., Vol. 67, pp. 2523 2528.

Farkas, J. (2004). *Charged particle and photon interactions with matter*, In: Mozumder, A. & Hatano, Y. (eds): Food Irradiation, Marcel Dekker, New York, pp. 785–812.

Farkas, J. (2006). *Irradiation for better foods*, Trends in Food Sci. & Tech., Vol. 17, pp. 148–152.

FDA (Food and Drug Administration), (2006a). http://www.cfsan.fda.gov/~dms/opa-torx.html

FDA (Food and Drug Administration), (2006b). http://www.cfsan.fda.gov/~dms/opa2pmnc.html#iid5.

Follett, P. A. (2004). *Irradiation to control insects in fruits and vegetables for export from Hawaii*, Radiation Phys. & Chem., Vol. 71, pp. 161-164.

Fox, J. A. (2002). *Influences on purchase of irradiated foods.* Food Tech., Vol. 56, no. 11, pp. 34–37.

Frankhauser, R. L.; Monroe, S. S. & Noel, J. S. (2002). *Epidemiologic and molecular trends of Norwalk-like viruses associated with outbreaks of gastroenteritis in the United State,* J. Infect. Dis., Vol. 86, no. 1, pp. 1-7.

Galan, I.; Garcia, M. L. & Selgas, M. D. (2010). *Effects of irradiation on hamburgers enriched with folic acid,* Meat Sci., Vol. 84, pp. 437-443.

Goetz, J. & Weisser, H. (2002). *Permeation of aroma compounds through plastic films under high pressure: in-situ measuring method,* Innovative Food Sci. & Emerg-ing Tech., Vol. 3, pp. 25-31.

Goulas, A. E.; Riganakos, K. A.; Badeka, A. & Kontominas, M. G. (2002). *Effect of ionizing radiation on the physic chemical and mechanical properties of commercial monolayer flexible plastics packaging materials,* Food Additives Contaminants, Vol. 19, pp. 1190-1199.

Goulash, A. E.; Riganakos, K. A. & Konotminas, M. G. (2004). *Effect of ionizing radiation on physicochemical and mechanical properties of commercial monolayer and multilayer semirigid plastic packaging materials,* Radiation phys. & chem., Vol. 69, pp. 411-417.

Guinebretiere, M.H.; Girardin, H.; Dargaignaratz, C.; Carlin, F. & Nguyen-The, C. (2003). *Contamination flows of Bacillus cereus and spore-forming aerobic bacteria in a cooked, pasteurized and chilled zucchini puree processing line,* International J. Food Microb., Vol. 82, no. 3, pp. 223-232.

Guinebretiere, M. H. & Nguyen-The, C. (2003). *Sources of Bacillus cereus contamination in a pasteurized zucchini puree processing line, differentiated by two PCR-based methods,* Fems Microb. Ecology, Vol. 43, no. 2, pp. 207-215.

Guy, J. H. (2011). *Phytosanitary Applications of Irradiation,* Comprehensive Reviews in Food Sci. & Food Safety. Vol. 10, no. 2, pp. 143–151.

Hallman, G. J. (2001). *Irradiation as a quarantine treatment,* pp. 113-130. In: R, Molins (ed). Food irradiation principles & applications, Wiley-Int. sci., pp. 113-130.

Hallman, G. J. (2011). *Phytosanitary Applications of Irradiation,* Comprehensive Reviews in Food Science and Food Safety, vol. 10, no. 2, pp. 143-151.

Haack, R. A. (2006). *Exotic bark- and wood-boring beetles in the United States: recent establishments and interceptions,* Can. J. Res., Vol. 36, no. 2, pp. 269–88.

Hammad, A. A.; Abo-elnour, S. A. & Salah, A. (2006). *Use of irradiation to ensure hygienic quality of minimally processed vegetables and fruits,* IAEA-TECDOC-1530. pp. 106-129.

Han, J.; Gomes-Feitosa, C. L.; Castell-Perez, E.; Moreira, R. G. & Silva, P. F (2004). *Quality of packaged romaine lettuce hearts exposed to low-dose electron beam irradiation,* Lehensmittel Wissenschaft und Technologie, Vol. 37, pp. 705-715.

Hatha, A. A. M.; Maqbool, T. K. & Kumar, S. S. (2003). *Microbial quality of shrimp products of export trade produced from aquacultured shrimp,* International J. Food Microb., Vol. 82, no. 3, pp. 213-221.

Heather, N. W. & Hallman, G. J. (2008). *Pest management and phytosanitary trade barriers.* Wallingford, UK: CABI., p. 257.

Hendrichs, J. M.; Vreysen, J. B.; Enkerlin, W. R. & Cayol. J. P. (2005). *Strategic options in using sterile insects for area-wide integrated pest management,* pp. 563-600 *In* Dyck, V. A.; Hendrichs, J. & Robinson, A. S. [eds.], Sterile Insect Technique. Principles and

Practice in Area-Wide Integrated Pest Management. Springer, Dordrecht, The Netherlands, 787 pp.

Hosseinzadeh, A. & Shayesteh, N. (2011). *Application of gamma radiation for controlling the red flour beetle Tribolium castaneum* Herbst (Coleoptera: Tenebrionidae). African J. Agri. Res., Vol. 6, no. 16, pp. 3877-3882.

Hvizdzak, A. L.; Beamer, S.; Jaczynski, J. & Matak, K. E. (2010). *Use of Electron Beam Radiation for the Reduction of Salmonella enterica Serovars Typhimurium and Tennessee in Peanut Butter*. J. Food Protec., Vol. 73, no. 2, pp. 353-357.

IAEA (International Atomic Energy Agency), (2001). *Consumer acceptance and market development of irradiated food in Asia and the Pacific*, Vienna, Austria, p. 98.

IAEA (International Atomic Energy Agency), (2002). *Dosimetry for food irradiation*, Vienna, Austria, p. 168.

IAEA (International Atomic Energy Agency), (2009). *Irradiation to ensure the safety and quality of prepared meals*, Vienna, Austria, p. 375.

International Plant Protection Convention (IPPC), (2009). *Regulation of wood packaging material in international trade*. Rome: Food and Agriculture Organization (FAO).

Jeong-Ok, L.; Seong, A. L.; Mi-Seon, K.; Hye-Rim, H.; Kyoung-Hee, K.; Jong-Pil, C. & Hong-Sun, Y. (2008). *The effects of low-dose electron beam irradiation on quality characteristics of stored apricots*. J. Korean Society Food Sci. & Nutrition, Vol. 37, pp. 934-941.

Johnsson, L. & Dutta, P. C. (2006). *Determination of phytosterol oxides in some food products by using an optimized transesterification method*, Food Chem., Vol. 97, pp. 606-613.

Johnson, M. & Estes-Reynolds, A. (2004). *Consumer Acceptance of Electron-Beam Irradiated Ready-to-Eat Poultry Meats*, J. Food Process Preserv., Vol. 28, pp. 302-319.

Jung, H. H. (2007). *Packaging for nonthermal processing of food*. Chapter 5: Packaging for high-pressure processing, irradiation, and pulsed electric field processing. Blackwell Pub. : IFT Press, pp. 67-86.

Junqueira-Gonc-alves, M. P.; Maria, J. G.; Ximena, V.; Carolina, M. D.; Paulina, A. & Joseph, Miltz. (2011). *Perception and view of consumers on food irradiation and the radura symbol*, Radiation Physics & Chem., Vol. 80, pp. 119–122.

Ju-Woon, L.; Jae-Hun, K.; Jang-Ho, K.; Sang-Hee, O.; Ji-Hyun, S.; Cheon-Jei, K.; Sung-Hee, C. & Myung-Woo, B. (2005). *Application of gamma irradiation for the microbiological safety of fried-frozen cheese ball*, J. Korean Society of Food Sci. & Nutrition, Vol. 34, pp. 729-733.

Kale, D. D. (1994). *Effect of irradiation on polymeric packaging films*, Report submitted to department of atomic energy, under board for research in nuclear sciences sponsored project, 1992-1994.

Kilcast, D. (1994). *Effect of irradiation on vitamins*, Food Chemi., Vol. 49, pp. 157-164. Kim, M. J.; Lee, J. W.; Seo, J. H.; Song, H. P.; Yook, H. S.; Choi, J. M. & Byun, M. W, (2003). *Safety evaluation on mutagenicity of white layer cake containing gamma irradiated egg white*, J. Korean Society Food Sci. & Nutrition, Vol. 32, pp. 1172-1173.

Klassen, W. & Curtis. C. F. (2005). *History of the sterile insect technique*, pp. 3-36, *In* Dyck, V. A.; Hendrichs, J. & Robinson, A. S. The Sterile Insect Technique: Principles and Practice in Area-Wide Integrated Pest Management. Springer, Dordrecht, The Netherlands, p. 787.

Koopmans, M. & Duizer, E. (2004). *Food borne viruses: an emerging problem*, J. Food Microb., Vol. 90, pp. 23-24.

Kottapalli, B.; Wolf-Hall, C. E. & Schwarz, P. (2006). *Effect of electron-beam irradiation on the safety and quality of Fusarium-infected malting barley*, Int. J. Food Microb., Vol. 110, no. 3, pp. 224-231.

Kume, T.; Furuta, M.; Todoriki, S.; Uenoyama, N. & Kobayashi, Y. (2009). *Status of food irradiation in the world*, Radiation Physics and Chemistry, Vol. 78, no. 3, pp. 222-226.

Landgraf, M.; Gaularte, L.; Martins, C.; Cestari, A.; Nunes, T.; Aragon, L.; Destro, M.; Behrens, J.; Vizeu, D. & Hutzler, B. (2006). *Use of irradiation to improve the microbiological safety of minimally processed fruits and vegetables*, IAEA-TECDOC-1530, pp. 41-59.

Lee, J. W.; Park, K. S.; Kim, J. G.; Oh, S. H.; Lee, Y. S.; Kim, J. H. & Byun, M. W. (2005). *Combined effects of gamma irradiation and rosemary extract on the shelf-life of a ready-to-eat hamburger teak*. Radiation Physics & Chem. Vol. 72, no. 1, pp. 49-56.

Levanduski, L. & Jaczynski, J. (2008). *Increased resistance of Escherichia coli O157: H7 to electron beam following repetitive irradiation at sub-lethal doses*, International J. Food Microb., Vol. 121, no. 3, pp. 328-334.

Lim, S.; Jung, J. & Kim, D. (2007). *The effect of gamma radiation on the expression of the virulence genes of Salmonella Typhimurium and Vibrio spp*, Radiation Physics & Chem., Vol. 76, no. 11, pp. 1763-1766.

Loaharanu, P. (1994). *Food irradiation in developing countries: a practical alternative*, IAEA Bull., Vol. 1, pp. 30-35.

Lynch, M. F.; Tauxe, R. V. & Hedberg, C. W. (2009). *The growing burden of food borne outbreaks due to contaminated fresh produce: risks and opportunities*, Epidemiology & Infection, Vol. 137, no. 3, pp. 307-315.

Marcotte, M. (2005). *Effect of irradiation on spices, herbs and seasonings-comparison with ethylene oxide fumigation*, www.foodirradiation.com.

Mehta, K. (1992). *Process qualification for electron-beam sterilization*, Med. Device & Diagnostic Ind., Vol. 14, no. 6, pp. 122-134.

Miller, R. D. (2005). *Electronic irradiation of foods: an Introduction to the Technology*, Springer. NY., p.295.

Miller, A. & Chadwick, K. H. (1989). *Dosimetry for the approval of food irradiation processes*, Radiat. Phys. & Chem., Vol. 34, pp. 999–1004.

Mitchell, G. E.; McLauchlan, R. L.; Isaacs, A. R.; Williams, D. J. & Nottingham, S. M. (1992). *Effect of low dose irradiation on composition of tropical fruits and vegetables*. J. Food Comp. Anal., Vol. 5, pp. 291-311.

Morehouse, K. M. & Komolprasert, V. (2004). *Irradiation of Food and Packaging*, An Overview, With Permission from ACS: ACS Symposium Series 875, *Irradiation of Food and Packaging*, chapter 1, pp. 1-11.

Moreno M. A.; Castell-Perez, M. E.; Gomes, C.; Da Silva P. F.; Kim, J.; & Moreira, R. G. (2008). *Treatment of cultivated high bush blueberries (Vaccinium corymbosum L.) with electron beam irradiation: Dosimetry and product quality*, J. Food Process Engineering, Vol. 31, pp. 155-172.

Niemira, B. A.; Sommers, C. H. & Fan, X. (2002). *Suspending lettuce type influences recoverability and radiation sensitivity of Escherichia coli O157:H7*, J. Food Prot., Vol. 65, pp. 1388-1393.

Niemira, B. A.; Sommers, C. H. & Boyd, G. (2003). Effect of freezing, irradiation, and frozen storage on survival of *Salmonella* in concentrated orange juice. J. Food Prot., Vol. 66, pp. 1916-1919.

Niemira, B. A. & Lonczynski, K, A. (2006). *Nalidixic acid resistance influences sensitivity to ionizing radiation among Salmonella isolates*, J. Food Prot., Vol. 69, pp. 1587-1593.

Norhana, M. N. W.; Poole, S. E.; Deeth, H. C. & Dykes, G. A. (2010). *Prevalence, persistence and control of Salmonella and Listeria in shrimp and shrimp products*: a review. Food Control, Vol. 21, no. 4, pp. 343-361.

NRC (Nuclear Regulation Commission), (2009). *Fact sheet on commercial irradiators*, Http//.www.nrc,gos/

Nunes, T, P.; Martins, C. G.; Behrens, J. H.; Souza, K. L. O.; Genovese, M. I.; Destro, M. T. & Landgraf, M. (2008). *Radioresistance of Salmonella species and Listeria monocytogenes on minimally processed arugula (Eruca sativa Mill.): Effect of irradiation on flavonoid content and acceptability of irradiated produce*, J. Agri. & Food Chem., Vol. 56, no. 4, pp. 1264–1268.

O'Bryan, C. A.; Crandall, P. G.; Ricke, S. C. & Olson, D. G. (2008). *Impact of irradiation on the safety and quality of poultry and meat products*: A review, Critical Reviews in Food Sci. & Nutrition, Vol. 48, pp. 442-457.

Ozen, B. F. & Floros, J. D. (2001). *Effects of emerging food processing techniques on the packaging materials*. Trends in Food Sci. & Technol., Vol. 12, no. 2, pp. 60-67.

Pentimalli, M. D.; Capitani, A.; Ferrando, D.; Fern, P.; Ragni, A. & Segre, L. (2000). *Gamma irradiation of food packaging materials: an NMR study*, Polymer., Vol. 41, pp. 2871-2881.

Patterson, M.; Connoly, m. & Darby, D. (2006). *Effect of gamma irradiation on the microbiological quality of seeds and seeds sprouts. In use of irradiation to ensure the hygienic of fresh, pre-cut fruits vegetables and other minimally processed food of plant origin,*Vienna: International Atomic Energy, p. 536.

Pinu, F. R. Yeasmin, S.; Bar, M. L. & Rahman, M. M. (2007). *Microbiological conditions of frozen shrimp in different food market of Dhaka city*, Food science and technology research, Vol. 13, no. 4, pp. 362-365.

Prakash, A.; Chen, P. C.; Pilling, R. L.; Johnson, N. & Foley, D. (2007). 1% calcium chloride treatment in combination with gamma irradiation improves microbial and physicochemical properties of diced tomatoes. Foodborne Pathog. Dis., Vol. 4, pp. 89-98.

Prendergast, D.M.; Crowley, K. M.; McDowell, D. A. & Sheridan, J. J. (2009). *Survival of Escherichia coli O157:H7 and non-pathogenic E. coli on irradiated and non-irradiated beef surfaces*, Meat Sci., Vol. 83, no. 3, pp. 468-473.

Rajkowski, K. T. & Fan, X. T. (2008). *Microbial quality of fresh-cut iceberg lettuce washed in warm or cold water and irradiated in a modified atmosphere package*, J. Food Safety, Vol. 28, pp. 248-260.

Romero, M. G.; Mendonca, A. F.; Ahn, D. U. & Wesley, I. V. (2005). *Influence of dietary vitamin E on behavior of Listeria monocytogenes and color stability in ground turkey meat following electron beam irradiation*, J. Food Prot., Vol. 68, pp. 1159-1164.

Sádecká, J.; Kolek, E.; Petka, J. & Kovác, M. (2005). *Impact of gamma-irradiation on microbial decontamination and organoleptic quality of oregano* (Origanum vulgare L.). In: Proceedings of Euro Food Chem. XIII, Hamburg: pp. 590–594.

SCF (Scientific Committee on Food), (2003). *Revision of the opinion of the scientific committee on food on the irradiation of food*, European commission health and consumer protection directorate general, SCF/CS/NF/IRR/24.

Scott Smith, J.; & Suresh, P. (2004). *Irradiation and food safety*, Food tech., irradiation & food safety. Vol. 58, no. 11, pp. 48-55.

Sommers, C. H. & Boyd, G. (2006). *Variations in the radiation sensitivity of foodborne pathogens associated with complex ready-to-eat food products*. Radiation Physics & Chem., Vol. 75, no. 7, pp. 773- 778.

Song, H. P.; Kim, B.; Choe, J. H.; Jung, S.; Kim, K. S.; Kim, D. H. & Jo, C. (2009). *Improvement of foaming ability of egg white product by irradiation and its application*, Radiation physics & Chem., Vol. 78, no. 3, pp. 217-221.

Stefanova, R. Vasilev, N. V. & Spassov, S. L. (2010). *Irradiation of food, current legislation framework, and detection of irradiated foods*, Food Analytical Methods, Vol. 3, pp. 225-252.

Stephen, G. S. & Downing-Matibag, T. (2009). *Consumer acceptance of food irradiation: a test of the recreancy theorem*. International J. Consumer Studies, Vol. 33, pp. 1470-6423.

Stewart, E. M. (2009). *Effect of gamma irradiation on the quality of ready meals and their meat components. In: Irradiation to Ensure the Safety and Quality of Prepared Meals*, Results of the coordinated research project organized by the joint FAO/IAEA division of nuclear techniques in food and agriculture (2002-2006). IAEA, Vienna, pp. 313-342.

Stoffers, N.; Linssen, H.; Josef, P. H.; Franz, R. & Welle, F. (2004). *Migration and sensory evaluation of irradiated polymers*, Radiation Physics & Chem., Vol. 71, no. 1, pp. 205-208.

Suresh, P.; Leslie, A. & Braby, L. (2005). *Electron beam technology for food irradiation*, The International Review of Food Science and Technology (Winter 2004/2005). An Official Publication of the International Union of Food Science and Technology (IUFoST).

Thayer, D. W. & Boyd G. (1999). *Irradiation and modified atmosphere packaging for the control of Listeria monosytogenes on turkey meat*, J. Food Prot., Vol. 62, no. 10, pp. 1136-1142.

The Institute of Food Science and Technology (IFST), (2006). *The use of irradiation for food quality and safety*. info@ifst.org Web: www.ifst.org

Tomlins, K. (2008). *Food safety and quality management*, Food Africa. http://foodafrica.nri.org/safety/safetydiscussions1.html.

Tsiotsias, A.; Savvaidis, I.; Vassila, A.; Kontominas, M. & Kotzekidou, R. (2002). *Control of Listeria monocytogenes by low-dose irradiation in combination with refrigeration in the soft whey cheese' Anthotyros'*, Food Microb., Vol. 19, pp. 117-126.

Twaroski, M.; Bartaseh, L.; Layla, I. & Bailey, A. B. (2006). *The Regulation of Food Contact Substances in the United Sates*, In Chemical Migration and Food Contact Materials, edited by Watson, D; Barnes, K. & Sinclair, R., pp. 17-42. Cambridge, UK: Woodhead Publishing Limited.

Variyar, P. S.; Rao, B. Y. K.; Alur, M. D. & Thomas, P. (2000). *Effect of gamma irradiation on migration of additive in laminated flexible plastic pouches*, J. Polym. Mater., Vol. 17, pp. 87-92.

Venugopal, V. (2005). *Seafood processing: adding value through quick freezing, reportable packaging and cook-chilling and other methods*, Venugopal, V. (ed), CRC Press, Taylor and Francis group, Boca Raton, FL., pp. 281-318.

Vojdani, J. D.; Beuchat, L. R. & Tauxe, R. V. (2008). *Juice-associated outbreaks of human illness in the United States, 1995 through 2005*. J. Food Prot., Vol. 71, no. 2, pp. 356-364.

Wang, Q. (2001). *Effect of ionizing radiation on tensile properties of zein films*, Food packaging of the Annual Meeting of IFT, New Orleans, Louisiana, USA.

Welle, F.; Mauer, A. and Franz, R. (2002). *Migration and sensory changes of packaging materials caused by ionising radiation*, Radiation Physics and Chem., Vol. 63, no. 3-6, pp. 841-844.

World Health Organization (WHO), (2005). www.who.Int/media centre/fsctsheets/.

World Trade Organization (WTO), (2010). Available from: http://www.wto. org/english/res_e/statis_e/world_region_export_08_e.pdf.

Zhu, M. J.; Mendonca, A.; Ismail, H. A. & Ahn, D. U. (2009). *Fate of Listeria monocytogenes in ready-to-eat turkey breast rolls formulated with antimicrobials following electron-beam irradiation*, Poultry Sci., Vol. 88, pp. 205-213.

Part 2

Control for Food Processing and Vision System

3

Processing and Utilization of Legumes in the Tropics

Fasoyiro Subuola[1], Yudi Widodo[2] and Taiwo Kehinde[3]
[1]*Institute of Agricultural Research and Training, Ibadan*
[2]*Indonesian Legumes and Tuber Crops Reserach Institute,*
[3]*Department of Food technology, Obafemi Awolowo University, Ile-Ife,*
[2]*Indonesia*
[1,3]*Nigeria*

1. Introduction

Legumes belong to the family *Leguminosae*. In the tropics, they are the next important food crop after cereals (37). They are sources of low-cost dietary vegetable proteins and minerals when compared with animal products such as meat, fish and egg (8). Indigenous legumes therefore are an important source of affordable alternative protein to poor resource people in many tropical countries (23) especially in Africa and Asia where they are predominantly consumed. In the developing countries, research attention is being paid to better utilization of legumes in addressing protein malnutrition and food security issues. Legumes can be classified as:

1. Pulses or grain legumes which are various peas and seeds that are low in fat content
2. Oilseeds such as soybean and groundnut
3. Forage leguminous crops such as *Mucuna pruriens*, *Psopocarpur tetragonolobus* (winged bean).
4. Swollen root or tuberous root consumed as vegetable or fresh salad such as *Pachyrrhisus erosus, P. tuberosus* so called as Yam bean. Yam bean in Indonesia is widely planted by farmers for cash crop with better income. The other species such as *Flemingia grahamiana, F. procumbent, Psoralea esculenta and Pueraria lobata* are still under-utilized.

The highlight of this chapter is on the pulses and the leguminous oilseeds. Different types of legumes grown are consumed in different tropical regions in the world. Legume growing areas in Tropical Africa include Nigeria, Senegal, Togo, Cameroun and Cote d'Ivoire and in Tropical Asia include Indonesia and India (9). Some legumes are commonly used as commercial food crops such as cowpea in West Africa while some are lesser known, neglected or underutilized outside their indigenous areas. Table 1 shows some of the legumes grown in the tropics. Underutilization can be due to the hard- to cook phenomenon in legumes and lack of information on potential food uses. Pigeon pea, African yam bean, lima bean and bambara groundnut are neglected or underutilized crops in many parts of tropical Africa.

Common name	Botanical names	Other names	Areas available / consumed
Cowpea	*Vigna unguiculata*		Asia, Tropical Africa, West Indies
Blackeyed pea	*Vigna sinensis*	Catjan cowpea, Hindu cowpea, Kaffir bean	Asia, Africa, West Indies
Soybean	*Glycine max*		America, Asia, Africa
Groundnut	*Arachis hypogaea*	peanut	Tropical Africa, Central and South America
Pigeon pea	*Cajanus cajan*	Red gram, Congo bean	West Africa, East Africa, Pakistan, Middle East, Asia
Lentils	*Lens esculenta, Lens culnaris*	Split pea, red dhal	Central America, India, North Africa West Asia
Mung bean	*Phaseolus aureus*		East Asia, East Africa
African yambean	*Stenostylis stenocarpa*		West and East Africa
Lima bean	*Phaseolus lunatus*	Sieve bean, butter bean	Central America, Africa, Tropical Africa
Faba bean	*Vicia faba*	Broad bean, horse bean, windsor bean	Africa
Kidney bean	*Phaseolus vulgaris*	Navy bean, pinto bean, snap bean, black bean, haricot bean, pea bean	East Africa, Latin America
Chickpea	*Cicer ariethinum*	Lathyrus pea, grass pea, Khesari pea, Chickling pea	India, Pakistan
Lathyrus pea	*Lathyrus sativus*		
Bambarra groundnut	*Vigna subterranea*		Tropical Africa
Jack bean	*Canvalia ensiformis*		
Winged bean	*Psophacarpus tetragonolobus*		Tropical Asia, South East Asia

Reference (23)

Table 1. Common legumes grown in the tropics

2. Nutritional, health and economic Importance of legumes

Legumes are rich in protein and their chemical composition varies depending on variety, species and region (24). Table 2 shows the chemical composition of some tropical legumes. The protein content of legumes is twice or triple that of cereals depending on the type of the legume. The protein of legumes though adequate in essential amino acid lysine is however deficient in sulphur containing amino acids methionine and cystine (19). Legumes, however, form good supplements for cereals which are lacking in essential amino acid lysine. Improved nutritional quality can therefore be achieved by combining legumes with cereals.

Most legumes are low sources of fat with the exception of soybean and groundnut. Legumes are also good sources of different minerals such as calcium and phosphorus (24). The bioavailability of these minerals can be improved through processing. Legumes contain anti-nutritional factors such as lectins, saponin, haemagglutin, protease inhibitor, oxalate, goitrogen, phytates, trypsin inhibitor and tannin (8). These compounds reduce protein digestibility and availability. Some anti-nutritional factors in legumes have been reported to have health benefits. Tannin, a polyphenolic compound is reported to possess antioxidative activity (4). Raw legumes have higher content of anti-nutritional factors but can be eliminated or reduced by processing. Legumes are also good sources of carbohydrates, minerals, dietary fibres and water soluble vitamins which are important in human health. Dietary fibre consists of indigestible polymers which are made up of cellulose, hemi-cellulose, pectin and lignin. They provide bulk in natural food and are resistant to hydrolysis by enzymes in the alimentary tract (18). Dietary fibre is important in aiding absorption of water from the digestive track. It also has health benefits such as lowering of blood pressure and serum cholesterol, protection against cardiovascular diseases, diabetes, obesity and colon cancer (36). Legumes also have complex sugars such as raffinose and starchyose which are responsible for flatulence. Legumes are important both in human and animal nutrition especially in tropical Africa where they are more consumed (11). Legumes are processed into various semi- finished and finished products (29). Retailed legume products serve as a means of economic empowerment for individuals which also help to boost the national economy of some countries.

Legumes	Protein	Fat	Carbohydrate	Fibre	Ash
Soybean	37-41	18-21	30-40	4-6	4-5
Cowpea	22-26	1-2	60-65	4-5	3-4
Groundnut	20-33	42-48	22-25	3-4	2-3
Hyacinth bean	24-28	1-2	65-70	7-9	4-5
Common bean	20-27	1-2	60-65	4-5	4-5
Pigeon pea	15-29	1-3	60-66	5-10	3-4
Lima bean	19-25	1-2	70-75	4-6	3-5
Winged bean	30-40	15-20	35-45	6-7	3-5
Bambara groundnut	16-18	6-8	50-57	3-6	3-4

Reference (9)

Table 2. Chemical composition of some legumes (g/100g)

3. Legume processing

One major way of utilizing legumes is through food processing. Food processing involves techniques of converting raw materials into semi-finished and finished products that can be consumed or stored (23). Food can be processed at different levels including home-based food processing and at industrial level. Industrial food processing could be at the cottage level or on a large scale. The advantages of legume processing include:

- transformation of raw produce into edible forms
- improving digestibility of foods

- improving the nutritional quality of foods
- reducing and eliminating anti-nutritional factors
- improving consumer appeal and acceptability of foods
- destruction of food enzymes causing food spoilage thus extending shelf-life
- deactivation of spoilage and pathogenic microorganisms in the food products
- serving as a means of income generation

3.1 Unit operations in legume processing

Common legumes grown in the tropics include cowpea, soybean, pigeon pea, African yambean, bambara groundnut, kidney bean, lima bean etc (7, 14). Right from harvesting of the pods from the field, the seeds of these legumes pass through common post-harvest processes to obtain the dried seeds (9). The dried seeds are further processed into semi-finished or finished products through several processing steps called unit operations. Different unit operations are intended to fulfill different purposes (33).

3.2 Primary unit operations in the processing of legumes

3.2.1 Sundrying

Raw mature grains at harvest are at about 20% moisture content and are subject to spoilage unless dried. They come in long husk or pods which are removed by hands or mechanically. In some areas, grain legumes are steeped in water for hours (2-8 hours) before sun-drying. Seeds are dried on raised platform. In some other cases, grains are treated with oil before drying. The purpose of steeping and oil treatment is to aid dehusking process.

3.2.2 Husking

This process is also called hulling. Husking can be done by dry method or the wet method. Traditionally in African and Asian countries, the dry method involves pounding of the dried grains in mortar with pestles or in hand- operated wooden or stone sheller. Improved power-operated shellers have been designed and abrasive hulling machines have also been developed to improve the hulling process. Wet grinding process for husking involves soaking of the grains before drying. Improvement in husking process has been done through conditioning techniques through moisture adjustment to allow easy husking.

3.2.3 Winnowing

The separated husks are removed from the cotyledons by winnowing. Winnowing can done manually which is time consuming and laborious. Improved abrasive hulling machines which separate husk form cotyledons have been developed.

3.2.4 Seperation

This process is used to remove or separate whole grains from split, broken and powdery ones. It is done manually using sieves or mechanically with machines designed with a sieving device. Sieving manually is laborious and time consuming.

3.2.5 Storage

Proper post-harvest handling of legumes will prevent both qualitative and quantitative losses. It is important that legumes should be dried to safe moisture level of 12-14% to ensure good storage. Table 3 shows the safe moisture content of some tropical legumes. Dried seeds with high moisture content have increased the rate of mold attack and infestation. Dehulled seeds are often dried to a safe moisture level. Milled seeds are packaged and stored under dried conditions that will not allow absorption of moisture thereby leading to spoilage (33).

Legumes	Safe moisture content (%)
Broad bean, cowpea, kidney bean,white bean	15.0
Lentil, pea	14.0
Groundnut (shelled)	7.0
Soybean	13.0

Reference (22)

Table 3. Safe moisture content of some legumes

3.3 Secondary processing of legumes

3.3.1 Sorting and cleaning before use

Legumes are sorted and cleaned to remove dirts, stones, chaff, broken and spoilt seeds and other foreign materials. Sorting is done by hand sorting which is laborious and time consuming or through mechanical or electronic sorting device. Cleaning can be done by dry or wet methods. Dry cleaning is intended for grain legumes meant for storage purpose. Wet cleaning is usually done by washing with water.

3.3.2 Soaking

Different seeds are soaked in water for different periods of time. Soaking in water allows the seeds to absorb water, to decrease and eliminate anti-nutritional factors in legumes. However, soaking for long periods of time has been found to reduce nutritional quality of legumes through leaching of nutrients into the soak water (35).

3.3.3 Blanching

Blanching is a mild heat treatment of seeds. Legumes are usually blanched by soaking in hot water or boiled in water for few minutes. This process destroys food enzymes and some anti-nutritional factors in the legumes. Blanching can also aid the dehulling process.

3.3.4 Boiling / cooking

This process improves the appeal and sensory properties of legume. Boiling is usually at 100^0 C for some minutes. It tenderizes the seeds through water absorption. Traditionally, cooking of beans can be done using firewood. Pressure cooking pots allows legumes to be cooked under pressure and it reduces cooking time. This process eliminates heat labile anti-nutritional factors such as trypsin inhibitors (10).

3.3.5 Roasting

Legumes are roasted on the open frying pan in the presence or absence of salts or ash. Roasting improves the taste and edibility of legumes. It is important also in reducing and eliminating anti-nutritional factors. Roasted legumes are characterized by unique flavours which can increase their sensory appeal.

3.3.6 Fermenting

The process increases the digestibility of plant proteins and also reduces the anti-nutritional factors. Fermentation enhances flavour, colour and texture of legumes. Changes in these attributes are major stimuli in development of legume fermented products. It reduces heat stable anti-nutritional factors such as phytate. Fermented legumes are consumed as condiments e.g fermented locust bean (*iru*).

3.3.7 Germinating

Germination enhances desired qualities such as improved digestibility, reduced anti-nutrients like trypsin inhibitors (3). It improves nutritional quality of the proteins by hydrolyzing them into absorbable polypeptides and essential amino acids. Germinated or malted legumes are eaten in form of sprouts and are better than ungerminated ones. Sprouting improves the availability of vitamins B and C. It also reduces polyphenols content. Chicken pea and broad beans are commonly germinated before eating, cooking or use in salad dressing.

3.3.8 Milling

Dehulled legumes may be wet-milled or dry-milled. Milling is a size reduction process of the seeds into smaller particle forms. Wet-milling of seeds will produce a paste while dry-milling results in flour production. Different types of equipment have been designed for milling for household or industrial purpose. Wet milled legume may be mixed with other ingredients and steamed in leaves to produce pudding (*moinmoin)* or fried in hot oil to obtain bean cake (*akara*). The rehydrated flour maybe used to obtain these products

3.3.9 Sieving

Sieving removes unwanted materials from whole ground legume seeds (dry or wet). Example of wet sieving is in the filtration of ground soybean paste in the production of soymilk. The sieving process removes the unwanted residue called *okara*. For the dry-milled legume flour, sieving helps to achieve different ranges of particle sizes. Wet sieving can be done using cheese-cloth or muslin cloth while dry sieving can be done with different kinds of local or standard sieves. Some milling equipment have sieving devices incorporated into the design.

3.3.10 Frying

Several legumes are wet milled, mixed with other ingredients in preparing different local or oriental dishes. Frying improves the appeal and eating quality of legumes. It also improves digestibility and reduces anti-nutritional factors (1).

3.3.11 Canning

This is a sophisticated technology of packaging cooked beans in cans. The packaged beans are usually in brine, sugar or tomato purees. This technology allows for all year round availability of the product and for food preservation. Legumes processed in this form are however expensive.

4. Household utilization of some tropical legumes

4.1 Cowpea

Cooked beans: This can be in the form of cooked whole beans or cooked dehulled beans. Whole beans take longer period of time to cook than dehulled beans. Whole beans are boiled for about 45 to 60 minutes on the cooking stove or gas cooker depending on the hardness of the hull at household level. It is eaten whole or mashed. It used in or may be eaten alone or in combination with other food products like bread, *gari,* boiled yam with vegetable soup or fish meat sauce. Cooked dehulled beans reduce flatulence and is an excellent meal for both children and adults. The whole cooked bean can also be made into bean porridge by adding other ingredients such as palm oil, salt, pepper, onion and spices. Cooked beans prepared for income generating purpose are usually cooked with firewood which imparts a characteristic flavour. This, however, has its occupational hazards to processors. Long term effect of wood smoke in contact with eyes has its health and cost implications (17). Modern cooking methods also involve the use of locally fabricated cooking gas equipment at the commercial level.

Bean soup: In this food preparation, beans are washed, soaked, dehulled , boiled, mashed and sieved. The sieved beans is then cooked with palm oil along with other ingredients such as pepper, spices and seasoning with or without fresh or dried fish to taste to produce *gbegiri*. It is eaten with reconstituted yam flour product *amala.*

Bean cake and pudding: beans are washed, soaked, dehulled and milled into paste. In making the bean cake, the paste is mixed to a fluffy texture by trapping in air. Other ingredient such as onion and pepper are milled with the dehulled beans and the paste is fried with oil. Among the Yorubas of Nigeria , this product is called *akara* while the steamed pudding is called *moinmoin*. The pudding however is mixed with other ingredients that include vegetable oil. Traditionally, the mixture is packaged in leaves and steamed. Steaming of the pudding however today may be done in stainless steel cups. Some local processors use polyethylene bags in steaming the paste. Use of polyethylene bag is however being discouraged due to leaching of the chemicals in the package into the product which may lead to future health complications. Bean cakes and pudding are excellent diets that are usually consumed with fermented maize gruel *ogi, bread, gari, eko, or just on its own*. Bean cake and pudding are usually consumed as a breakfast meal, but they can also be consumed during lunch and supper too. They are considered as light meals.

4.2 Soybean

This has been known as an excellent source of protein, fat and minerals especially calcium. Soybean also has its unique characteristics in that it can be processed into a number or variety of products. Many economically challenged families in Nigeria utilize soybean processing as a means of income generation for household as well as ensure food security.

Soymilk: This is a popular soybean product rich in protein, fat and minerals. It is usually processed by soaking soybean in water, followed by milling, sieving, boiling and adding ingredients such as sugar and desired flavours to taste. A common hindrance or limitation to soymilk consumption is the beany flavour. However, research efforts have been conducted to reduce the beany flavour and obtain a better tasting and acceptable product. Soy-corn milk is another product from a blend of fresh sweet corn and soymilk to improve the nutritional quality of soymilk (31).

Soy cheese: This in the Orient is called *tofu*. In Nigeria, the local name is *soya-wara* or *soy-warankasi*. It is a highly digestible product that is good for people suffering from lactose intolerance. Locally, it is processed by first preparing soymilk and further precipitating the milk with a coagulant. Different cheap locally sourced coagulants have been used in soy cheese processing. This includes the enzyme based *Calotropis procera* leave water extract or acidic based lime juice, lemon juice, fermented maize water liquor (32, 28). Fermented maize water liquor is the most common type. Some local processors also use alum. Calcium salts are not usually used due to the cost implication.

Tempeh: This is a soy product that orginated form Indonesia. It is made from whole soybean seeds which are soaked, dehulled and partly cooked. Spores of *Rhizopus oligoporus*, used a s fermenting culture is mixed with the seeds. The seeds are spread thinly on a tray and allowed to ferment for 24 to 36 hours at 30^0 C. Good *tempeh* is characterized by proper knitting together to have a firm texture. This can be cut, soaked in brine or salty sauce and then fried. *Tempeh* has also been processed from other types of beans or mixture with whole grains. Figure 1 shows the pod of climbing beans, the mature seeds and the processed beans made into *tempeh*.

Fig. 1. Common climbing beans from young pod to matured seeds which can be used as vegetable, old seeds may be processed into *tempeh* mixed or supplemented with soybean.

Soy yoghurt: Yoghurt is a fermented milk product produced from mixed culture of *Lactobacillus bulgaricus and Streptococcus thermophillus* (6). Soy yoghurt is processed from soybean which is quite cheaper than yoghurt from milk. It is a good source of protein and minerals.

Soy sauce: This is a condiment common in East and South East Asia. It is processed by fermenting soybean seeds with two molds of *Aspergillus oryzae* and *Aspergillus soaje* in the presence of salt and water. The fermentation process yields a product called *Moromi* which is pressed to obtain a liquid called soy sauce. Soy sauce is also called *Miso* which may also be prepared from rice or barley.

Natto: This product is traditional to the Japanese. Soybean seeds are soaked in water for 12 to 20 hours. The seeds are fermented with *Bacillus subtilis* at 40^0 C for 24 hours. The product is cooled and aged in the refrigerator for a week.

4.3 Groundnut

Boiled and roasted groundnut: In some West African countries, groundnuts are cooked with the pods to get the cooked/ boiled groundnuts while shelled or unshelled groundnut are usually roasted. The shelled groundnut can be roasted in the presence or absence of salt. This can be consumed directly. Roasted groundnuts can also be grounded into powder and used in the preparation of sauce or as ingredients in other food dishes.

Peanut butter: This is usually used as sandwich spread. Groundnuts also called peanuts are dry roasted and ground into a smooth paste. Stabilizers in form of partial or complete hydrogenated vegetable oil, sweetener, spices, emulsifier and salt are also added.

4.4 Fava beans

Roasted fava beans: fava beans are cleaned, roasted at about 200^0 C for 20 minutes, cooled and packaged. The roasted beans can be consumed directly as snack or used as raw ingredient in gruel preparation

4.5 Chickpea

This can be processed into split legume called *kikk* in Ethiopia. This is used in producing a traditional sauce. This can be used in eating with local staples like *injera,* a cereal based product.

4.6 African locust bean

Dawadawa: Fermented African locust bean is called dawadawa or Iru . Dawadawa is generally processed from fermented oilseed called African locust bean (2, 27). This product is a traditional Nigerian condiment. The seeds are cooked,dehulled, spread thinly in containers usually calabash lined with leaves and fermented for 24-36 hours. African locust bean seeds are very hard to cook. Traditionally they are cooked overnight over firewood. Dawadawa has characteristic ammoniacal smell with its unique flavour in dishes. Dawadawa is used as natural seasonings in preparing soups, stews and traditional delicacies. Dawadawa has been processed also from other legumes such as soybean, bambara groundnuts and pigeon pea seeds (2, 25).Table 4 shows fermented products from different legumes.

Legume	Condiment	Predominant micro organisms	Optimum fermenting conditions
Soybean	Soy sauce or miso	*Aspergillus oryzae* *Aspergillus soaje*	-
Soybean	*dadawa*	*Bacillus subtilis, B. licheniformis*	3 days 35^0 C
African locust bean *(Parkia biglobosa)*	*dadawa*	*Bacillus spp, Staphyloccous aureus*	3 days, 35^0 C, pH 7-9
Bambara groundnut	seasoning	*Staphylococcus sp, Streptococcus sp ,Enterococcus sp*	3 days, 37^0 C, pH 7.8
Pigeon pea	seasoning	-	3 days,37 ^{0}C, pH 7-
African oil bean (*Pentraclethra macrophylla*)	*ogiri*	*Bacillus spp, Staphylococcus sp, Micrococcus sp*	3 -5 days, 30-33 ^{0}C, pH 5-8.7

References (2, 25)

Table 4. Fermented legumes and oilseeds used in preparing traditional condiments

5. Industrial products from legumes

5.1 Flours

Legumes are multi-purpose crops. At the household, cottage and large scale level, flours have been processed from different types of legumes. This has added to household convenience. Due to changing trends in consumer demands for more convenient products. Research studies have been geared towards developing innovative products from legumes. Many people working outside of their homes desire foods that can be easily prepared. Cowpea, soybean, pigeon and African yam bean seeds have been processed into flours (5, 16, 25). The common unit operations involved in flour production include washing, soaking, dehulling, drying , milling, sieving and packaging. Flours have been developed into different household recipes such cake, cookies, kokoro (20, 30) with comparable sensory attributes with products from fleshly prepared legumes. Composite flours have also been developed from cereals and tuber crops mixed with legume flours. In Africa, cowpea is the most popular legume (9). Cowpea flour is usually rehydrated and utilized in formulations as desired.

5.2 Vegetable oils

Vegetable oils are usually produced from soybean and groundnut more at the industrial level than at the household level. These oils contribute to gross domestic products and foreign exchange earnings in vegetable oil producing countries (21). Groundnut has about 42-48% oil content and soybean has about 18-21% (21) which are extracted locally by means of mechanical presses. Refining of expressed oil in Nigeria is still rudimentary. Cake from the expressed seeds is used as animal feed and that of groundnut is fried to produce *kulikuli* a snack commonly eaten by children. Cake from melon seeds are used in producing *robo* a fried snack.

5.3 Legume protein isolates

These are concentrates which have versatile functionalities (24). Soy protein isolate is a common isolate. It has high protein content of about 90%. It is made of defatted soy meal by removing most of the fats and the carbohydrates. Soy protein isolate is usually combined with other food ingredients such as minerals, vitamins and flavours in preparation of soy protein shake powder. Protein isolates have also been developed from a variety of legumes such as pinto and navy beans (34).

6. Limiting factors to household utilization of legumes

The hard- to -cook phenomenon due to the hard testa of some of the legume has led to long cooking times and utilization of more fuel during preparation. This led to under-utilization of legumes (12). The presence of anti-nutritional factors such as phytates which affect digestion and cause flatulence have also limited consumption of legumes by people. Research studies have shown that removal of the outer testa will reduce some of these effects. Dehulling process can be laborious and time consuming for hard testa seeds especially for cottage or large processing. Research efforts should however be geared towards improving the utilization of legumes to achieve improved nutritional status.

7. Legume processing and Income generation

In a developing country like Nigeria , legume processing into products like soymilk, soy cheese, cowpea cake and puddings are common income generating activities. The processing is usually carried out by women. This food processing activity plays a vital role in the survival and sustenance of their household and in meeting domestic financial obligations. However, these products are usually prepared under poor sanitary conditions. These processors need to be trained on improved processing methods and food safety practices (17). Processing techniques at the household or cottage level for processing needs to be upgraded. This to enhance productivity in terms of yield and quality to ensure food security, income generation and food safety especially in the developing countries. The following highlighted points below are important areas of focus for improvement.

7.1 Need for improved appropriate processing facilities

Equipment utilized in household and local processing needs to be upgraded to allow for increased productivity, reduced drudgery and efficient time management.

7.1.1 Improved product quality and safety

Most household and cottage processing activities in the tropics are usually carried out without product quality control. There is no standardization of product quality. Many local processors also produce under non-hygienic conditions. The products from different process batches have slight differences in product quality. This is due to lack of measurement of quality parameters during processing. Most operations are done using subjective judgments e.g hand feeling to estimate temperature, ingredients added to the taste of the processor (not according to a standard formula) etc. Local processors need to be sensitized and trained on the benefit of ensuring consistent quality products for increased income generation.

7.1.2 Improved shelf-life of freshly processed legume products

Dried low moisture semi-finished legume products like flours have excellent keeping qualities at ambient condition when stored away from moisture. However, the freshly prepared products such as soymilk, soy-cheese, bean pudding and cakes have short shelf-life of about a day or two at ambient condition. Dried and fried soy-cheese samples have better shelf-life of about a week in comparison with the freshly prepared products. Research studies need to be conducted be conducted on developing low cost techniques for extending the shelf-life of these products. Refrigeration, freezing and sophisticated preservative techniques which have high cost implication are usually not adopted by local processors.

7.2 Need for improving nutritional quality of local staples

Legumes are important foods in addressing protein - energy malnutrition concerns in developing countries. They contain the essential amino acid lysine which is deficient in cereals. On the other hand, cereals contain amino acid methionine which is limiting in legumes. Legumes can be supplemented with cereals to give a balance of amino acid called protein complementation. Flours are important improved products from legumes. They can be utilized in fortification of local staples to form composite flours like maize –soybean flour, cassava-soybean flour, maize-cowpea flour, cassava-cowpea flour, maize- pigeon pea flour, cassava-pigeon pea flours. The flours have been used in developing different types of recipes such as snacks and different nutritionally improved diets. There is still the need to develop novel food recipes from such composite flours to increase nutrient diversification.

7.3 Need for value addition through proper packaging

Dried legume seeds in the tropics are usually sold in the market places using different measuring weights which can indicate inconsistency in size. Proper packaging , labeling and branding are important considerations that local processors can add to boost sales for supermarket and export i.e pre-packaging. It minimizes time wastage, ensures uniform packed weights, eliminates access by rodents in the store, facilitates quick shopping but requires trust on the part of the buyers or consumers.

7.4 Need to explore the processing of underutilized legumes

Food security has been described as the availability, affordability and accessibility to nutritious, wholesome and safe food at all times (13). However, in Africa, many people are living below the poverty line of a dollar a day. Neglected legumes that are not commonly consumed due to the hard-to-cook phenomenon and lack of adequate knowledge on ways of utilization are usually regarded as underutilized legumes. Such include pigeon pea, African yam bean, bambara groundnut, lima bean etc. There is the need to improve the dehulling process of legumes with hard testa. To combat food crisis and poverty in the world, intensive research studies on the underutilized crops as a means to mitigate against global food insecurity need to be conducted. Many underutilized legumes grow well in tropical and sub-tropical countries in Africa, but the impeding need is the capacity to carry out intensive coordinated research studies in these countries in order to improve their utilization as food for improved nutritional status.

8. References

[1] Abd El-Moniem G.M., Honke J. and `Bednarsaka A. (2000). Effect of frying various legumes under optimum conditions on amino acids, in vitro protein digestibility, phytate and oligosaccharides. *Journal of Science of Food and Agriculture*, 80:57-62.

[2] Achi O. K. (2005). Traditional fermented protein condiments in Nigeria. *African Journal of Biotechnology*, 4: 1612-1621.

[3] Ahmed F.A., Albel Rahim E.A., Osama M.A., Volker A.E and Corinna L. (1995). The changes in protein patterns during one week germination of some legumes seeds and roots. *Food Chemistry*, 52: 433-43.

[4] Amarowicz R. and Pegg R.B (2008). Legumes as a source of natural antioxidants. *European Journal of Lipid Science and Technology*, 110:865-878.

[5] Ashaye O.A, Fasoyiro S.B and Kehinde R.O (2000). Effect of processing on quality of *Ogi* fortified with full fat cowpea flour. *Moor Journal of Agricultural Research*, 1: 115-122.

[6] Ashaye O.A., Taiwo L.B., Fasoyiro S.B. and Akinnagbe C.A (2001). Compositional and shelf-life properties of soy-yoghurt using two starter cultures. *Nutrition and Food Science*, 31: 247-250.

[7] Apata D.F. and Ologhobo A.D (1994). Biochemical evaluation of some Nigerian legumes seeds. *Food Chemistry*, 49:333-338.

[8] Apata D.F. and Ologhobo A.D. (1997). Trypsin inhibitor and the other anti-nutritional factors in tropical legume seeds. *Tropical Science*, 37:52-59.

[9] Borget M. (1992). Food Legumes. Technical Centre for Agricultural and Rural Cooperation, Wageningen, The Netherlands.

[10] Bishoi S. and Khetarpaul N. (1993). Effect of domestic processing and cooking methods on in-vitro starch digestibility of different pea cultivars (*Pisum sativum*). *Food Chemistry*, 47:177-182.

[11] Burkill, H. M. (1995). The Useful Plants of West Tropical Africa. Second Edition, Volume 3, Families J-L. Royal Botanical Garden, Kew, United Kingdom.

[12] El-Taby Shehata A.M. (1992). Hard-to-cook phenomenon in legumes. *Food Reviews International*, 8:191-221.

[13] FAO (1997). Guide for the conduct of the constraints analysis component. Special Programme for the Food Security, Handbook Series, SPFS/Doc/18. Rome, Italy.

[14] Fasoyiro S.B., Ajibade S.R. , Omole A.J. , Adeniyan O.N. and Farinde E.O. (2006). Proximate, mineral and anti-nutritional factors of some underutilized grain legumes in South- Western Nigeria. *Nutrition and Food Science*, 38:18-23.

[15] Fasoyiro S. B., Obatolu V.A, Ashaye O.A., Adeojo F.A. and Ogunleti D.O. (2009). Chemical and Sensory qualities of pigeon pea (*Cajanus cajan*) developed into a local spice dawadawa. *Nigerian Food Journal*, 27: 150-159.

[16] Fasoyiro, S.B, S.R Akande, K.A Arowora, O.O Sodeko, P.O Sulaiman, C.O Olapade and C.E Odiri (2010). Physico-chemical and sensory properties of pigeon pea (*Cajanus cajan*) Flours. *African Journal of Food Science*, 4:120-126.

[17] Fasoyiro, S.B., Obatolu, V.A., Ashaye O.A., and Lawal B.O. (2010). Knowledge Assessment, Improved Storage Techniques and Training of Local Processors and Vendors of Soy Products on Food Safety Practices in South West Nigeria. *Journal of Food and Agricultural Information*, 11: 340-350.

[18] Fennema O.R. (1996). Food Chemistry. Third edition. Taylor and Francis, Newyork. Pg 218-220.

[19] Friedman M. (1996). Nutritional value of proteins from different food sources: a review. *Journal of Agricultural and Food Chemistry*, 44: 6-21.

[20] Granito M., Valero Y. and Zambran R. (2010). Baked product development based on fermented legumes and cereals for school children . *Archivos Latinoamericanos de Nutritcion*, 60:85-92.

[21] Hawkes C. (2006). Uneven dietary development: linking the policies of globalization with the nutrition transition, obesity, and diet-related chronic diseases. *Globalization and Health* 2:4.

[22] Hayma D. (2003). The storage of tropical agricultural products. Fourth edition. STOAS Digigraf. Wageningen, The Netherlands.

[23] Ihekoronye A.I. and Ngoddy P.O. (1985). Integrated Food Science and Technology for Tropics. Macmillan Publishers Ltd. London. 284p.

[24] Liu, K. (1997). Soybeans, Chemistry, Technology and Utilization. Chapman and Hall, USA. Pg 532.

[25] Obatolu V.A., Fasoyiro S.B. and Ogunsumi L.O (2007). Functional properties of yam beans . *Journal of Food Processing and Preservation*, 31:240-249.

[26] Obatolu V.A. (2008) Effect of different coagulant on yields and quality of tofu from soymilk. *European Food Research Technology*, 226: 467-472.

[27] Odunfa S.A (1983). Carbohydrate changes in fermenting locust bean during preparation. *Plant Foods Human Nutrition*, 32: 1-10.

[28] Okoruwa A.E. (1999). Nutritional value and uses of food legumes in Africa. Paper presented at the legume breeding workshop International Institute of Tropical Agriculture, Ibadan, Nigeria, 27 September-8 October, 1999.

[29] Ojimelukwe, P.C. (2009). Sourcing and Processing of Legumes. In: Nigerian Agro Raw Materials Development, Volume 1: Some Industrial Crops and Salient Issues (P. A. Onwualu, S. C. Obasi and U. J. Ukpabi editors). Raw Materials Research and Development Council, Abuja.

[30] Omueti O. and Morton I.D. (1996). Development by extrusion of soyabari snack sticks: a nutritionally improved soya-maize product based on the Nigerian snack (Kokoro). *International Journal of Food Sciences and Nutrition*, 47, 5-13.

[31] Omueti O., Oguntona E.B., Jaiyeola O. and Ashaye O.A. (2000). Nutritional evaluation of home-level prepared soy-corn milk- a protein beverage. *Nutrition and Food Science*, 30:128-132.

[32] Omueti O. and Jaiyeola O. (2006). Effects of chemical and plant coagulants on yield and some quality attributes of tofu. *Nutrition and Food Science*, 36: 169-176.

[33] Potter N. N. and Hotchkiss J.H. (1998). Food Science. 5th edition. Springer USA. 608p.

[34] Seyam A.A., Banank O.J. and Breen M.D. (1983). Protein isolates from navy and pinto beans: their uses in mararoni products. *Journal of Agricultural Food and Chemistry*, 31:499-502.

[35] Taiwo K.A. (1998). The potential of cowpea as human food in Nigeria. *Food Review International*, 14:351-370.

[36] Ubom D.E. (2007). Nutrition, health and our environment. Sendina limited, Nigeria. Pg 140.

[37] Uzoechina O.B. (2009). Nutrient and anti-nutrients potentials of brown pigeon pea (*Cajanus cajan* var bicolor) seed flours. *Nigerian Food Journal*, 27: 10-16.

4

Processing of Sweet Corn

Mariusz Szymanek
University of Life Sciences in Lublin
Poland

1. Introduction

1.1 Sweet corn kernel structure, chemical composition, and sensory qualities

Sweet corn cobs constituting raw material for processing must be characterized by the highest quality of kernels. Kernel quality is defined not just by the chemical and sensory properties, but also by the mechanical parameters of kernels. This appears to fully justify joint consideration of all those properties. Sweet corn is probably a mutant of fodder corn (Orłowski, 2000). Significant differences between the two are related more to the genetics than to the structure of the kernels. As emphasized by Salunkhe and Kadam (1998), the structure of the kernels is strongly related to thegenetic modification and to ripeness.

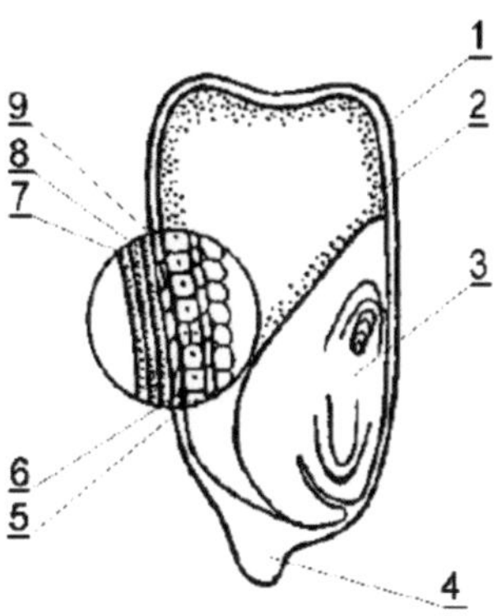

Fig. 1. Schematic of longitudinal cross section of a sweet corn kernel: 1 - pericarp, 2 - endosperm, 3 - germ, 4 - pedicel, 5 - aleurone layer, 6 - tube cells, 7 - epicarp, 8 - mesocarp, 9 - cross cells (Salunkhe and Kadam, 1998).

Sweet corn kernel is built of the pedicel, the pericarp, the germ, and the parenchyma (Fig. 1). The pedicel is a hard and fibrous remnant of the tissue that joins the kernel to the cob core (Szpaar and Dregiew, 1999). During kernel shearing, a part of the pedicel remains on the kernel, which has a negative effect on its nutritional value. Kernel shape is described as flattened and wedge-like, with the tip much broader than the base end by which the kernel is attached to the cob core. The kernel is deep set on an abbreviated shoot (rachis) forming the cob, and is covered with a thin pericarp. The pericarp is a component of the seed coat tissue and forms the outer layer of the kernel. The thickness of the layer determines the kernel skin level of tenderness. This feature is important in the estimation of kernel quality for processing. As compared to other varieties, sweet corn is characterized by lower thickness of the epicarp, on average 25-30 μm (Ito and Brewbaker, 1991). The epicarp is composed of a single layer of pericarp, in the cavities of which single cells of the mesocarp are located. It also includes one or two layers of cross-cells and one or more tube cells adherent to the pericarp. The germ, located obliquely at the base of the kernel, is large and constitutes about 15% of the volume or 11.5-14% of the weight of the kernel. Germ size, however, is cultivar-related and may constitute 5% of the mass of the whole kernel (Puangnak, 1998). In turn, the parenchyma is the largest element of the kernel. It is in the parenchyma that the sugars, starch, and water-soluble polysaccharides are accumulated in. In the phase of consumption ripeness, the consistency of kernels is soft, delicate, creamy, and the taste is sweet and fragrant (Reyes and Varseveld, 1982). The kernel seed coat has colours from pale yellow to orange, often also with a violet tint and shiny. In the genotype of sweet corn, apart from the genotype of sweet corn, except gene *su* (sugary) gene determining the sweet taste and tenderness of the kernel, also other genes have been discovered, responsible for enhancing the sweetness and the taste and utility qualities - the gene *se* (sugary enhancement) and the gene *sh2* (shrunken 2). Genetic combinations of the genotypes *su* and *sh2* helped in the creation of very sweet cultivars (Simonne *et al.*, 1999). In the phase of full ripeness, sweet corn has wrinkled kernels, almost completely filled with vitreous parenchyma, mostly white or yellow in colour. The reserve substance of the parenchyma is composed of amylodexstrins which are responsible for the sweet taste. The kernel has a bulbous shape, oval, wedge-shaped or angular, a smooth or wrinkled surface, and white, yellow, read or brownish colouring. It is enclosed in a cover formed from fused pericarp and seed coat, beneath which there is a layer of aleurone cells, the parenchyma, and the germ. From the moment of pollination till harvest the cob of sweet corn undergoes numerous physical and chemical transformations which largely affect the taste and the quality of kernels. The taste is especially strongly affected by the transformations of sugars (Wong, 1994). With respect to the content of sugars, three types of sweet corn cultivars are distinguished: normally sweet cultivars, type *su* (sugary), with sugar content of 4-6%, cultivars with increased sugar content, type *se* (sugary enhancement) - 6-8%, and very sweet cultivars, type *sh2* (shrunken 2) - 8-12% (Warzecha, 2003). Apart from the content of sugars, fresh mass of sweet corn kernels contains 2.1-4.5% of proteins, 3-20% of starch, 1.1-2.7% of fats, 0.9–1.9% of cellulose, 9–12 mg of vitamin C, small amounts of vitamins A, B1, B2, PP, and mineral components such as: sodium, potassium, magnesium, calcium, phosphorus, iron, selenium, copper, nickel and chromium (Felczyński *et. al.*, 1999; Lee and McCoon, 1981). The chemical composition of the kernels is related to the weather conditions, ripeness, and method of storage (Salunkhe and Kadam, 1998). According to data from the USDA (Hardenburg and Watada, 1986), the nutritional value of sweet corn kernels is related to the content of water (72.7%) and to the total content of solid parts (27.3%). Solid parts include hydrocarbons (81%), proteins (13%), lipids (3.5%), and others (2.5%). Starch is the dominant hydrocarbon component. Sweet corn has the highest nutritional value in the phase of milk ripeness. With progressing phase of ripeness, in the transition to the phase of wax ripeness the content of sugars decreases, accompanied by an increase in the content of starch (Suk and Sang, 1999). In 100 g of kernels there is about 3.03 g of saccharine, 0.34 g of glucose and

0.31 g of fructose. The content of saccharine increases, and that of reducing sugars decreases as the kernels reach the optimum ripeness . The content of proteins in the kernels decreases from the surface towards the centre of the kernel. The content of proteins, free aminoacids, water-soluble and insoluble hydrocarbons, increases up to the phase of wax ripeness, and then gradually decreases (Azanza *et. al.*, 1996). The amounts of the particular components in various cultivars and in various phases of ripeness variable. In comparison to other cereals, sweet corn is relatively rich in oil. Approximately 90% of the oil is accumulated in the germ. Sweet corn is more tasty than other corn species, thanks to its high content of water-soluble polysaccharides. That component imparts to the kernels their tender and creamy character. The most important parameters that affect the sensory quality of the kernels include sweetness, texture, and taste (Wong and Swiader, 1995). Sweetness depends on the content of sugars, while texture depends on a number of factors, such as tenderness of the seed coat, moisture, content of water-soluble polysaccharides. Taste, in turn, is frequently associated with the content of DMS (dimethyl sulphide). Decrease in kernel quality related to loss of taste and aroma after the harvest is a problem for the processing industry. The loss of taste in fresh or frozen state of the kernels is caused by enzyme activity (Collins and Biles, 1996). Fresh kernels are characterized by faint aroma or its total absence. Wade (1981) states that cut kernels have three characteristic types of aroma. Two of these are similar to the aroma of fresh green vegetables, and the third is described by that author as a methol-type aroma. With progressing ripeness of sweet corn, the level of DMS in the kernels decreases, which is a serious problem for the processing industry due to the related considerable loss of taste of processed sweet corn products.

The consumption quality of fresh sweet corn largely depends on the content of sugars and water-soluble hydrocarbons in the kernels (Evensen and Boyer, 1986). The quality of sweet corn kernels can be determined in many ways. The basic discriminant of kernels for the processing industry is kernel hardness and taste. One of the most important factors determining the quality of kernels for processing is the use of cultivars characterized by uniform ripening. The choice of cultivar affects not only the yield of kernels cut off cobs, but also the taste quality of the kernels. Other quality factors include the colour, sweetness, and tenderness of the kernel cover. With ripening, the cover becomes harder and harder. The quality of sweet corn is correlated to the content of sugars. The transformation of sugars into starch is also related to decreasing moisture content of the kernels.

1.2 Characterization of sweet corn varieties

The choice of variety is one of the more important factors the determine whether sweet corn production is a success. Population (fixed) varieties have been largely replaced in cultivation by hybrid (heterotic) varieties, created by man in the first half of the 20th century as new forms of the crop plant. Hybrid varieties spread less, produce bigger and evenly ripening cobs, and are higher yielding compared to the population varieties. They are characterized by high sensory qualities and are suitable for direct consumption and for the processing industry alike. They meet the requirements of the fruit and vegetable processing industry in terms of having very delicate kernel skin and kernels easy to separate from the cobs in whose mass the kernels constitute 30–40%. Primary differences among the hybrid varieties include the duration of their vegetation period, content of sugars, and suitability for various uses. The productive value of the varieties is determined primarily by their yield capacity and the earliness of their ripening.

Producers involved in sweet corn growing for direct consumption are interested in obtaining a large number of well kernelled cobs. Those producing sweet corn for industrial processing expect a high yield of material suitable for processing. In terms of the length of the vegetation period the following groups of corn varieties are distinguished:

- early varieties (70–80 days),
- medium early varieties (85–90 days),
- late varieties (95–110 days).

Also important is the division of varieties with respect to the content of sugars in kernels at the phase of harvest ripeness. With reference to their genetic features, they are classified as:

- normally sweet, with the gene „*su-1" (sugary)*,
- with increased content of sugars, with the gene „se" or „se+" *(sugary enhanced)*,
- very sweet, with the gene „*sh-2"(shrunken 2)*.

Variety selection is an important consideration in sweet corn production and includes factors such as sweetness, days to maturity, seed color, size, yield potential, and tolerance to pests. The Cooperative Extension Service can provide a list of varieties recommended for each region.

Genotype	Sweetness	Conversion of sugars to starch	Isolate form
Normal sugary (su)	Moderately sweet	Rapid	(sh2) vaieties
Sugary enhanced (se), (se+)	Sweeter than (su) less sweet than (sh2)	Not as rapid as (su)	(sh2) vaieties
Super sweet or shrunken (sh2)	Very sweet	Very slow	(su), (se), (sh +) vaieties

Table 1. Sweet Corn Genotypes (http://attra.ncat.org/attra-pub/PDF/sweetcorn.pdf)

Modern sweet corn varieties are classified as: "normal sugary" (su); "sugary enhanced" (se) and (se+); and "shrunken" (sh2), also called "super sweet." These differ in flavor and tenderness, and in the rate at which starches are converted to sugar. In general, (se) lines yield the best, followed by (sh2), and finally (su). Cross-pollination of sweet corn with other kinds of corn or with some other sweet corn genotypes can result in starchy-tasting kernels. Generally, a minimal isolation distance of 250 feet between those varieties or types is recommended; 700 feet, however, is preferred for more complete isolation. Table 1 summarizes the general characteristics of sweet corn genotypes, including isolation requirements. The normally sweet varieties contain 4–6% of sugars in fresh kernel mass, those with higher sugar content from 6 to 8%, and the very sweet varieties, sometimes also called super sweet or extra sweet, from 8 to 12%. Another criterion of division or classification of cucltivars is the colour of their kernels which can be yellow (the largest group of varieties), white, yellow-white *(bicolor)* and red (Wong, 1994).

2. Uses of sweet corn

Sweet corn kernels can be consumed both as fresh produce and in processed forms. In practice, most frequently three basic directions of their utilization are distinguished:

- direct consumption - cobs harvested at milk ripeness of seed, for so-called freshproduce market;
- fruit-and-vegetable processing industry - cobs harvested at late-milt ripens of seed, for pickles and frozen foods;
- industrial processing - cobs harvested at full ripeness of seed, for flour, meal, etc.

3. Sweet corn processing technology

3.1 Harvest of sweet corn cobs

Sweet corn for processing is harvested at a relatively immature stage as compared to field corn. Processing of corn is used to increase its shelf life but as a consequence, a significant loss of nutrients may occur via heat degradation or leaching (Scott and Eldridge, 2005). Sweet corn for processing is picked at different stages of maturity depending on the way it is to be processed. The corn for freezing is harvested at about the same stage as that for fresh market, while the corn for whole kernel pack and cream-style is harvested at a slightly later stage of maturity. For whole kernel canning and freezing, optimum kernel moisture ranges from 70 to 76%. For cream-style canning corn, optimum kernel moisture is about 66%. Experience showed that it correlated very closely with the moisture percentage and with postharvest grade evaluation Olson (2000). There are many reasons why crops should be harvested at optimal maturity for their specific end uses. An accurate determination of the sweet corn maturity for harvest can ensure the best possible crop yield and quality. The optimum moisture for harvesting shrunken sweet corn for freezing and canning is no less than 76% and no more than 79%. This compares to the range for standard sweet corn of 70 -72%. Because the shrunken sweet corn loses only about ¼ percent of moisture per 24-hr period at the 76% level as compared to 1% per 24-hr period for standard sweet corn, the harvest window for shrunken corn harvest for a processing line and results in fewer bypassed fields due to planting, mistiming, or weather delay (Marshall and Tracy, 2003). However, according to Warzecha (2003), it is easier to mechanise the harvest of standard sweet corn than that of shrunken sweet corn. The standard sweet corn compares to shrunken varieties mature to longer. Sweet corn has a very short period of optimum harvest maturity, and its quality changes rapidly close to and following the peak. Ears harvested immature will have a small diameter, a poor cob fill, and kernels that are watery and lack sweetness. At optimum harvest maturity, the kernels are plump, sweet, milky, tender, and nearly of maximum sizes. After optimum harvest maturity has been reached, the eating quality of sweet corn begins to decrease rapidly, while the husk appearance changes very little. Overmature corn is rather starchy than sweet, tough, and the kernels are often dented (Motes *et al.*, 2007). However, according to Kumari *et al.* (2007), the unfavourable correlation coefficients between the sugar content and grain weight suggested that it is difficult to obtain high-yielding sweet corn hybrids of good quality.

Results of sweet corn ear size measured at different harvest date are presented in Table 2.

Particular	Harvest date				LSD
	1st	2nd	3rd	4th	$\alpha = 0.05$
Moisture content (%)	77.41^{a} (0.95)	75.62^{ba} (0.88)	72.31^{c} (1.05)	69.83^{d} (1.09)	2.05
Yield ($t \cdot ha^{-1}$)	18.64^{a} (1.15)	17.98^{ba} (1.21)	16.31^{c} (0.98)	15.88^{dc} (1.11)	1.20
Length (cm)	22.21 (2.09)				-
Max. diameter (cm)	4.94 (0.98)				-
Number of kernels per row (pcs)	28.05 (1.57)				-
Number of kernel rows (pcs)	14.72 (1.54)				-
Bulk density ($kg \cdot m^{-3}$)	612.21^{a} (9.12)	619.54^{ba} (8.65)	624.36^{cb} (10.21)	634.54^{d} (8.86)	9.36

Numbers in the same line followed by the same letter are not significantly different at $p<0.05$.

Table 2. The mean values of kernels moisture content, yield ears, ear length, ear diameter, number of kernels per row, number of kernel rows and bulk density with standard deviation in parenthesis

The mean size of 100 husked ears measured at a first harvest date are: length 22.21±2.09 cm and max. diameter 4.94±0.98 cm. The yield decreased from 18.64 to 15.88 $t \cdot ha^{-1}$, when the moisture content of kernels decreased from 77.41 to 69.83%. The similar decreasing of moisture content with increasing harvest maturity has been reported by Wong, (1994).

At all harvest date (moisture contents), an increase of deformation with an increase in applied forces was observed (Table 3).

Particular	Harvest date				LSD
	1st	2nd	3rd	4th	$\alpha = 0.05$
Compression force (N)	29.48^{a} (1.78)	35.54^{ba} (1.32)	42.71^{c} (1.54)	49.56^{dc} (1.23)	7.11
Shear force (N)	8.21^{a} (0.28)	10.41^{ab} (0.32)	12.34^{cb} (0.41)	15.21^{dc} (0.37)	3.21
Puncture force (N)	9.11^{a} (0.18)	12.65^{ba} (0.19)	15.28^{cb} (0.17)	17.23^{dc} (0.21)	4.56

Numbers in the same line followed by the same letter are not significantly different at $p<0.05$.

Table 3. The means values of compression, shear and puncture force with standard deviation in parenthesis

The hull rupture is marked by an audible "click", and a sudden decrease of the force occurs. The point marked by the abrupt force decrease is often called the bio-yield point and the loading was stopped once this point was reached. The measured parameters were the rupture force, when the kernel hull undergoes failure during compression, shear and puncture, the deformation up to the rupture point.

The force required for the hull rupture increase as the moisture content decreased. At the moisture content ranging from 77.41 to 69.83% the compression force increased from 29.48 to 49.56 N, the shear force increased from 8.21 to 15.21 N and the penetration force increased from 9.11 to 17.23 N. Burton (1982) reported that the average puncture tensile strength forces increased with later harvest date.

During the period when sweet corn ears are suitable for harvesting and kernel moisture decreases total of sugars decreased from 6.24 to 5.11% and starch increased from 14.49 to 22.19% (Table 4).

Particular	Harvest date				LSD
	1st	2nd	3rd	4th	$\alpha = 0.05$
Total sugars (%)	6.24[a] (0.24)	5.92[ba] (0.21)	5.54[c] (0.19)	5.11[d] (0.18)	0.41
Starch (%)	14.49[a] (0.22)	16.21[ba] (0.24)	18.70[c] (0.21)	22.19[d] (0.23)	2.24

Numbers in the same line followed by the same letter are not significantly different at $p<0.05$.

Table 4. The means values of total sugars and starch level with standard deviation in parenthesis

It was observed that the harvest date affect total sugars and starch level. The mean values of total sugars and starch only between 1st and 2nd harvest date weren't significantly different. (Table 2). Similar trend have been reported by Suk and Sang (1999).

The processing recovery and corn cut yield increased (from 41.14 to 50.02%, and 7.67 to 7.94 $t \cdot ha^{-1}$, respectively) for different harvest date (Table 5).

Particular	Harvest date				LSD
	1st	2nd	3rd	4th	$\alpha = 0.05$
Processing recovery (%)	41.14[a] (3.48)	43.13[ba] (3.98)	48.25[cb] (4.11)	50.02[dc] (4.02)	6.86
Corn cut yield ($t \cdot ha^{-1}$)	7.67[a] (0.13)	7.75[ba] (0.11)	7.86[cb] (0.10)	7.94[dc] (0.12)	0.13
Bulk density ($kg \cdot m^{-3}$)	585.51[a] (9.42)	592.31[ba] (9.12)	601.74[cb] (9.06)	609.11[dc] (9.21)	12.36

Numbers in the same line followed by the same letter are not significantly different at $p<0.05$.

Table 5. The means values of processing recovery, corn cut yield and bulk density with standard deviation in parenthesis

The explanation for these increase of recovery could be found in the decline of moisture level and increases of starch content. Michalsky (1986) found that lower average moisture content and higher starch level make easy to mechanical cutting the kernel off the cob and reduce losses of kernel flesh. This is why the sweet corn for whole kernel canning is harvest at lower moisture content than for frozen – style corn. Although corn cut yield was the highest at the lowest moisture content (69.83%) it was observed that some single kernels begun to wrinkle. A similar result was reported by Olson (2000), who found that that the highest quality cut corn from most of the standard sweet corn hybrids would be obtained at a kernel moisture level of 72 to 73%. At 74 to 75% moisture, the flavor and taste were good but kernel size and uniformity, color, and cut-corn yield of the standard sweet hybrids might be below par. At 70 to 71%, a critical dividing point, yield was higher but the cut corn would appear to be older (large; darker yellow kernels) and might be tougher.

3.2 Kernels removing from the cobs

3.2.1 Cutting method for removal of kernels from raw sweet corn cobs

The food processing industry currently applies machines utilizing rotary cutter heads for severing corn kernels (Kessler and Harry, 1998) (Fig. 2).

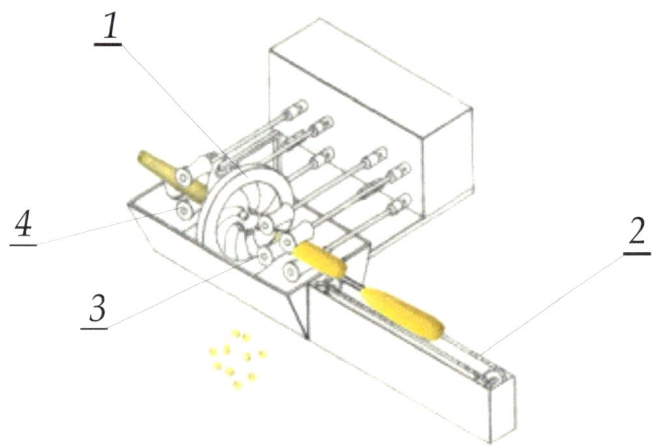

Fig. 2. Schematic of sweet corn kernel cutter: 1 - knife head, 2 - feeder of cobs, 3 –,rollers of copying system, 4 - removing rollers of cob cores (Robertson and Farkas, 1982).

The action of separating sweet corn kernels from the cob for consumption purposes is performed with the use of special cutters, the operation of which is described by many authors. To obtain kernels of high quality, the working elements of such machines should be carefully adjusted so that all the kernels are cut off as close to the cob core as possible, but without cutting off the cob husks whose presence among the kernels acquired worsens the quality of the product. Until recently it was recommended that kernels should be cut off at 2/3 of their length with the kernel germs remaining on the cob core, as predominant in corn production were varieties with long cob husks. New hybrid varieties of corn are free of that defect, and to increase the amount of material acquired cutter knifes are set to the maximum length of kernels detached. Care must be taken that the cut through the kernel be smooth, without tearing the seed cover, and set so that no thick cob husks are among the kernels detached. This requires frequent sharpening of the cutter knives. Adjustment of the working elements of the kernel cutter should be made suitably to changes in the dimensions and properties of the material processed. Corn cobs should oriented with their narrower end towards the cutter head, as in that position the cutter knives adapt better to correct detachment of kernels. During kernel cutter operation, it is necessary to systematically check all the moving elements of the machine, and to clean and lubricate as required. This has an immense effect on the quality of kernels detached, as with the cutter head dirty the positions of cutter knives take longer to adjust to the changing cob diameter, and them may result in increased amount of incorrectly cut off kernels.

The mechanical removal of sweet corn kernels from the cob by cutting leads to waste, loss of nutritional value, and loss of yield (Hanna *et al.*, 1988). This means that it is not possible to obtain all the kernels of corn. The cutting operation severs the upper part of the kernel from the lower part, which remains on the cob. Thus, the hull of the kernel is broken open and a part of each kernel is wasted (approximately 20% remains on the cob), including much or the entire germ of the kernel. Furthermore, during subsequent wet-processing, including washing and blanching of the severed kernels, some of the corn meat is leached from its pouch and lost (Kunicki, 2003). However the proportion of cut-off kernels is strictly related to the moisture product processing characteristics. The successful detachment of sweet corn kernels as intact kernels promises large advantages in terms of yield and effluent over cut kernels during processing, but it does require low strength kernel attachment to the cob. The study by Niedziółka and Szymanek (2004) showed that blanching prior to the cutting process, instead of blanching after kernel cutting, resulted in a 12% increase in yield of cut kernels and quality. However, a similar study by Trongpanich *et. al.* (2002) did not show a statistically significant difference (0.05), while Michalsky (1986) reported that by delaying the time of harvest until the moisture content decreased, it is possible to increase the detached part of the kernel.

The studies of Szymanek (2011) showed that the cut kernel yield increased with the increase in cutter head angular speed. The change of speed from 167.5 to 293.2 rad $\cdot$ s^{-1} increased approximately 51% for the Jubilee variety, 29% for the Boston variety, and 54% for the Spirit variety. The differences recorded between the varieties were statistically significant. An increase of cutter head angular speed from 167.5 up to 293.2 rad $\cdot$ s^{-1} resulted in a statistically significant decrease of inferior kernel share. The highest reduction in inferior kernels in the investigated speed range was obtained for the Spirit variety, at about 64%, whereas the lowest was for Boston, at about 44%. An increase in the amount of kernels cut, along with a change of a cutter head angular speed, also decreased sugar content loss. The most substantial decline in sugar losses within the studied range of cutter head speed was observed for the Jubilee variety, with about 48%, while the lowest for Spirit and Boston was about 33%. A change in a cutter head angular speed generally resulted in a statistically significant effect on the differences in sugar content. Increasing cutter head angular speed in the range of 167.5 to 293.2 rad $\cdot$ s^{-1} induced a decrease in kernel mass losses subject to the plant variety. The changes were in the range from 51 to 28%. Spirit variety had the highest reduction in sugar losses.

The losses occurring in the production process have negative effects not only in economical terms, but also pose considerable environmental pollution problems. The loss of kernels could, however, be reduced by means of the respective choice of sweet corn varieties. Robertson *et al.* (1980) think that solutions to this problem are being developed by food engineers and plant breeders by modifying the mechanical separation method and by modifying the raw product processing characteristics. However, claims that reduction of the kernel moisture could be achieved by delaying the cob harvesting dates, which, in turn, reduces loss during the kernel cutting process. In addition, blanching the cobs prior to attempting the cutting process (Stewart *et al.,* 1997) and selecting the optimum operation parameters of the kernel cutter devices also reduces loss (Niedziółka and Szymanek, 2006). The method proposed by Robertson and Farkas (1982), whereby the cobs are halved along their length and the kernel is subsequently detached on reciprocally moving conveyor belts can practically reduce kernel loss to almost zero. However, this method is less efficient and consumes too much energy and has not so far been applied on industrial scale. In recent years, the growing concern about applying freezing methods, including cryo-liquids, in the food processing industry has also been reflected by the growth in manufacturing of frozen products , including sweet corn (frozen cobs or kernel). In highly technologically advanced countries, about 30% (20% in Poland) of the production output of sweet corn is designated for freezing and about 40% for canned food (60% in Poland) (Waligóra, 2006). Freezing in boiling liquids means very high convective heat-transfer coefficients, considerable temperatures and consequently very short freezing times. Unlike in classic (slow) freezing, the cryo-liquid freezing method results in a high quality of the product processed. Kernel frozen in temperatures of -40C fit long-term storage requirements without any significant change in taste and technological properties (Matheus *et al.,* 2004). The growing demand for frozen products brings about a need to develop technologies that can provide storage of fresh vegetables and fruit, while their physical properties undergo no essential changes.

3.2.2 Shelling method for removal of kernels from frozen sweet corn cobs

In the shelling method for removal of kernels from frozen sweet corn cob the corn cobs prior to shelling are subjecting to blanching and freezing by spraying in liquid nitrogen.

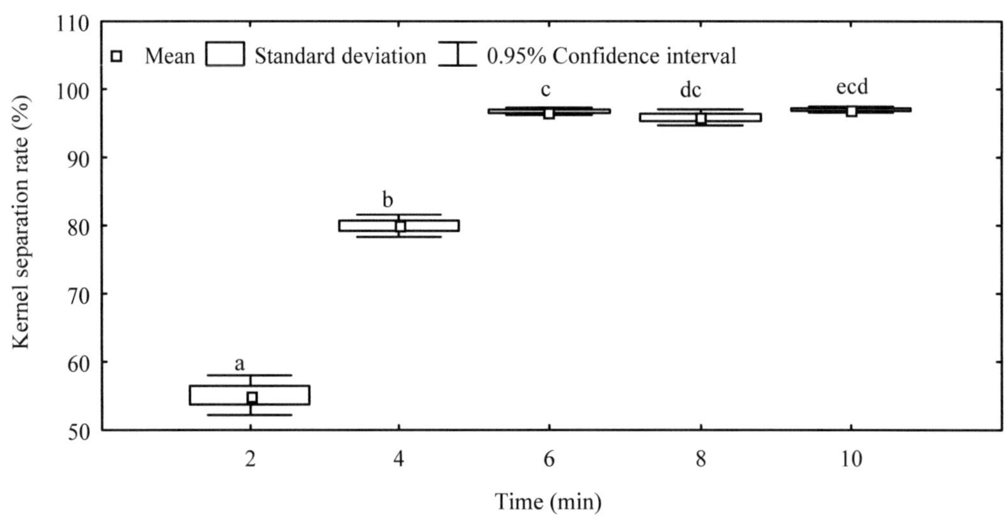

Numbers in the same letter are not significantly different at $p<0.05$.

Fig. 3. The effect of freezing time on kernels separation rate

The mean values of kernel separation rate of cobs for 2, 4, 6, 8 and 10 minutes freezing time were: 55.1%; 79.9%, 96.7%, 95.8% and 97.0% respectively (Fig. 3).

In the range from 2 to 6 minutes freezing time we can observe the statistically significant increases of kernels separation rate and then from 6 to 10 minutes the not significant changes of kernels separation rate. This might be due to lack of uniformity in kernels freezing which results in different kernels hardness and not complete kernels removal.

The advantage of this method is the reduction of waste and increased yield of corn. During the analyses of kernels separation (Szymanek, 2011), it was confirmed that 86.3% of intact kernels removed after 2 and 4 min of freezing contained adhering cob related tissue, and 13.7% of kernels were without such tissue. However, for intact kernel freezing for 6, 8 and 10 min, the relations were opposite and amounted to, respectively, 8.9 and 91.1%. This means that the optimal freezing time influences not only the quantity but also the quality.

Robertson *et al.* (1980), in a comparison between cut and intact kernels, found that intact kernels have more adhering cob-related tissues which is perceived as a defect of intact kernels

In the range from 2 to 10 minutes were observed the decrease of kernels damage from 7.3 to 3.9%. The share of damaged kernels showed the same tendency as changes of kernels separate rate. The proportions of damaged kernels for 6, 8 and 10 minutes freezing time although different in value are not significantly different. Decreasing of kernels damages together with lengthen of freezing time might be due to increasing of hardness of whole kernels which results in kernels being more resistant to mechanical action of shelling unit. The mechanism of damages formation can be compared to threshing dried cob. Nguyen (1986) reported that when shelling dried cob, only the linkage between kernel and corn cob is broken When threshing fresh cob, we have to break down two linkages: kernel - corn cob and kernel - kernel. This causes a considerable amount of broken kernels. Similar situation might occur when freezing kernels.

4. Summary

The use value of sweet corn, resulting from its high nutritional values, taste qualities, and extensive possibilities of application, fully justifies the sense of increasing the area of its variation. The search for new and more efficient methods of harvest at simultaneous assurance of favorable economic effects and high quality requirements for the sweet corn cobs and kernel produced, become a necessity. Sweet corn cobs harvested for the processing industry are subjected to machining consisting in the kernel cutting off from cob core. Since a considerable part of sugars is cumulated in the lower part of the kernel, it is recommended to cut kernels off cobs as close to the core as possible. The irregular shape of kernels and their low content of dry mass (approx. 27%) are the reason for frequent mechanical damage to kernels. Kernels, especially those located at extreme parts of the cob, differ in their size and hardness. Also the shapes of cobs (cylindrical or tapered) and their variable size (variety-related) make the detachment of kernels more difficult. Hence the process of kernel cutting off from cob cores is a major problem for the processing industry.

The study has shown that sweet corn kernels can be removed from the cob by using shelling method when there are first subject to rapid freezing by spraying liquid nitrogen. The freezing time of 2, 4 and 6 minutes affects significantly the increase of hardness and separation rate and decrease of damage and kernels losses. Starting from 6 minutes freezing time, the time of 8 and 10 minutes had no significance effect on average values of analysed parameters. The visual observation of kernels showed that for freezing time of 2 and 4 minutes 86.3% of intact kernels had and 13.7% had not adhering tissue. In contrast, after 6, 8 and 10 minutes freezing time only 8.9% of intact kernels had and about 91.1% had not adhering tissue.

5. References

Azanza, F.; Tadmor, Y.; Klein, B.P. (1996). QTL influencing chemical and sensory characteristics of eating quality in sweet corn. *Genome,*Vol. 39, 40-50.

Burton, L.V. (1982). The measurement of maturity of country gentlemen corn. *Canner*, Vol. 54, 27–29.

Collins, J.K.; Biles, C.L. (1996). Flavour qualities of frozen sweet corn are effected by genotype and blanching. *J. Sci. Food and Agric.*, Vol. 72(4), 425-429.

Evensen, K.B.; Boyer, C.D. (1986). Carbohydrate and Sensory Quality of Fresh and Stored Sweet Corn. *J. Amer. Soc. Hort.Sci.*, Vol. 111(5), 734-738.

Felczyński, K.; Bąkowski, J.; Michalik, H. (1999). Czynniki wpływające na jakość plonu i wartość odżywczą kukurydzy cukrowej. *Ogrodnictwo*, Vol. 3, 18-22.

Hanna, H.Y.; Story, R.N.; Adams, A.J. (1988). Effects of sweet corn production practices on yield and other characteristics. ASHS-SR Meeting, New Orleans, LA, Feb *HortScience*, Vol. 23(5), 824.

Hardenburg, R.E., Watada, A.E. (1986). The Commercial Storage of Fruits, Vegetables, and Florist and Nursery Stocks. U.S. Dept. Agric. *Handbook*, Vol. 66.

Ito, G.M.; Brewbaker, J.L. (1991). Genetic analysis of pericarp thickness in progenies of eight corn hybrids. *J.Am.Soc.Hort.Sci.*,Vol. 116(6), 1072-1077.

Kessler, Jr.; Harry, T. (1998). Machine for cutting krnels from ears of corn. USA Patent 5830060.

Kumari, J.; Gadag, R.N.; Jha G.K. (2007). Genetic analysis and correlation in sweet corn (*Zea mays*) for quality traits, field emergence and grain yield. *Indian J.Agric. Sci.*, Vol. 77(9), 613-615.

Kunicki, E. (2003). Uprawa kukurydzy cukrowej. Kraków, Polska, Plantpress.

Lee, Y.C.; McCoon, M. (1981). Lipoxygenase and off –flavor development in some frozen foods. *Korean J. Food Sci. Technol.*, Vol. 13, 53.

Marshall, S.W.; Tracy, W.F. (2003). Sweet corn. In: RAMSTAD P.E., WHITE P. (eds.): Corn Chemistry and Technology. American Association of Cereal Chemists, Minneapolis, 537-569.

Matheus, A.O.R.; Martinez, N.M.; Bertorelli, L.O. & Venanzi, F. (2004). Adaptability of sweet corn ears to a frozen process. *Archivos Latinoamericanos de Nutricion*, Vol. 54(4), 438-43.

Michalsky, F. (1986). Sweet corn – Vegetable of future? *Mais,* Vol. 2, 40-43.

Motes, J.E.; Roberts, W.;Cartwright, B. (2007). HLA-6021-Sweet corn production. Available at http://osufacts.okstate.edu.

Nguyen, Q.L. (1986). Sledonanie kvality praûce zberaca kukurice. Proceedings of the International Conference. Grain harvest, 161–163. Nitra-Czechoslovakia

Niedziółka, I.; Szymanek, M. (2004). Wpływ blanszowania na wybrane właściwości mechaniczne ziarna kukurydzy cukrowej. *Acta Agrophysica*, 4(2), 449-457.

Niedziółka, I.; Szymanek, M. (2006). Effects of some working parameters of corn cutter on cutting process. *Inżynieria Rolnicza*, Vol. 6, 81-89.

Olson K. (2000). Northland foods: planing the end. *International Food and Agribusiness. Management Review*, Vol. 3, 423-432.

Orłowski, M. (2000). Polowa uprawa warzyw. Kukurydza cukrowa, 383-386.

Puangnak, W. (1998). Effect of hybrid, maturity and kernel structure on lipid content, composition and aroma development in sweet corn, M.S. thesis, University of Maryland, College Park.

Reyes, F.G.; Varseveld, G.W. (1982). Sugar composition and flavor quality of high sugar (shrunken) and normal sweet corn. *J. Food Sci.*, Vol. 47, 753-755.

Robertson, G.H.; Guadagni, D.G.; Lazar, M.E. (1980). Flavor and texture of preserved intact sweet corn: Comparison with cut sweet corn and storage tests. *J. Food Sci.* Vol. *45*, 221–223.

Robertson, G.H.; Farkas, D.F. (1982). Apparatus for removing corn from cob. USA Patent 4318415.

Salunkhe, D.K.; Kadam, S.S. (1998). Handbook of vegetable science and technology; Production, composition, storage and processing. Marcel Dekker, Inc.

Scott, C.E.; Eldridge, A.L. (2005). Comparison of carotenoid content in fresh, frozen and canned corn. *J. Food Comp. Anal.*, Vol. 18, 551-559.

Simonne, E.; Simonne, A.; Boozer, R. (1999). Yield, ear characteristics, and consumer acceptance of selected white sweet corn varieties in the southeastern United States. *Hort. Technology*, Vol. 9(2), 289-293.

Suk, S.L.; Sang, H.Y. (1999). Sugars, soluble solids and flavor of sweet, super sweet and waxy corns during grain filling. *Korean J. Crop Sci.*, Vol. 44(3), 267-272.

Stewart, B.; Chevis, P.; Perera, C. (1997). Peroxidase isoforms of corn kernels and corn on the cob: Preparation and characteristics. *Lebensmittel Wissenschaft und Technologie*, Vol. 30, 192-201.

Szpaar, D.; Dregiew, D. (1999). Kukuruza. Uczebno–prakticzeskoje rukovodstvo po vyraszczyvaniu kukuruzy. Minsk.

Szymanek, M. (2011). Studies on shelling of frozen sweet corn. *J.Food Proc. Eng.*, Vol. 34(3), 716–727.

Trongpanich, K.; Stonsoavapak, S.; Hengsawadi, D.; Lowitoon, N.A. (2002). A comparative study on pretreatment processes of canned whole kernel sweet corn. *Kasetsart J.Nat. Sci.*, Vol. 36 (1), 63–68.

Wade, J.H. (1981). Characterization of aroma components of corn. Ph. D. thesis, University of Georgia, Athens.

Waligóra, H. (2006). Harvest and utilization of sweet corn. *Kukurydza*, Vol. 2(28), 26-27.

Warzecha, R. (2003). Słodki smak kukurydzy. *Owoce warzywa kwiaty*, Vol. 6, 20-21.

Wong, A.D. (1994). A study of kernel composition affecting the quality of shrunken2 sweet corn. Disseration Abstracts International, 55(1) 10 Order no. DA9416451.

Wong, A.D.; Swiader, J.M. (1995). Nitrogen and sulfur fertilization influences aromatic flavor components in shrunken2 sweet corn kernels. *J. Am. Soc. Hort. Sci.*, 120(5), 771-777. http://attra.ncat.org/attra-pub/PDF/sweetcorn.pdf

5

Design of a Simple Fuzzy Logic Control for Food Processing

Abdul Rahman O. Alghannam
Department of Agriculture Systems Engineering,
College of Agricultural and Food Sciences, King Faisal University, Al-Hassa
Saudi Arabia

1. Introduction

The main objectives of the food process control are to maintain food safety, quality assurance, less processing time, and high production at minimum cost (Linko And Linko (1998). In the food industry, end-products must achieve a compromise between several properties, including sensory, sanitary and technological properties. Among the latter, sensory and sanitary properties are essential because they influence consumer choice and preference. Advanced process control techniques have been widely applied in the chemical, petrochemical and forest-based industries, after the apparently first computer-aided process control system was installed in 1959 in an oil refinery in Port Arthur, Texas (Johnson (1996), Anderson et al. 1994). In general, computerized control systems in the food industry have been recently comprehensively discussed Mittal (1997). There is no doubt that advanced, intelligent control techniques such as model-based, expert, neuro-fuzzy and hybrid control systems would offer particular advantages also in food and allied processes (Caro and Morgan(1991). Investments in automation, robotics, and advanced control techniques are likely to result in marked savings in costs, increased productivity, improved and more consistent product quality, and increased safety8. (Linko And Linko (1998).

2. Fuzzy logic control for food processing

There is a collection of papers that are written in the topic of "fuzzy logic and the quality control of the food" classified by Perrot et al. (2006) . Only forty of which are on the subject of Supervision Decision help system control. Most of them are classical applications of the Takagi–Sugeno controller, such as Linko et al. (1992) for extrusion cooking, Zhang and Litchfield (1993) for drying, Norback (1994) for cheese-making, Alvarez et al. (1999) for controlling isomerizes hop pellet production, Honda et al. (1998) for controlling the sake brewing process, and O'Connor et al. (2002)for controlling the brewing process. Guillaume et al. (2001) optimized a fuzzy rule basis using a genetic algorithm to establish a decision support system for the cheese-making process. Davidson et al.(1999) developed a fuzzy control system for continuous cross flow in which he used a fuzzy arithmetic that estimates the browning of peanut roasting. Perrot et al. (2000) proposes a fuzzy logic approach to control the quality of the biscuits in an industrial tunnel oven, Voos et al. (1998)develops a fuzzy control of a drying process in the sugar industry based on operator experience, and

Curt et al. (2002) develops five Takagi–Sugeno modules to control the quality of the sausage during ripening. It is also used to perform supervisory tasks such as Acosta-Lazo et al. (2001)for the supervision of a sugar factory. Perrot et al. (2004) developed a decision help system to control the cheese ripening process, integrating the uncertainty of human measurements. Petermeier et al. (2002) used a hybrid approach to develop a model of the fouling behavior of an arbitrary heat treatment device for milk. This is developed by combining deterministic differential equations with cognitive elements for the unknown parts of the knowledge model. These authors emphasize the relevance of this open field of research in the context of food processes and the interest of fuzzy symbolic representation of expert reasoning. Nevertheless, they call into question the optimality of the approaches developed on the basis of imperfect and incomplete expert knowledge.

More papers emphasized on the application of control techniques such as Kupongsak and Tan (2006) who applied fuzzy set and neural network techniques to determine food process control set points for producing products of certain desirable sensory quality. The results demonstrated a great potential of the fuzzy set concept and neural network techniques in sensory quality-based food process control. Soares et al. (2010) applied high performance nonlinear fuzzy controllers for a soft real-time operation of a drying machine. All the criteria evaluation used for controller's performance analysis for several steps tracking tasks has showed much better performance of the fuzzy logic controller. The absolute errors were lower than 8,85 % for Fuzzy Logic Controller, about three times lower than the experimental results. Omid (2011) Designed an expert system for sorting pistachio nuts through decision tree and fuzzy logic classifier The correct classification rate and root mean square error RMSE for the training set were 99.52% and 0.07, and for the test set were 95.56% and 0.21, respectively. These encouraging results as well as the robustness of the FIS based expert system makes the approach ideal for automated inspection systems. A prototype-automated system for visual inspection of muffins was developed by Zaid et al. (2000). The automated system was able to correctly classify 96% of regarded and 79% of ungraded muffins. The algorithm procedure classified muffins to an accuracy of greater than 88%, compared with 20_30% variations in quality decisions amongst inspectors. Podržaj and Jenko (2010) found out that temperature control based on fuzzy logic is suitable for processes in which a high degree of precision is required. Research conducted by Venayagamoorthy et al. (2003) has shown that the fuzzy logic controller augmenting the conventional P and PI controller does control more efficiently the industrial food processing plant with respect to set point tracking and disturbance rejection. Turing a predictive fuzzy logic controller for resin manufacturing the predictive FLC scheme is found to be highly useful and satisfactory in controlling an exothermic process. Nagarajan et al. (1998).

3. Food processing control

Some unique aspects of food processing problems:

Managing the properties of food starting from the input stage with the aim of controlling them is not an easy task for several reasons (Perrot et al. (2006), Welti et al. 2002):

- There are many parameters in food industry that must be taken into consideration in parallel. A single sensory property like color or texture can be linked individually to several dimensions recorded by the human brain.

- The food industry works with non-uniform, variable raw materials that, when processed, should shaped into a product that satisfies a fixed standard.
- The process control of foods are highly non-linear and variables are coupled.
- Little data are available in traditional manufacturing plants that produce, for example, sausage or cheese and this situation is applies to most food processing industry.
- In addition to the temperature changes during a heating or cooling process, there are biochemical (nutrient, color, flavor, etc.) or microbial changes that should be considered.
- The moisture in food is constantly fluctuating either loss or gain throughout the process which can affect the flavor, texture, nutrients concentration and other properties.
- Other properties of foods such as density, thermal and electrical conductivity, specific heat, viscosity, permeability, and effective moisture diffusivity are often a function of composition, temperature, and moisture content, and therefore keep changing during the process.
- The system is also quite non-homogeneous. Such detailed input data are not available.
- Often, irregular shapes are present.

4. Design of fuzzy logic control system

In most process control problems, it is relatively easy to design a PID controller, merging fuzzy rules into the system produces many extra design alternatives, and despite the availability of publications on fuzzy control on food processing, there are few general guidelines for setting the parameters of a simple and practical fuzzy controller.

The approach here is based on a three step design procedure, that builds on PID control Jantzen (1998):

1. Start with a PID controller.
2. Insert an equivalent, linear fuzzy controller.
3. Make it gradually nonlinear.

5. Start with a PID controller

The term control in engineering refers to a discipline whose main interest is to solve problems of regulating and controlling the behavior of physical system. In food process, both an open and close loop control configuration are applied. A bottle washing machine performing predefined sequence of operations without any information "with no concern" regarding the results of its operation is an example of an open loop control system. The bottle washing machine mentioned above as an open-loop system would operate in a close-loop mode if it were equipped with a measuring device capable of generating signal related to the degree of cleanness of the bottles being washed.

For decades, food process engineers have adopted control strategies that have been introduced by control engineering. The control engineering is based on the foundation of feedback and feed forward theory and linear system analysis. It is basically an interconnection of components forming a system configuration that will provide a desired system response (Dorf and Bishop, 2010). The feedback control acts when a deviation from the set point occurs. It uses the difference of the controlled variable between the set point and the actual one to control the actuators of the process in the following four modes (Kreider and Rabl, 1994):

- Two position (On/Off)
- Proportional
- Integral
- Derivative.

The distinct advantage of the feedback control system is the ability to adjust the transient response and steady-state performance. A schematic diagram of the feedback theory is depicted in figure 1

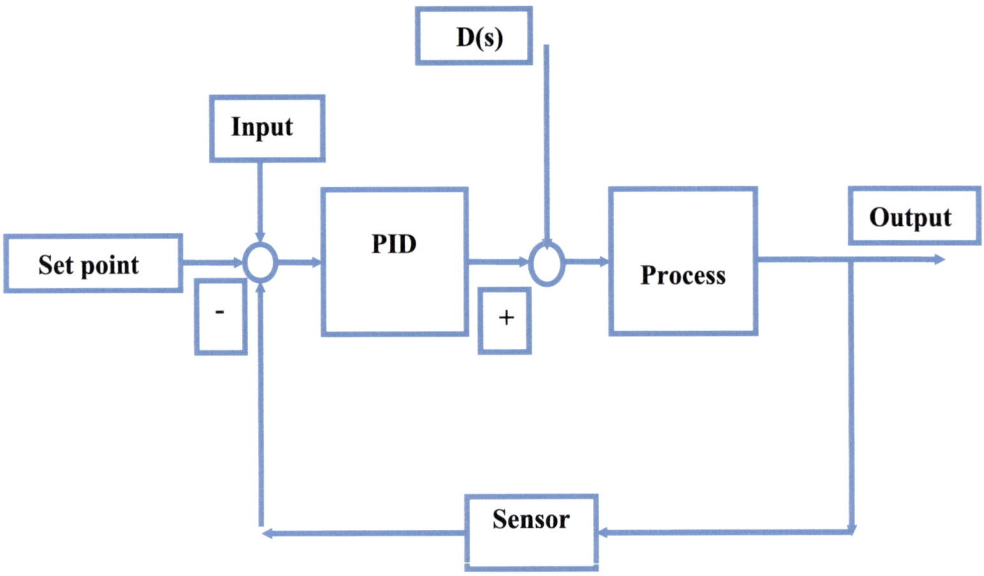

Fig. 1. A schematic drawing of PID control system

5.1 On/Off controller

The two-positions control applies to an actuator or a relay that is either fully open or fully closed. When the controlled variable drops below the lower decided limit, the actuator opens fully and remains open until the controlled variable reaches the upper limit. The upper and lower limits are sometimes adjustable. The two-position control is the least expensive method of automatic control and convenient for the use in systems with large time constants (Kreider and Rabl, 1994).

5.2 Design of PID controller

The Proportional, Integral, and Derivative are usually used in a variety of combinations with one another to achieve the right control process. A schematic drawing of a process feedback system with a PID controller is depicted in figure1.

The proportional control corrects the controlled variable in proportion to the difference between the controlled variables and the set point. The error is calculated as follow:

$$e = T_{set} - T_{sensed} \quad (1)$$

Integral control is often added to proportional control to eliminate the error inherent in proportional-only control. The integral component has the effect of continuing to increase or decrease the output as long as any offset continues to exist (Smith and Corripio, 1997; Kreider and Rabl, 1994).

Derivative control is often called the rate action, or pre-act and is used to anticipate where the process is heading by looking at the time rate of change of the error, and its derivative. In other words, it gives the controller the capability to "look ahead" by calculating the derivative of the error (Smith and Corripio, 1997).

6. Measuring device (sensors)

The use of advanced instrumentation and sensors in the food industry has led to continuing improvement in food quality control, safety and process optimization. Some of the basic measurement devices used in process control of foods are pressure, temperature level, and flow sensors. Other measurement devices are used such as color vision, speed of sound, viscometers texture sensors, chemo-sensors, biosensors, immune-sensors, electronic noses and tongues, sensors for food flavor and freshness: electric Noses, tongues and testers In situ freshness monitor of frying oil (resonant viscosity probe) , knife-type meat freshness tester (glucose profiling biosensor). Figure 3 shows an overall diagram of food process control systems that illustrates some of the possible variables, parameters, actuators, sensors. (Kress-Rogers and Brimelow (2002))

In most food process control applications, standard "off-the-shelf" devices are used to obtain the desires system performance. These devices commonly called industrial controllers. The manner in which the controller produces the control signal in response to the controller error is referred to as a control algorithm or control law. The most common control algorithms implemented in industrial controllers are the two position or on/off control , proportional Integral derivative control (PID), and fuzzy logic control. PID controllers are affordable, robust, fairly easy to use, tune and maintain, and generally commercially available. Figure2 Shows an overall diagram of food process control systems that illustrates some of the possible variables, parameters, actuators, sensors.

7. Insert an equivalent, linear fuzzy controller

Fuzzy logic deals with uncertainty. This technique which uses the mathematical theory of fuzzy sets simulates the process of normal human reasoning by allowing computer to behave less precisely and logically than conventional computers do. Some of the drawback, however, of the on/off switching system used in food process engineering are; they are incapable to maintain a set point temperatures accurately due to the non-linearity of this system and it is hard to design a controller to maintain a fixed process variable. Whereas the weakness of the Proportional Integral Derivative (PID) controller is its inability to implement human thinking. Most food related processes are multivariable, time-varying and non-linear. Non-linear processes are difficult to predict with conventional control systems designed for linear processes but cases involving several process variables have been generally dealt with by multi-loop controllers running several independent PID-loops concurrently (Linko And Linko (1998). However, the advantages of the fuzzy logic control systems are its easiness to use and maintain and its affordability. Fuzzy logic can be used for controlling a process that is too nonlinear or too ill-understood to use conventional control design. Also fuzzy logic enables control engineers to easily implement control strategies used by human operators. Other advanced control system are the hybrid systems that encompasses decision tree, neural networks(NN), evolutionary algorithms, and expert systems. Omid (2011)

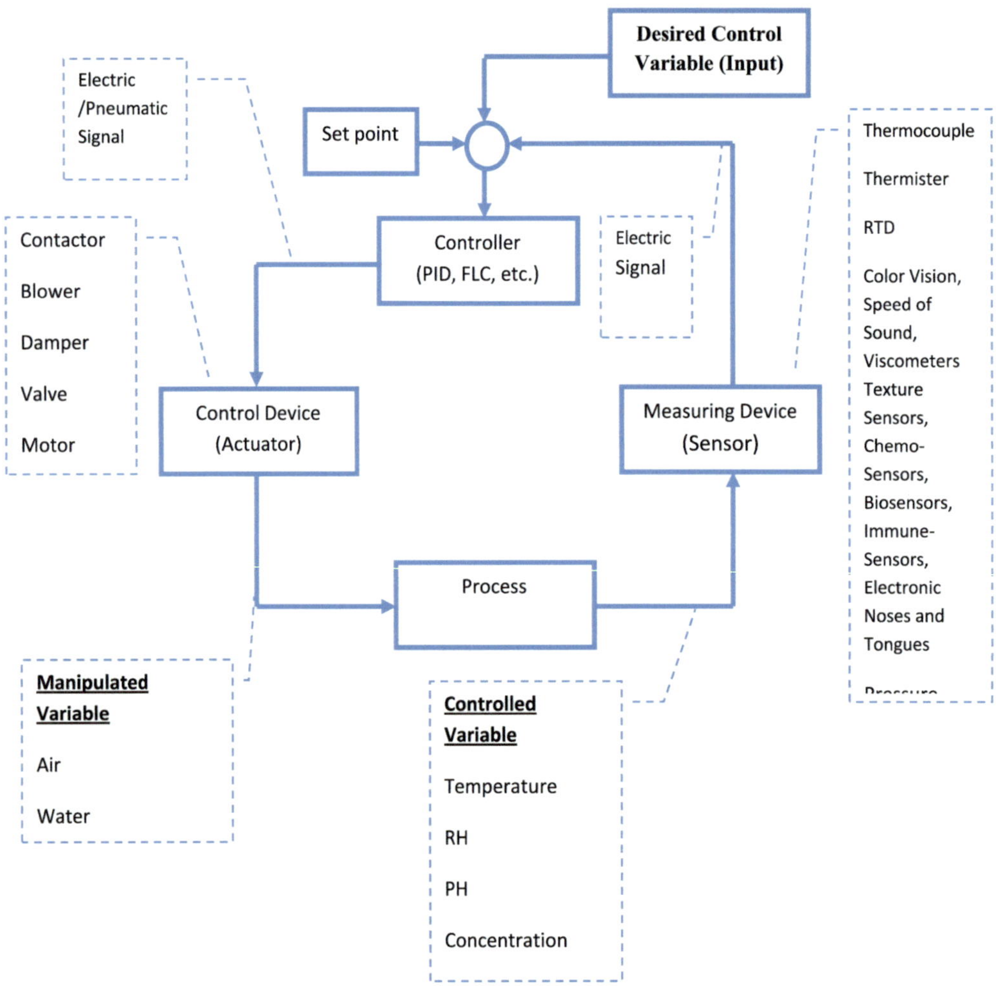

Fig. 2. An overall diagram of food process control systems that illustrates some of the possible variables, parameters, actuators, sensors.

On the other hand, fuzzy logic is simple to use if incorporated with analog-to-digital (D/A) converters, and micro controllers. This can easily be upgraded by changing rules to improve performance or add new features to the system . In many cases, fuzzy control can be used to improve existing controller systems by adding an extra layer of intelligence to the current control method. Although fuzzy logic control systems are still young in the food process engineering, the fields of applications of fuzzy systems are very broad. Those applications include: Pattern recognition and classification, modeling of classification control systems fault diagnosis operation research, decision support systems Omid (2011). Therefore, more precise systems are needed for the many applications in the field. Furthermore, fuzzy logic systems could be used as an alternative controller for the most food process plants. With the fast growing computer applications in food engineering, there is a need to test and evaluate more advance controllers to reach the best and most affordable food process control systems.

Fuzzy logic is basically a multi-valued logic that allows transitional values to be between the normal two valued evaluations like Yes/No, True/False, and Black/White. Instead, phrases like "very light" or "pretty heavy " can be formulated mathematically and dealt with by computers. Fuzzy controllers have three transitional steps; an input step, a processing step, and an output step. The input step, which is fuzzy matching, maps sensor output or the error or other inputs to the proper membership functions by calculating the degree of membership; the processing step triggers each appropriate rule and produce a result for each, then join the results of the rules together; and finally a crisp control value of the result is obtained through the output step (Bauer et al. 1998; Yen and Langari, 1999).

The control system block from figure1 is further detailed as follows to include the fuzzy logic control system in figure3.

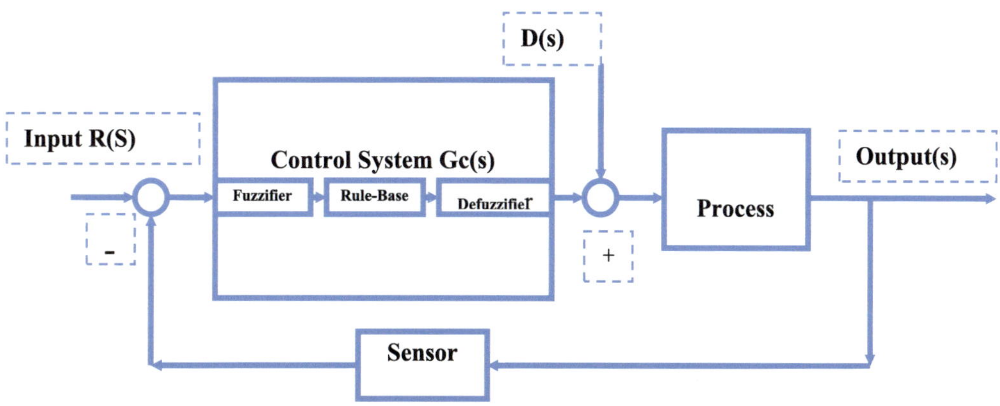

Fig. 3. A schematic drawing of feedback control system with fuzzy logic controller

A block diagram of a fuzzy control system is shown in Figure3. The fuzzy controller is composed of the following three elements:

a. A fuzzification interface, which converts controller inputs into information that the inference mechanism can easily use to activate and apply rules. shapes of membership functions are the triangular, trapezoidal and bell, but the shape is generally less important than the number of curves and placement. From 3 to 7 curves are generally enough to cover the intended range of an input value (the "universe of discourse")(Bauer et al. 1998; Yen and Langari, 1999).

b. A rule-base (a set of If-Then rules), which contains a fuzzy logic quantification of the expert's linguistic description of how to achieve good control. In other word, the rule base is derived from an "inference engine" or "fuzzy inference" module, which emulates the expert's decision making in interpreting and applying knowledge about how best to control the plant.

The processing stage is basically a group of logic rules in the form of IF-THEN statements, where the IF part is called the "antecedent" and the THEN part is called the "consequent"(Yen and Langari, 1999). For example the rule of a thermostat works as follow:

IF the temperature is "cold" THEN turn the heater to "high"

This rule basically implements the truth-value of the temperature input, which is cold to create a result in the fuzzy set to switch the heater to high. The results of all the rules are joined together using one of the defuzzification methods to finally come up with the crisp composite output. Sometimes, membership functions are formulated by "hedges". Examples of hedges include "more", "less", "about", "close to", "approximately", "very", "slightly", "too", "extremely", and "somewhat". These phrases may have precise definitions and mathematical representation. For example, "very", squares membership functions and since the membership magnitudes are always below 1, this reduces the membership function (Bauer et al. 1998; Yen and Langari, 1999).

Very:

$$\mu_{very\ A}(x) = [\mu_A(x)]^2 \tag{2}$$

"Extremely" cubes the values to give more reduction:

$$\mu_{extremely\ A}(x) = [\mu_A(x)]^3 \tag{3}$$

and "somewhat" broadens the function by taking the square root;

$$\mu_{somewhat\ A}(x) = \sqrt{\mu_A(x)} \tag{4}$$

Membership functions can be joined together using a number of logical operators. For example, AND (conjunction) uses the minimum value of all the antecedents. However, the OR and NOT use the maximum value and the complementary value respectively. There are other different operators used to define the result of a rule, however, the most commonly used method to calculate the output is the "max-min" inference method (Yen and Langari, 1999; Bauer et al. 1998).Those rules can be implemented using hardware or in software.

Jantzen (1998) has suggested some sources of control rules :

- *Expert Experience and Control Engineering Knowledge:* The most common approach to establishing such a collection of rules of thumb, is to question experts or operators using a carefully organized questionnaire.
- *Based on the Operator and Control Action: Fuzzy-if then* rules can be deduced from observations of an operator's control actions or a log book. The rules express input-output relationships.
- *Based on A fuzzy model of the Process* A linguistic rule base may be viewed as an inverse model of the controlled process. Thus the fuzzy control rules might be obtained by inverting a fuzzy model of the process. This method is restricted to relatively low order systems, but it provides an explicit solution assuming that fuzzy models of the open and closed loop systems are available. Another approach is *a fuzzy Identification* or fuzzy model-based control.
- *Based on Learning* The self-organizing controller is an example of a controller that finds the rules itself. Neural networks is another possibility
- A defuzzification interface, which converts the conclusions reached by the inference mechanism into the inputs to the plant. In other words, results of the fuzzy rules are defuzzified using one of the defuzzification techniques to give a final crisp value to be sent as the control parameter. Among the defuzzification techniques are (Bauer et al. 1998;Yen and Langari, 1999):
- Mean of Maximum (MOM)

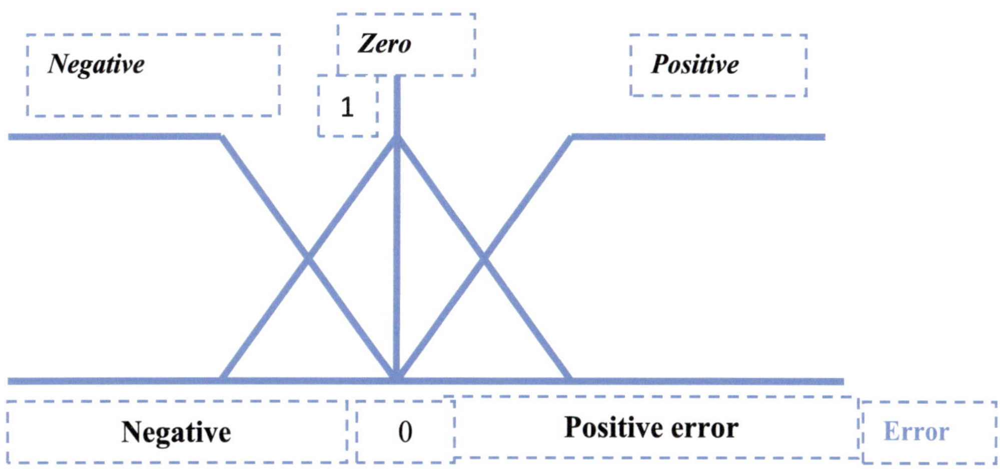

Fig. 4. Input membership functions

$$MOM(A) = \frac{\sum_{y^* \in P} y^*}{|P|} \quad (5)$$

$$P = \{y^* \mid \mu_A(y^*) = \sup_y \mu_A(y)\} \quad (6)$$

P = the set of output values with highest possibility degree in A

Center of Area (COA) or centroid method

$$COA(A) = \frac{\sum_x \mu_A(x) \times x}{\sum_x \mu_A(x)} \tag{7}$$

$\mu_A(x)$ = Weight for value x.

- Center Of Gravity Method for Singleton (COGS)

$$y = \frac{\sum_{k=1}^{m} c_i w_i}{\sum_{k=1}^{m} w_i} \tag{8}$$

c_i = The center of gravity

w_i = The degree of match with the input data

a. Membership functions

Example of triangular membership functions for error input with 50% overlap:

A triangular output membership functions with 50% overlap:

b. Optimization in food process control

In recent years, many efforts have been directed to the optimization and efficient control of food processing. Most bioprocesses have highly nonlinear dynamics, and constraints are frequently present on both the state and the control variables. Thus, efficient and robust dynamic optimization methods are needed in order to successfully obtain their optimal operating policies. Balsa et al. (1998)

Optimization can be defined as the process of finding the conditions that give the optimum (maximum or minimum) value of a function of certain decision variables subject to restrictions or constraints that are imposed (Edgar and Himmelblau. (1989)). Optimization may be the process of maximizing a desired quantity or minimizing an undesired one. The conditions (values of the processing variables) that produce the desired optimum value are called optimum conditions while the best of all feasible designs is called optimal design. In its most general meaning, optimization is the effort and process of making a decision, a design, or a system as perfect, effective, or functional as possible. Optimization for a system may mean the design of system parameters or the modification of its structure to minimize the total cost of the system's products under boundary conditions associated with available materials, financial resources, protection of the environment, and governmental regulation, taking into account the safety, operability, reliability, availability, and maintainability of the system. Optimizers or decision makers use optimization in the design of systems and processes, in the production and in systems operation. Some examples of the optimization use are: selection of processes or size of equipment, equipment items and their arrangement, operation conditions (temperature, pressure, flow rate, chemical composition of each stream in the system), equipment combination in specific processes to increase the overall system availability, etc.(Tzia and Liadakis (2003).

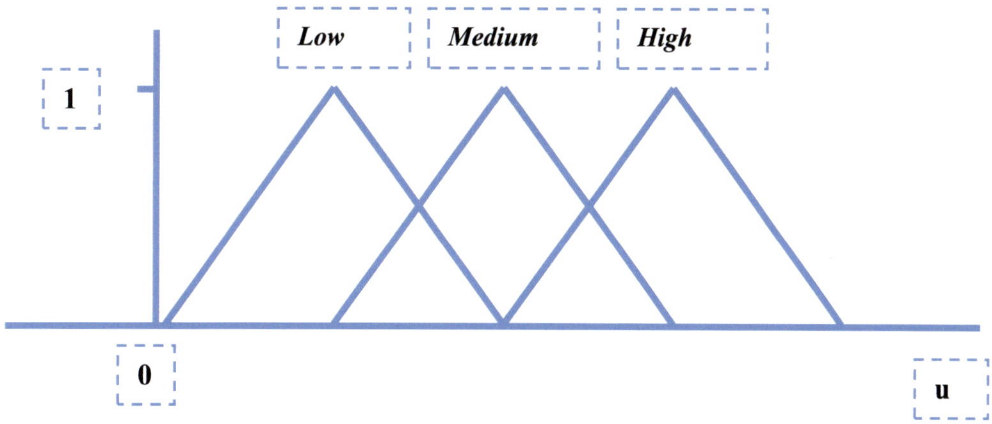

Fig. 5. The output membership functions

c. Performance Indices:

In order to determine the control parameters whether PID or fuzzy logic parameters a method of optimization should be applied to shorten time and get the best output (Edwards and Canning, 1997). Some simple tuning methods have been used for this optimization such as, Ziegler-Nichols method. This method uses small amount of information about the process to tune the system (Chipperfield and Fleming, 1993). Several general performance indices exist in the literature. These indices use the error as an indicator of the system deviation. The system is considered to have achieved the optimum output, when these indices reach the minimum value (Dorf and Bishop, 2010).

7.1 General indices

Integral of square of error:

$$ISE = \int_0^t e^2(t)\, dt \tag{9}$$

t : Settling time or could be an arbitrary steady state time

Another performance index is the integral of absolute error IAE:

$$IAE = \int_0^t \left| e \right| dt \tag{10}$$

This criterion is suitable for computer simulation.

Other general performance indices are also used for optimization such as ITAE, and ITSE (Dorf and Bishop, 2010). During pistachio classification, Omid (2011) compared root mean square error (RMSE), Mean absolute error (MAE), relative absolute error (RAE) and correct classification rate (CCR) as Performance indices. Balsa et al. (1998) Evaluated and compared the solution of the dynamic optimization of three bioprocesses, including a hybrid stochastic-deterministic method, where they found a significant advantages over other approaches.

8. Make it gradually nonlinear

Linear processes have the important property of superposition whereas nonlinear models do not. Superposition means that the response of the system to a sum of inputs is the same as the sum of responses to the individual inputs .These properties do not hold for nonlinear models .In this respect, it is important to recognize the fact that most food process systems are nonlinear.

There are three sources of nonlinearity in a fuzzy controller.

i. The Rule Base: The position, shape and number of fuzzy sets as well as nonlinear input scaling cause nonlinear transformations. The rules often express a nonlinear control strategy.
ii. The inference Engine :If the connectives AND and OR are implemented as for example MIN and MAX respectively, they are nonlinear.
iii. The defuzzification: Several defuzzification methods are nonlinear.

With these design choices the control surface degenerates to diagonal plane. A flexible fuzzy controller, that allows these choices, is two controllers in one. When linear it has transfer function and the usual methods regarding tuning and stability of the closed loop system apply.

It is possible to construct Aa rule base with linear input-output mapping (siler & Ying, 1989; Mizumoto, 1992; Qiao & Mizumoto; 1996) the following checklist summarizes the general design choices for achieving a fuzzy rule base equivalent to summation:

- Use triangular input sets that cross at μ=0.5;
- Use the algebraic product (*) for the end connective.
- The rule base must be the complete and combination of all input families;
- Use output singletons, positions determined by the sum of the peak positions of the input sets;
- Use Center Of Gravity Method for Singleton (COGS) defuzzification.

9. Summary

In this paper, the basics of the design of fuzzy logic control for food processing have been introduced. Due to its simplicity, it is a powerful method for managing the properties of food and solve its problems such as single sensory property like color or texture , non-uniformity, variability of raw materials, non-linearity, coupling of its variables continuous temperature and moisture change during a heating or cooling process, biochemical or microbial changes, density, thermal and electrical conductivity, specific heat, viscosity, permeability, effective moisture diffusivity.

Review of papers written in fuzzy logic control for food processing as well as steps for designing simple fuzzy logic control were introduced. It seems reasonable to start the controller design with PID controller; then tune a PID controller; replace it with a linear fuzzy controller; transfer gains; make the fuzzy controller nonlinear; then fine-tune it after the following steps have been achieved:

- Start with a PID controller.
  - Proportional
  - Integral
  - Derivative.
- Insert an equivalent, linear fuzzy controller.
  - Fuzzification
  - Rule-base (a set of If-Then rules)
  - Defuzzification
  - Membership functions
  - Optimization
  - Performance Indices
- Make it gradually nonlinear.

10. References

Acosta-Lazo G.G., Alonso-Gonzales C.J., Pulido-Junquera B. (2001) Knowledge based diagnosis of a sugar process with teknolid, International Sugar J. (103) 44–51.

Alvarez E., Cancela M.A., Correa J.M., Navaza J.M., Riverol C. (1999) Fuzzy logic control for the isomerized hop pellets production, J. Food Engrg. (39) 145–150.

Anderson, J., Backs, T., Van Loon, J. and King, M., 1994,Getting the most out of advanced process control, Chem. Engineering, 101(3): 78-89.

Balsa-Canto E., Alonso A. A., Banga J. R. (1998) .Dynamic Optimization Of Bioprocesses: Deterministic And Stochastic Strategies. ACoFoP IV (Automatic Control of Food & Biological Processes), Göteborg, Sweden, 21-23 Septe mber 1998. Correspondence to: Dr. Julio R. Banga, IIM-CSIC. Eduardo Cabello 6, 36208 Vigo, SPAIN. E-mail:julio@iim.csic.es

Bauer, P., S. Nouak, and R Winkler. 1998. Introduction to the Fuzzy Logic Course. An internet source document.
Http://www.flll.uni-Linz.ac.at/pdw/fuzzy/introduction.html

Caro, R. H. and Morgan, W. E., 1991, Trends in process control and instrumentation, Food Technol., 45(7): 62-66.

Chipperfield, A.J. and P. J. Fleming. 1993. MATLAB toolbox and applications for control. Peter Peregrinus Ltd. Six Hills Way, Stevenage, Herts. SG1 2AY, United Kingdom.

Curt C., Hossenlopp J., Perrot N., Trystram G.(2002) Dry sausage ripening control. Integration of sensory related properties, Food Control (13) 151–159.

Davidson V. J., Brown R.B., Landman J. J. (1999) Fuzzy control system for peanut roasting, J. Food Engrg. (41) 141–146.

Dorf R. C. and R. H. Bishop. 2010 .Modern Control Systems. Eighth Edition. Addison Wesley Longman. Inc. 2725 Sand Hill Road, Menlo Park, CA 94025.

Edgar TF., Himmelblau DM. 1989. Optimization of Chemical Processes. Singapore. Mc-Graw-Hill Book Co., Chapters 1–9, p. 3–438.

Edwards D. and H. T. Choi. 1997. Use of Fuzzy Logic to Calculate the Statistical Properties of Strange Attractors in Chaotic Systems. Fuzzy Sets and Systems 88(1997) 205-217.

Guillaume S., Charnomordic B. (2001)Knowledge discovery for control purposes in food industry databases, Fuzzy Sets and Systems (122) 487–497.

Honda H., Hanai T., Katayama A., Kobayashi T.H.T. (1998)Temperature control of Ginjo sake mashing process by automatic fuzzy modeling using fuzzy neural networks, J. Fermentation Bioengrg. (85) 107–112.

Jantzen J. (1998)Design Of Fuzzy Controllers Technical University of Denmark, Department of Automation, Bldg 326, DK-2800 Lyngby, DENMARK. Tech. report no 98-E 864 (design), 19 Aug 1998.

Johnson, D., 1996, Distributed control: A view of what' s new, Control Engineering, 43(12): 78-82.

Kreider, J. and A. Rabl. 1994. Heating and Cooling Buildings. McGraw-Hill, Inc. Princeton Road, S-1. Hightstown, NJ 08520.

Kress-Rogers E. and Brimelow C.(2002) Instrumentation and Sensors for the Food Industry

Kupongsak S., Tan J.(2006) Application of fuzzy set and neural network techniques in determining food process control set points Fuzzy Sets and Systems (157)1169 - 1178

Linko S. And Linko P. 1998. Developments In Monitoring And Control Of Food Processes. Trans. Chem. E, Vol 76, Part C.

Linko P., Uemura K., Zhu Y., Eerikainen T., Application of neural modeling in fuzzy extrusion control, Trans. I. Chem. E 70 (1992) 131–137.

Mittal, G. S., 1997, Computerized Control Systems in the Food Industry (Marcel Dekker, Inc., New York).

Nagarajan R., Kumar R. N., Halim R. A., Rosli A. (1998) A Predictive Fuzzy Logic Controller For Resin Manufacturing Computers ind. Engineering Vol. 34, No. 2, pp. 493-500.

Norback J.P. (1994) Natural language computer control of crucial steps in cheese making, Artificial Neural Networks Dairy Industry 49 (2) 119–122.

O'Connor B., Riverol C., Kelleher P., Plant N., Bevan R., Hinchy E., D'Arcy J., (2002) Integration of fuzzy logic based control procedures in brewing, Food Control (13) 23–31.

Omid M. (2011) Design of an expert system for sorting pistachio nuts through decision tree and fuzzy logic classifier. Expert Systems with Applications (38) 4339–4347

Perrot N., Agioux L., Ioannou I., Mauris G., Corrieu G., Trystram G., (2004) Decision support system design using the operator skill to control cheese ripening—Application of the fuzzy symbolic approach, J. Food Engrg. (64) 321–333.

Perrot N., Ioannoub I., Allaisc I., Curtc C., Hossenloppc J., Trystramc G. 2006. Fuzzy concepts applied to food product quality control: A review. Fuzzy Sets and Systems 157 (2006) 1145 - 1154

Perrot N., Trystram G., F. Guely, Chevrie F., Schoesetters N., Dugre E. (2000) Feed-back quality control in the baking industry using fuzzy sets, J. Food Process Engrg. (23) 249–279.

Petermeier H., Benning R., Delgado A., Kulozik U., Hinrichs J., Becker T. (2002) Hybrid model of the fouling process in tubular heat exchangers for the dairy industry, J. Food Engrg. (55) 9–17.

Podržaj P., Jenko M. (2010) A fuzzy logic-controlled thermal process for simultaneous pasteurization and cooking of soft-boiled eggs. Chemo-metrics and Intelligent Laboratory Systems (102) 1–7

Rao SS., Engineering Optimization. 1996. Theory and Practice, 3rd ed. New York: John Wiley & Sons, Inc.

Smith, C. and A. Corripio. 1997. Principles and Practice of Automatic Process control. Second Edition. John Wiley &Sons, Inc. New York

Soares M. P., Nicola C. B., Augusto J. F., Neto. F. Nonlinear Fuzzy Tracking Real-time-based Control of Drying Parameters. 2010 World Academy of Science, Engineering and Technology (71) 187-201

Taguchi G., Phadke MS. (1989) Quality engineering through design optimization. In: D Khosrow, ed. Quality Control, Robust Design, and the Taguchi Method. Pacific Grove, CA: Wadsworth & Brooks/Cole, Advanced Books & Software, pp. 77–96.

Tzia C. and Liadakis G. 2003.Extraction Optimization in Food Engineering Marcel Dekker, Inc.

Venayagamoorthy G. K., Naidoo D., Govender P.(2003) An Industrial Food Processing Plant Automation Using A Hybrid of PI and Fuzzy Logic.Control. The IEEE International Conferenceo n Fuzzy Systems 1059-1062

Voos H., Litz L., Konig H. (1998) Fuzzy control of a drying process in sugar industry, 6th European Congress Intell. Tech. Soft Comput. EUFIT '98, vol. 3 pp. 1476–1480.

Welti J., Barbosa-Cánovas G., Miguel J. A. 2002 Engineering and Food for the 21st Century. 2002 by CRC Press.

Yen J. and R. Langari. 1999. Fuzzy Logic -Intelligence, Control, and Information. Prentice-Hall, Inc. Upper Saddle River, New Jersey 07458.

Zaid Abdullah, M., Abdul Aziz, S., Dos-Mohamed, AM., 2000. Quality inspection of bakery products using a color-based machine vision system. Journal of Food Quality 23 (1), 39-50.

Zhang Q., Litchfield J. (1993) Fuzzy logic control for a continuous cross flow grain dryer, J. Food Process Engrg. (16) 59–77.

6

Machine Vision to Determine Agricultural Crop Maturity

Wan Ishak Wan Ismail and Mohd Hudzari Razali
Universiti Putra Malaysia
Malaysia

1. Introduction

Agriculture in Malaysia is still one of the biggest enterprise and the most important sector in Malaysia's economy. In the last 35 years, Malaysia has emerged as the number one producer of palm oil. In year 2010, Malaysia accounts for about 55% of world palm oil production and about 62% of world exports. The Malaysian oil palm industry continues to contribute significantly to the country's economic development and foreign exchange earnings. However, the Malaysian oil palm industry needs more emphasis on its research and development to meet the world challenge and to maintain Malaysia as the top world producer of palm oil. The major problem faced by the oil palm plantation is shortage of labor, especially in harvesting of oil palm FFB (MPOIP, 2007).

In 1994, the Palm Oil Research Institute of Malaysia (PORIM) established four classes of ordinary oil palm FFB belonging to Elaeis Guineensis species. The classes of FFB arranged in the ascending degree of ripeness are the unripe, under-ripe, ripe and the over-ripe categories. The unripe bunch has purplish black colored fruits, covering more than 90% of the bunch surface. Meanwhile, fruits belonging to underripe and ripe bunches appear reddish orange or purplish red and reddish orange respectively. Finally, the bunch belonging to the over-ripe class with more than 80% of the fruits in the bunch appears darkish red. In year 2000 the authors used camera vision to categories six oil palm FFB. Camera vision was used to investigate the relationship between oil content in oil palm fruit and its surface color distribution. Colorimeter was used to determine the correlation between mesocarp oil content and color values. The above research indicate that there is a direct relationship between the color of the oil palm fruits and its grade and quality (Wan Ishak, et al., 2000). The advantage of predicting the oil palm fruit maturity in real time oil palm plantation will help harvesters to immediately harvest the oil palm FFB at the correct maturity stage. Harvesting oil palm fresh fruit bunches (FFB) at the right stage of ripeness is critical to ensure optimum quality and quantity of oil production, and thus profitability to the industry. Camera vision techniques can be implemented to automate the assessment and increase its consistency.

2. Colour vision

Color is the most important indicator for the farmer to determine the fresh fruit bunches (FFB) oil palm fruit maturity in harvesting process. The development of vision system will replace the human eye for matured FFB recognition. In real plantation environment, the variations of the daylight caused changing the light intensity that effect on automatic recognition process. Colour is a property of waves (light waves) and the ability to see light as colour is determined by changes in the frequency or wavelength of light. The eyes interpret the incoming light into three colours and finally the three colours are received by the brain and re-interpreted as the complete spectrum in the form of a colour circle. Human perception of colour is a function of the response of three types of colour, that are blue, green and red (Gonzalez and Woods, 2002). Colour is considered a fundamental physical property of agriculture products and foods. There are common indicators used to recognize ripeness of the agricultural products and thus determine the best time to harvest the products. With the aid of modern technology, automated device with intelligent computing functions has been proposed to be used to replace human naked eye in deciding fruit maturity for harvesting time.

3. Machine vision system

Vision is the most powerful sense. It provides us with a remarkable amount of information of our surroundings and enables us to interact intelligently with the environment. Machine vision applications for automated inspection and sorting of fruits and vegetables have been studied and reported. Machine vision is a technology that employs a computer and camera to analyse and interpret images in a manner resembling human vision. The camera is equivalent to the human eye and the computer is equivalent to the human brain. Vision system is a new field of research in the agricultural sector. In agricultural applications, especially for fruit handling, we cannot detect fruit quality just by its shape or pattern. This is because a fruit may have a different shape and pattern but of the same level of quality. To solve this problem, the vision system should be able to analyse the colour of the object or fruit. Colour vision systems have been found more effective in colour inspection. A colour camera output can be decoded into three images to represent the red, green and blue (RGB) components of the full image. The three components of the colour image can be recombined in software or hardware to produce intensity, saturation and hue images, which can be more convenient for subsequent processing. Besides imaging objects in the visible (VIS) color region, some machine vision systems are also able to inspect these objects in light invisible to humans, such as ultraviolet (UV), near-infrared (NIR) and infrared (IR). The information received from objects in invisible light regions can be very useful in determining preharvest fruit maturity, disease or stress states and very useful in determining plant and vegetable variety. It is also useful in detecting postharvest quality and safety, such as defects, composition, functional properties, diseases and contamination of plants, grains and nuts, vegetables and fruits, and animal products

4. Oil Palm FFB ripeness

Figure 1 shows images of oil palm fresh fruit bunches of different degrees of ripeness from the five years old tree of Oil Palm Pasifera Variety. The top left is unripe, top right is under ripe, bottom right is ripe and bottom left is overripe. It can be seen from these images that the colour of each group is highly non-uniform which can be cluster to determine the maturity stages of fruit. The Pasifera (D X P) variety contains more oil in mesocarp layer with thin kernel compared to dura and tenera varieties.

Fig. 1. Four different classes of oil palm FFB ripeness

Optimum ripeness of oil palm fresh fruit bunches (FFB) is very crucial to the oil palm purchasing centers and palm oil mills. Presently the oil palm FFB are graded manually using visual human judgement of fruit bunch grader. The definition of matured colour oil palm FFB differs between planters and sellers due to the fact that human eyes perceive colours differently. The right stage of ripeness is critical to ensure optimum quality and quantity of oil production. The critical issue in harvesting the oil palm FFB is to harvest at the right stage of ripeness for ensuring that the maximum quantity and quality of palm oil is present for extraction of oil at the palm oil mills. If the oil palm FFB is harvested too early, the bunch will be too young and may not reach optimum oil content when processed while if it is harvested at overripe stage, the oil will contain high acid which will decrease the oil quality. Over ripe oil palm FFB is indicated by the number of loose fruits being detached from the bunches. The number of loose fruit detached from the bunches will increase as the oil palm FFB getting older or overipe. Uncollected loose fruits account 3-5% of bunch weight which resulted on further reduction in the oil extraction rate and profits. The time of harvest for the matured oil palm FFB is determined by the colour of the fruits by the naked eyes. The skin color of the fruit bunch will change from black to reddish and orange during ripening process. The change in the mesocarp color is due to the accumulation of the carotene pigments, which also corresponds to the oil content of the mesocarp when analysed. A fruit bunch normally takes 20 to 22 weeks to ripen and those less than 17 weeks old could be classified as unripe or black bunches (Kaida and Zulkifli, 1992). The current practice of harvesting the oil palm FFB is when about 10 loose fruits detached from the bunch when matured and found fallen on the ground. The Malaysian Palm Oil Board (MPOB) manual on FFB grading defines a ripe bunch when the mesocarp colour is reddish orange and the bunch has 10 or more empty sockets of detached fruitlets; an under ripe bunch has yellowish orange with less than 10 empty sockets; and an unripe bunch has yellow mesocarp with no empty sockets. Thus, the conventional method of determining on the 10 number of loose fruits fallen on the ground before harvesting the FFB is not practical and uneconomical. Thus, it is ideal to have a machine or device with artificial vision system replacing human eye to recognize level of oil palm fruits ripeness in order to determine the right harvest time. This will prevent the lost of loose fruits being detached from the FFB before harvest.

5. Development of crop colour model

Application of machine vision systems in outdoor fields addressed special difficulties of an unstructured work environment. Variability in light source temperature causes changes in the colors of objects in the image, making incorrect digital image value. Most light sources are not 100% pure white but have a certain color temperature, expressed in Kelvin. For instance, the midday sunlight will be much closer to white than the more yellow early morning or late afternoon sunlight. The image sensor sensitivity (ISO) was set at normal 100 which was experimentally suitable for capturing the FFB image under oil palm canopy in plantation. This is where the concept of white balance comes in. If we can tell the camera which object in the room is white and supposed to come out white in the picture, the camera can calculate the difference between the current color temperature of that object and the correct color temperature of a white object and then shift all colors by that difference. The white balances of the cameras were to be set in this experiment. The standard white calibration CR-A74 plate was used to set the white balance.

The shutter speeds determine how long the film or sensor is exposed to light. Normally this is achieved by a mechanical shutter between the lens and the film or sensor which opens and closes for a time period determined by the shutter speed. Slow shutter speeds let more light strike in the image sensor, so an image is lighter, while fast shutter speeds let less light strike it, so an image is darker . The FFB images were monitored and tested on different shutter speeds in actual oil palm plantation. Shutter speed of 0.125 s gave a good view for an image in all variability of lighting intensity on the farm.

Extech instrument datalogging light meter 401036 was used to capture the lighting intensity that affected the image captured during data collection process. When record button was activated, the intensity value (in foot candles unit) was recorded automatically into datalogger. For this study, the intensity was recorded under the canopy of oil palm tree.

The visual basic programming language version 6 was the platform to develop the software analysis. It contains the various components of functions to make an Excel worksheet programming software and populate it with sample scientific data. The API (Application Programming Interface) called GDI32, which stands for Graphic Device Interface: 32-bit version was applied and the hue pixel value was determined by the total value (0-255) assigned for each colour bands of red, green and blue as calculated here:

$$\text{if } B \geq G;\ H^0 = 360^0 - \cos^{-1}[-0.5[(R-G) + (R-B) / [(R-G)^2 + (R-B)(G-B)]^{0.5} \times 255/360$$

$$\text{if } B<G;\ H^0 = \cos^{-1}[-0.5[(R-G) + (R-B) / [(R-G)^2 + (R-B)(G-B)]^{0.5} \times 255/360$$

where R is red component pixel value, G is green component pixel value and B is blue component pixel value.

The embedded histogram application in the software analysis is a function to determine the maximum R,G and B pixel value of captured FFB images. These R, G and B values will then be converted to hue value which is assigned as an average value for overall FFB images.

Nikon coolpix 4500 digital camera (Nikon, Japan) was used to capture the fruit image and store it digitally. It was equipped with a lens type Nikon 3x Tele Converter TC-E3ED. Minolta MPOB colorimeter CR 10 (Minolta, Japan) was used to determine the ripeness of oil palm fruit based on mesocarp surface color. This colorimeter equipment was used to validate the ripeness of the oil palm fruits after determining the mesocarp oil content using Soxhlet extraction process.

Firstly, the image was captured using Nikon digital camera and the pictures were automatically saved in its local memory. The images of fruits were captured with highest optical zoom with 155 mm focus length. The size of image is 640 x 480 pixels screen to suit with the size of image analysis. At the same time, the lighting intensity of environment was recorded. The steps were repeated with days interval from immature fruits until the fruits were overripe. The distances between cameras and lightmeter with fruits were fixed where the camera covered under palm tree canopy, within 2-3 m from oil palm tree.

Figure 2 shows the Developed Software For Colour Modelling to predict the Luminance and RGB pixel value of the images. When histogram analyses were performed, the maximum reading from graph will indicate the luminance value of the image as shown on the right side of the figure. The Luminance pixel value was determined by the total value (0-255) assigned for each colour bands of red, green and blue.

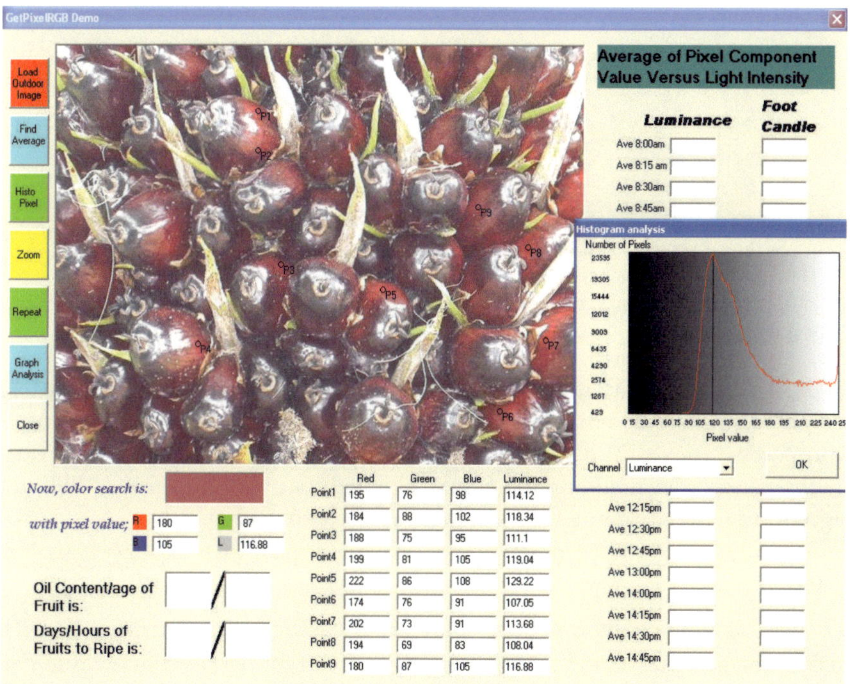

Fig. 2. Developed Software for Colour Modelling Analysis

6. Determination of oil content

The samples of fruits from different maturity stages were harvested from 5, 16 and 20 years old oil palm trees. The samples were weighed and chopped. The samples were dried at 70°C for a day to remove the water in the fruits. The dry nuts and mesocarp were weighed and blended before packed into filter papers (Whatman Cat No. 1001 150). The oil was extracted in Soxhlet extractor using chemical solvent, namely hexane. The remnant fibre and thimble were dried under 70°C for a day to remove the remaining hexane. The samples were weighed. All reading was inserted to automatic embedded calculation for oil to dry mesocarp ratio. Figure 3 shows the bunch analysis procedure to get oil extraction rate (OER). To determine the mesocarp oil content, the fruits were collected instead of whole bunch.

Fig. 3. Bunch Analysis Step Procedure for Oil Extraction

The hue pixel values of different fruits maturity were captured and analysed based on outdoor light condition of environment. The relationship of mesocarp oil content with optical properties for fruit color of hue was illustrated using trendline analysis of polynomial second order method. The results showed that the hue value of FFB image was highly significant in determining the oil content of oil palm fruit:

$$y = -0.0116x^2 + 5.2376x - 514.88; R^2 = 0.884$$

where y = mesocarp oil content, x = hue value, R^2 = coefficient of determination. The value of $R^2 = 0.884$ was acceptable and showed highly correlation between hue value and the oil content of FFB. The oil content of fruits increased with increasing hue value of fruit skin image. The mesocarp oil content will be maximum at an average of 75% which is similar to earlier findings (Mohd Hudzari, 2010). Figure 4 shows the graph of the relationship between the mesocarp oil content and their respective hue digital values.

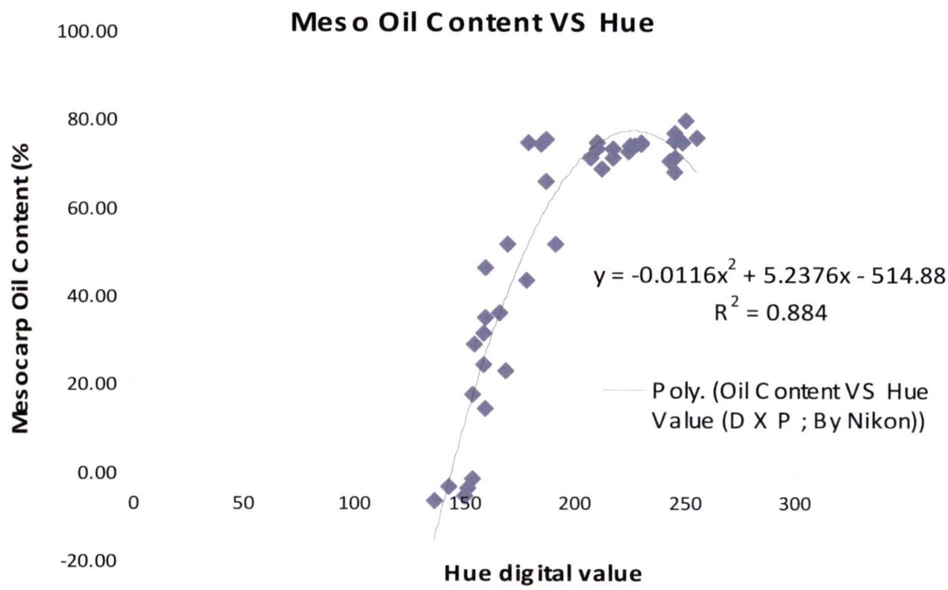

Fig. 4. The relationship of optical properties for fruit color versus mesocarp oil content.

For individual fruitlet, the hue value increases at overripe stage when it exhibits almost entirely reddish orange coloration. At overripe stage, the mesocarp oil content decreased or was stagnant due to reaction of free fatty acids (FFA) on FFB, which affected the quality and quantity of oil. Similar findings were reported on chemical changes on overripe olive oil which affect the quality and quantity of an oil.

7. Prediction of harvesting day

Figure 5 illustrates the graph to determine the number days of harvesting the oil palm FFB. The graph was developed through the experiments as described above.

The suitable days for harvesting were calculated based on the above equation of $y = -0.0116x^2 + 5.2376x - 514.88$. Let say the hue value from the camera is 200, so the oil content found using the above equation is: $y = -0.0116(200^2) + 5.2376(200) - 514.88 = 68.64$. The mesocarp oil content for hue of fruit color at 200 is 68.64%.

From the above graph of Figure 5, the matured fruit was found to have hue of 224 with 74.22% oil content which indicated the day for harvest. The unripe fruit with measured hue of 158 has an oil content of 24.44% and indicates a 63 days before the fruit began ripening. Triangulation method was used to determine the harvesting day for FFB which is illustrated by the equation below.

The day estimation to harvest the FFB model is described as below:

(J – L) / (K – L) = (M – 0) / (N – 0)

(J – L) / (K – L) = (M/N)

N = M / [(J – L) / (K – L)]

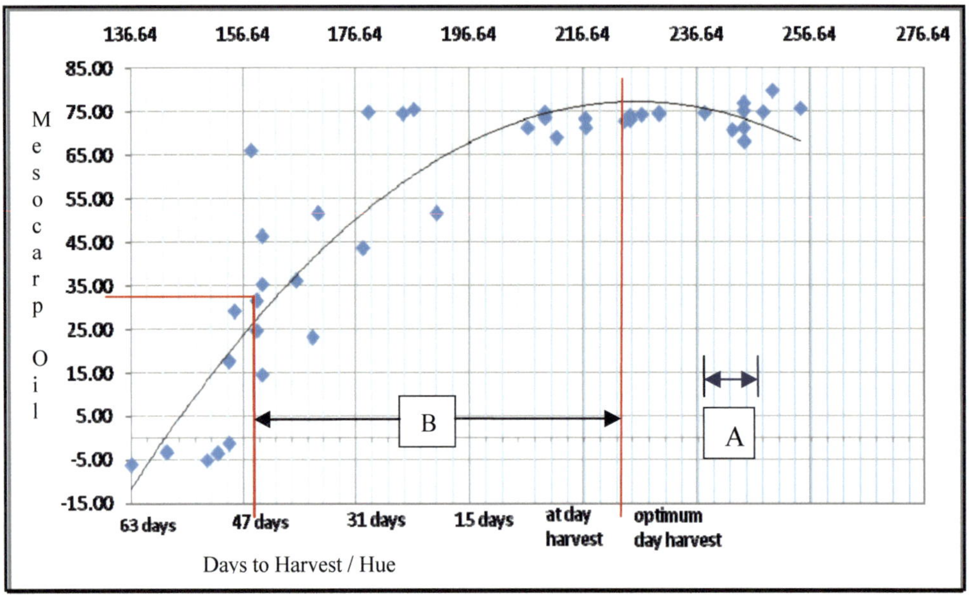

Fig. 5. The graph to determine the day of harvesting the FFB.

where *J* is the initial mesocarp oil content %, *K* mesocarp oil content %, *L* matured mesocarp oil content %, *J* and *L* are obtained by the chemical analysis in the laboratory, *M* is the initial day of observation, *N* is the day calculated and *O* is the day to harvest the FFB, which was set to 0. The suitable harvesting day is when the mesocarp oil content of FFB is at matured stage or ripe.

Using triangulation method, from 68.64%, the number of days when fruit needs to be harvested will be calculated as:

(24.44 - 74.22) / (68.64 - 74.22) = (63 – 0) / (x – 0)

(-49.78) / (-5.58) = (63) / (x)

x = 7.06 days

It means that the hue value of the FFB image of 200, the number of days required for harvesting is the next 7.06 days.

8. Real-time oil palm FFB maturity prediction model

The computer-implemented method for recognizing ripeness level of oil palm fruits uses optical properties of the oil palm fruit in determining ripeness with good accuracy. The method is computer-implemented and less likely to introduce any perceptive biases in contrast to human eye. This method provides a non-destructive approach to determine oil content of the oil palm fruits. The computer-implemented method of predicting oil content of oil palm fruits comprising the steps of providing a digital image of the oil palm fruits; acquiring average pixel value of RGB of the digital image; deriving average hue value of the digital image from the acquired average value of RGB; fitting the derived average hue value into an established model to obtain a predicted oil content of the oil palm fruits; or matching the derived average hue value to a database containing a range of preset hue value to obtain the predicted oil content; wherein each of the preset hue value is tagged with respective predetermined oil content. Subsequently, the digital image is processed to generate average pixel value of RGB. Particularly, a colour histogram preferably used as the approach to determine the maximum RGB and average RGB value as well as generate a RGB-colourspace. The histograms are used as the tool for graphing the frequency of intensity of red, blue, green and luminance of the image. Further, the RGB-colourspace is then transformed into a HSI colourspace which describes colour of the image into hue (average wavelength), saturation (amount of white), and intensity. The hue value acquired from the conversion to the HIS system is subjected to less influence of the light intensity of the images as intensity and hue are two separated information in the HIS system. This feature allows the image taken by any common available camera system can be straightly analyzed to acquire the predicted oil content without adjusting the light intensity before or after capturing of the images.

The above model of $y = -0.01x^2 + 5.2x - 514.8$; where y = oil content of the oil palm fruits, and x = hue value. Through substituting the derived hue value, the model give a predicted oil content of the oil palm fruits captured in the image. More preferably, the average hue value is of HIS coordinate system by conversion of the average RGB value.

To avoid potential error in the produced result due to lack of significant lighting, the system may further comprise a light sensing mean to detect light intensity of surrounding environment when acquiring the digital image of the oil palm fruits and prompt the user of insufficient light intensity upon detecting the light intensity of below 500cd. More preferably, the light intensity is in the range of 500 to 8000 cd. Any light intensity further below or higher may lower accuracy of the method in predicting the oil content. The above mentioned light intensity range of 500 to 8000 cd is the practical light intensity for any image taken under the palm tree canopy.

More specifically, the derived average hue value of the related digital image is placed into a plotted graph of average hue value against oil content that the graph is pre-established based on data collected from a series of experiments. Further, the location of the derived average hue value of the related digital image on the pre-established graph can be used to refer corresponding oil content of the mesocarp on the graph.

Apart from that, the method may include a step of acquiring predicted number of days left for optimum harvesting of the oil palm fruits by associating the predicted oil content to another model of relative ripeness of oil palm fruits correlated to oil content. To determine the best harvesting time, the method may use a triangulation method which can be deemed as the model of relative ripeness of oil palm fruits correlated to oil content. The second model is summarized as the following, $N = M (J-L/K-L)$, where N is the predicted day left, K is the predicted oil content in percentage based on derived average hue value, L and J are prefixed values for optimal oil content in percentage and initial oil content in percentage respectively. Entering the predicted oil content from the above mentioned two approaches into this model, the number of days left for optimum harvesting is predicted. The graph approach is used to generate the predicted oil content. More specifically, data collected from experiments are used to plot a graph of oil content against relative ripeness. Upon locating the position of the predicted oil content generated from the entered digital image on the plotted graph, the predicted relative ripeness can be referred from the graph, while the difference in between the predicted relative ripeness and the ripeness of optimal oil content in the graph is used to infer the days left for optimal harvest. With the generated information, users are able to harvest the oil palm fruits with optimal oil content thus obtaining greater profits. Furthermore, the predicted number of days left for optimal harvest is associated to the digital image. The method may have another step of categorizing the oil palm fruits into a group of unripe, ripe or overripe based on the derived average hue value and/or the relative ripeness acquired.

9. Conclusions

These experiments were conducted to determine and model an equation between the hue optical properties of the oil palm fruits at different stages of maturity after considering the affecting outdoor environment intensity in oil palm plantation. The maturity stages were confirmed by determining its mesocarp oil content. This study was carried out on selected immature fruits and monitored with days interval until one loose fruit found which indicated matured FFB. The factor of lighting intensity under canopy of oil palm plantation was not affected for image value which remains considerable using hue colorspace with setting camera parameters. The radiation from the sunlight was intercepted by the oil palm canopy by inter cropping systems for growth resources of solar radiation and caused the intensity decrease. The procedure in monitoring the image pixel value of different maturity stages for real oil palm fruit contributed. Developed system can be applied for maturity detection on other fruits in Malaysia. However, different fruits will have different characteristics for their properties.

10. References

[1] Kaida Khalid and Zulkifly Abbas (1992). A Microstrip Sensor for Determination of Harvesting Time for Oil Palm Fruits, Journal of Microwave Power and Electromagnetic Energy,Vol.27 No.1, 1992

[2] Malaysian Palm Oil Industry Performance (MPOIP) (2007). Global Oils & Fats Business Magazine Vol. 6, Issue 1.(Jan-March), 2007

[3] Mohd. Hudzari R. (2010). A New Technique for Predicting the Maturity Stage of Oil Palm (*Elaeis guineensis.*) Fresh Fruit Bunch. Unpublished PhD Dissertation, Universiti Putra Malaysia, Malaysia

[4] R. C. Gonzalez and R. E., Woods (2002). Digital Image Processing, Prentice Hall, New Jersey, 2002.

[5] Wan Ishak W. I., Mohd. Zohadie B., A.Malik A.H. (2000). Optical Properties for Mechanical Harvesting of Oil Palm. Journal of Oil Palm Research, Vol. 11, No. 2, pp. 38-45

Part 3

The Application of Nanotechnology in the Food Industry

7

Continuous Membrane Bioreactor (CMBR) to Produce Nanoparticles from Milk Components

Vanessa Nieto-Nieto[1], Silvia Amaya-Llano[2] and Lech Ozimek[1]
[1]*Department of Agricultural, Food and Nutritional Science, University of Alberta, Edmonton, Alberta,*
[2]*Programa de Posgrado en Alimentos del Centro de la República (PROPAC), Universidad Autónoma de Querétaro, Querétaro Qro.*
[1]*Canada*
[2]*México*

1. Introduction

Membrane technology has been used in the dairy industry as alternative to some unit operations since the early 1970s. The new applications of membranes were often due to the development of membrane science. The commercial availability of Nanofiltration membranes allowed the demonstration of some biological activities in milk and whey peptide sequences (Pouliot 2008). Nanotechnology is used in different stages of the membrane development and has gain great interest among the agro-food sector and it involves the manufacture, processing and applications of structure devices and systems controlling the shape and size of particles at the nanometer scale (Garcia et al., 2010). The size range that holds so much interest is typically from 100 nm down to the atomic level; approximately 0.2 nm, because in this range material can have different and enhanced properties compared with the same material at a large size. As the particle size decreases, a greater proportion of atoms are found at the surface compared to inside. For example, a particle size of 30 nm has a 5% of its atoms on the surface, at 10 nm 20%, and at 3 nm 50% of the atoms are in surface. Thus a nanoparticle (NP) has a much greater surface area per unit mass compared with larger particles, leading to grater reactivity. In addition to the surface area effect, quantum effect can also govern the properties of matter as size is reduced to the nanoscale, affecting the optical, electrical and magnetic behaviour of materials (Thassu et al., 2007).

The application of nanotechnology in the food industry covers many aspects, such as food safety, packaging material, disease treatment, delivery systems, bioavailabilty, and new tools for molecular and cellular biology and new material for pathogen detection. However the four major areas in food industry to benefit from nanotechnology are the development of new functional materials, micro and nanoscale processing, new product development and the design of nanotracers and nanosensors for food safety and biosecurity (Moraru et al., 2003).

It is the interest of this chapter to discuss the production of bioactive NPs from food commodities and to explain how nanotechnology in a continuos membrane bioreactor (CMBR) and in particular nanofiltration processes can be employed to alter food products.

The term "bioactive food component" refers to nonessential biomolecules that are present in foods and exhibit the capacity to modulate one or more metabolic processes, which results in the promotion of better health (Swanson, 2003).

A major strategy for the delivery of these components into food is through encapsulation, which consists in coating one or several components (core) within a secondary material (encapsulant). This is used to mask the color and taste of nutrients, and to protect sensitive nutrients during processing, storage, and transportation (Kuo, 2010).

Bioactive compounds are added into food looking to provide health benefits and protection as antioxidants and anti-aging agents, reducing the risk of cardiovascular diseases or with the intention to boost the nutritional content of the product.

The application of nanotechnology in the food industry is still in a developing stage, however in recent years; research has been carried out in order to formulate food-grade encapsulants to enable the delivery of desirable bioactive compounds through the food supply.

Due to their sub-cellular size, NPs offer promising means of improving the bioavailability of bioactive compounds, especially poorly soluble substances such as functional lipids (e.g. carotenoids, phytosterols, ω-3 fatty acids), natural antioxidants, and other compounds that are widely used as active ingredients in various food products. NPs can dramatically prolong compound residence time in the gastrointestinal tract by decreasing the influence of intestinal clearance mechanisms and increasing the surface available to interact with the biological support. They can also penetrate deeply into tissues through fine capillaries, cross the epithelial lining fenestration and are generally taken up efficiently by cells, thus allowing efficient delivery of active compounds to target sites in the body (Chen et al., 2006).

2. Nanomaterials

NP can be produced from a variety of different materials, from metals like gold and silver, organic solvents or some food derived ingredients. The use of food-grade ingredients is generally accepted by the regulatory agencies, has shown to exert good results as nanomaterials and would contribute to dissipate the fear among consumers given the fact that these polymers are part of the human diet.

Food-grade proteins and polysaccharides, such as whey protein, casein, gelatine, soy protein, zein, starch, cellulose, and various other hydrocolloids are used. In addition, they may contain other components, such as water, lipids, minerals and sugars (Jones & McClements, 2010).

2.1 Lipid-based nanoparticles

Lipid-based nanoparticles can encapsulate compounds with different solublities, but in particular these particles are useful carriers of poorly water soluble compounds. The term "lipid" is used in a broad sense to include phospholipids, triacylglycerides, fatty acids, steroids and waxes (Peters et al., 2011).

Different structures have been proposed using lipids as a main encapsulant constituent some of the most commonly discussed structures are: liposomes and solid lipid nanoparticles (SLN).

Liposomes are spherical, self-closed structures formed by one or more concentric lipid bi-layers with an encapsulated aqueous phase in the center and between the bi-layers (Pisal et al., 2010). They can accommodate hydrophilic, lipophilic and amphiphilic compounds in their aqueous and or lipid compartment. These structures are used as carrier systems for the protection of bioactive compounds by improving their solubility and bioavailability and preventing their unwanted interaction with other molecules (Mozafari, 2010).

The preparation of SLN is carried out at high temperatures, generally above the melting point of the lipid component used and further a cooling stage is required to solidify the lipid (Bunjes, 2010). Some disadvantages of SLN are the limited loading capacity of the bioactive compound and expulsion of the bioactive compound during storage. In order to overcome these drawbacks Nanostructured Lipid Carriers (NLC) were developed (Souto & Muller, 2006). NLC are composed of oily droplets embedded in a solid lipid matrix, this provides more stability with a controlled nanostructure that improves the bioactive compound loading and firmly incorporates the bioactive compound during storage (Shidhaye et al., 2008).

Lipid-based nanoparticles, have been used to contain different bioactive compounds such as vitamins (α-tocopherol) (Shukat & Relkin, 2011), β-carotene (Hentschel et al., 2008), and ω-3-fatty acids (Muchow et al., 2009).

2.2 Protein-based nanoparticles

Proteins have very good gelling and emulsifying properties and for this reason they are widely used as encapsulating material. Thanks to the consistent primary structure of proteins a wide variety of nutrients can be incorporated allowing them to form complexes with polysaccharides, lipids or other biopolymers. In order to improve bioavailability, the food industry is currently attempting to increase the circulation time of conventional nanocarriers in the gastrointestinal tract, notable by surface coating with protein (Chen et al., 2006).

Food proteins can undergo denaturation due to exposure to high temperatures or pressure. The resulting un-denaturated product can then re-assemble, building a new structure (Bengoechea et al., 2011). Or well if proteins are combined with polysaccharides, these materials can form biopolymeric nanostructured particles by complexation (Ron et al., 2010).

When a globular protein becomes denaturated its physical and chemical interaction change appreciably through exposure of nonpolar and sulphur containing groups that were originally present within the compact interior of the globular protein. Consequently, denaturated proteins have a greater tendency to aggregate, irreversibly, with each other through hydrophobic bonding and disulfide bond formation (Jones & McClements, 2010).

Nanoemulsions are dispersions of nanoscale droplets as the result of mixing two immiscible phases, made by the application of high shear. The rupture of droplets may be achieved by ultra-sonication or microfluidisation. The amount of surfactant requires to stabilise nanoemulsion is greater (Augustin & Hemar, 2009). To prevent the droplet from recombining into larger droplets a thin encapsulating layer is introduced to help stabilize the system; this layer is traditionally made up of proteins or phospholipids which act as surfactants (Sanguansri & Augustin. 2006).

A particular case of protein based NP are nanotubes derived from partially hydrolyzed α-lactalbumin. This nanostructure is very stable and strong and has a 8.7 nm cavity that can be used to contain bioactive compounds (Graveland-Bikker, 2006).

Milk proteins such as casein, α-lactalbumin and β-lactoglobuline, lactoferrin, bovine serum albumin, have been widely used in the development of nanostructures with very different applications. However their physicochemical properties facilitate their functionality as vehicle of bioactive compounds (Livney, 2010). However the use of protein in the building of nanostructures is not limited to milk proteins, vegetable proteins have been used as it is the case of pea protein (Donsi et al., 2010) and zein (Podaralla and Perumal et al.,2010).

2.3 Polysaccharide-based nanoparticles

Polysaccharides are used to improve texture in foods or as stabilizer agents in emulsions, and some of the most commonly used are: starch, xanthan gum, pectin, alginate and chitosan. In combination with proteins and lipids, nanostructed particles can be formed.

According to their structural characteristics, these polysaccharide-based nanoparticles are prepared mainly by four mechanisms: covalent crosslinking, ionic crosslinking, polyelectrolyte complexation, and self-assembly of hydrophobically modified polysaccharides, Liu et al. (2008) reviewed these four mechanisms in detail. Of these methods polyelectrtolyte complexetion is commonly used in the preparation of nanoparticles containing food derived bioactive compounds.

When proteins and polysaccharides carry an opposite charge, complex formation is driven by attractive electrostatic interactions between the two biopolymers (Augustin & Sanguansri, 2009). Attractive interactions and complexation begin at values slightly above the protein's isoelectric point; further pH reduction induces greater complexation (Jones et al., 2010).

It is important to establish the electrical characteristics of the polysaccharide molecules used, since electrostatic interactions may be used to assemble specific biopolymer structures. The electrical charge on polysaccharide depends on the nature of the ionic groups along the chain background, as well as solution conditions. Some polysaccharides are neutral (starch, cellulose) some anionic (alignate, carrageenan, xanthan, gum Arabic) and some cationic (chitosan) (Matalanis et al., 2011).

Sugar beet pectin was used to produce a nanostructured system for stabilization, protection and delivery of hydrophobic bioactive compounds. For this, vitamin D was bounded to β-lactoglobulin, later a complex between the carrier and the polysaccharide of the opposite charge was formed. This system has been proposed for enrichment of clear and non-fat beverage (Ron et al., 2010).

3. Nanotechnology and nanofiltration

The application of membrane technology in the nanoscale range has increased as a result of the continuous application of nanotechnology in the food sector. In the same way that nanoparticles are produced as an encapsulated systems based on food ingredients and designed to have a broad range of applications, nanofiltration can be utilized as a nanoscale process to produce food ingredients with added value.

In the membrane based separation process, the driving force for material transport through selective membranes is a pressure difference, thus these processes are called pressure-driven membrane process, and such is the case of microfiltration (MF) and ultrafiltration (UF). In the case of nanofiltration (NF) a pressure difference is not the only driving force, selective removal of ions based on charge is another feature of this technology (Muthukumarappan & Marella 2010). Thus the selective separation process results in nanometer scale products.

Nanofiltration membranes have apparent pore diameter between 0.5 and 5 nm and are capable of removing from water, both organic matter and ions of mineral salts (Goncharuk et al., 2011). In such way nanofiltration can separate components according to their molecular size without a change phase of the solvent; this avoids thermal damage and requires a lower consumption of energy (Darnoko & Cheryan, 2006).

Nanofiltration is currently used to supply high quality drinking water (Goncharuk et al., 2011), desalting milk, whey and other dairy fluids (Pouliot, 2008), additionally a wide range of value added components can be obtained after a nanofiltration treatment.

3.1 Membrane bioreactors in the nanoscale process

Biocatalytic membrane reactors combine selective mass transport with chemical reactions, and the selective removal of products from the reaction site increases the conversion of product-inhibited or thermodynamically unfavorable reactions. Membrane reactors using biological catalysts can be used in production, processing and treatment operations (Giorno & Drioli, 2000).

In a bioreactor, the conversion of raw materials into value added products is carried out by enzymatic hydrolysis; further, in combination with a NF membrane the separation of substrate and product will occurred; as the membrane will only allow the passage of certain components known as "permeate" (the product) and retains/rejects other components known as "retentate" (Muthukumarappan & Marella, 2010) .

The use of biocatalysts for large-scale production is an important application because it enables biotransformations to be integrated into productive reaction cycles. Biocatalysts (e.g. enzymes, microorganisms and antibodies) can be used: (1) suspended in solution and compartmentalized by a membrane in a reaction vessel or (2) immobilized within the membrane matrix itself. In the first method, the system might consist of a traditional stirred tank reactor combined with a NF membrane-separation unit. In the second method, the NF membrane acts as a support for the catalyst and as a separation unit (Giorno & Drioli, 2000).

Some of the advantages of the use of membrane bioreactor as a nanoscale process are that the catalytic/separation process (Giorno & Drioli, 2000; Rios et al., 2004):

- does not require additives
- is able to function at moderate temperature and pressure
- reduces the formation of by-products
- catalytic enzymes are extremely efficient and selective compared with chemical catalysts.

- the enzyme is retained and reused
- substrate/product inhibition is reduced
- the end product is free of enzyme
- one single step operation (reaction/separation).

However the use of NF membranes is not limited to the enzymatic membrane reactors. NF has been commonly used as a pre-concentration step in the milk processing operations in order to reduce the energy consumption in heat based treatments.

The whey that is obtained during cheese making with a very high mineral content, could be used for the manufacture of other valuable food ingredients such as whey protein concentrate and whey protein isolate, however the mineral content should be reduce to provide a higher quality. With NF the demineralization of whey can be achieve, since NF membranes are often used for separation of charged solutes (Suárez et al., 2009).

Thus, membrane technology at this scale can be used for (Vandanjon et al., 2007):

- Separation of high molecular weight component from low molecular weight components.
- Fractionation of products according to their molecular weight.
- Concentration of mixture or selected fractions
- Purification

Table 1. presents an overview of the NF processes that have been proposed in the food sector. NF has been used for concentration or purification of compounds with biological activity such as antioxidants, prebiotics and bioactive peptides. Alternatively NF has also been proposed for separation of valuable components such as lactose or undesirable components such as salts, biogenic amines, toxins or heavy metals.

3.2 Recovery of compounds with antioxidant activity

Concentration of bioactive compounds can be achieved by NF treatment, because this process does not require high temperatures in contrast to other concentration treatments such as evaporation. Thus, by using mild temperatures the functional properties of this temperature sensitive compounds are preserved (Benedetti et al., 2011).

Due to their beneficial effects on human health, Isoflavones were concentrated from soybean using a NF membrane. However in order to ensure a good yield and to avoid the fouling of the membrane due to the accumulation of fat globules in the membrane, defatted soy flour was used as raw material and the final results of this study showed that the nanofiltration process for the concentration of isoflavones is viable (Benedetti et al., 2011).

In order to separate low molecular weight, hydrophobic components, the nanofiltration membrane needs to meet certain characteristics that will be defined by the compound that is being recovered. Darnoko & Cheryan (2006) selected a hydrophobic membrane to separate β-carotene from the palm methyl esters, increasing the concentration in from 0.45 g/L to 1.88 g/L (34).

Flavonoids and polyphenolic compounds from *Sideritis scardica Griseb* were concentrated up to 3-4 times. The concentrated extracts preserve their high antioxidant activity and could be used as a source of concentrated biologically active material. Also the separation of flavonoids form low molecular polyphenols was possible in base of their molecular weight difference (Tylkowski et al., 2010).

Separated component	Component separated from	Membrane material	NF module	MWCO (Da)	T (°C)	Pressure (kPa)	Reference
Phenolic compounds	Aqueous mate (*Ilex paraguariensis*) extract	--------	Spiral wound	150-300	25	690	Negrão et al. 2011
Aroma and protein	Tuna cooking juice	Polyamide coated with polyesthersulfone	tubular	400	40	3500	Walha et al. 2011
Anthocyanins	Açai juice	Semi-aromatic polyamide layer on top of a polysulphone microporous support	Plate and frame	--------	35	1000 1500 2000 3000	Couto et al. 2011
		Aromatic polyamide					
		--------		~1000			
		Thin film		--------			
		Thin film					
		Poliestersulfone		~1000			
Fructooligo-saccharides	Sucrose from sugar cane molasses	--------	Spiral wound	400	95	4000	Kuhn et al. 2010
Alcohol	Wine	Polyamide	--------	300	45	500 1500	Takacs et al. 2010
				400		1000 2000	
Biogenic amines	Model solution	--------	Flat	1000	70	3500	Sabaté et al. 2008
Mineral salts	Whey	Aromatic polyamide	Spiral wound	300	--------	500 - 2400	Pan et al. 2011
Cyanobacterial toxins	Drinking water	Polyamide composite with a microporous supporting layer	Flat-sheet	200	--------	8000	Coral et al. 2011
				300			

Table 1. Nanofiltratrion processes proposed for concentration, purification or separation of relevant food ingredients.

Conidi et al. (2011) proposed an integrated membrane process based on the preliminary UF of enzymaticaly depictinised juice with a 100 kDa membrane to remove suspended solids followed by a NF step with a 450 Da membrane. The separation and concentration of polyphenols of the bergamot juice was possible. The retentate contained the phenolic compounds, which showed a high total antioxidant activity.

Crude rice bran is a by-product of rice milling, rich in phytochemicals with high nutritional value such as γ-oryzanol. A two-step nanofiltration system was set up for enrichment of this phytochemical in rice oil. The first membrane stage, produced the separation of glycerides and γ-oryzanol, promoting the oil enrichment in this phytochemical. In the second membrane stage the oil was refined to acceptable consumption levels of free fatty acids and its γ-oryzanol content was further enhanced. The antioxidant activity of the resulting product oil possessed a significantly higher antioxidant capacity than the feed oil (45.9, 19.3 μmol Trolox/g respectively). This results show that the product generated through NF membrane process has a superior nutritional value (Sereewatthanawut et al., 2011).

3.3 Oligosaccharides

Oligosaccharides (OS) are not digested in the human small intestine and stimulate the growth of bacterial flora, hence they are used as prebiotics (Botelho-Cuhna et al., 2010). The resulting commercial OS products contain plenty of side resultants of low molecular weight sugars such as glucose, fructose, sucrose, galactose and lactose, lowering the performance of the end product. Therefore, high purity of OS products is required (Feng et al., 2009). Thus NF seems to be a viable alternative for industrial scale purification of OS mixtures.

Many recent reports of the application of membrane technology to produce and concentrate OS have been published using different sources such as whey protein concentrate (Barile et al., 2009), caprine milk (Martinez-Ferez et al., 2008), rice husk (Vegas et al., 2008) and soybean waste water (Wang et al., 2009).

In a recent study, apple pomace samples were subjected to simultaneous saccharification and fermentation resulting in a mixture of lactic acid and OS. Lactic acid was further removed by ion exchange and the mixture was processed with two sequential steps using NF membranes in order to refine and concentrate the following OS: glucooligosaccharides, galactooligosaccharides, xylooligosaccharides and arabinooligosaccharides. More than 90% of low molecular weight compounds (residual lactic acid, arabinose and NaCl) were removed from the solution. On the other hand, just a limited part of the OS was lost in the permeate. As a result of the treatments, the mass fraction of the OS in the final product increased from 0.360 up to 0.677 kg/kg of non-volatile compounds. These results confirm that coupling two NF stages (discontinuous diafiltration followed by concentration) is a suitable alternative for obtaining OS concentrates. The refined product obtained in this work showed degrees of purity in the range reported for commercial OS and was assayed for its prebiotic potential and the results showed the promotion of beneficial bacterial growth (Gullón et al., 2011).

Eucalyptus wood-derived xylooligosaccharides were obtained using UF and NF membranes (1-50 kDa) for separation and concentration. The UF unit was used to separate OS from higher molecular weight products or to fractionate OS of different degree of polimerization. In addition a NF unit was used for concentrating liquors and/or for removing undesired low molecular weight compounds, such as monomeric sugars or phenolic compounds, increasing the xylooligosaccharide concentration 3.8 times (Gullón et al., 2008).

In other cases the separation process can be primarily intended to recover the low molecular weight sugars, such is the case of lactulose, which is very useful in the treatment of chronic constipation and is normally produced by isomerization of lactose using hydroxide/boric acid forming a complex. This complex can be split by changing conditions from basic to acidic and the borate is converted into NaCl and H_3BO_3. NF was used as an effective method for the desalination of lactulose syrup, thus the disaccharides (lactulose and lactose) were retained by the membrane in the presence of high concentrations of NaCl and H_3BO_3 in the syrup. In this way NF provides a commercial alternative to chromatography for concentration and purification (Zhang et al., 2011).

NF has been used in the winemaking for sugar control, by reducing the sugar concentration in must before fermentation with the purpose to reduce the alcohol content in wine. For this García-Martín et al. (2010) proposed a two successive NF steps for sugar removal. The musts obtained were mixed with untreated must or the retentate of the first NF stage, in order to reduce the alcoholic content by 2°. With this treatment other compound such as polyphenols, anthocyanins, catechins and tannins were partially removed and slight changes in color and aroma were observed.

A goats´ milk product enriched in oligosaccharides (more than 80%), lactose and salts-free was obtained following the two-stage tangential filtration process by Martinez-Ferez et al., (2006). Tubular ceramic membranes with molecular mass cut-offs of 50 (ultrafiltration) and 1 kDa (nanofiltration), respectively, were employed and 15 new oligosaccharide structures (4 neutral and 11 acidic), were identified. Goat´s milk is a suitable source of oligosaccharides for applications in human nutrition due to their prebiotic and anti-infective properties.

Nanofiltration membranes can be used for purifying galactooligosaccharides (GOS) products from the monosaccharide hydrolysis products, glucose and galactose. According to Gosling et al., (2010), up to 88% of the di- and oligosaccharides were recovered from a commercial GOS mixture using a nanofiltration membrane with only 19% of the monosaccharides remaining in the retentate stream (Goulas et al., 2003). Catarino et al., (2008) investigated the fractionation of saccharide mixtures with calcium using ultrafiltration and nanofiltration processes and reported that saccharide fractionation was enhanced in the presence of calcium. Feng et al., (2009) reported the separation of sugar solutions in total recycle mode operations using nanofiltration membranes and the process resulted in oligosaccharides yield of 70% and purity of 54%. These results show nanofiltration as an alternative process to industrial chromatography.

3.4 Bioactive peptides

Controlled enzymatic hydrolysis of proteins produces smaller peptidic fractions, which can exert a positive effect upon health. These fractions have been defined as bioactive peptides (Hartmann & Meisel, 2007) and NF processes are particularly useful for separation of peptides due to the suitable molecular weight cut-off and because of the electrochemical effects, which play an important role in the case of charged molecules (Saxena et al., 2009).

Fish protein hydrolysates were produced and then submitted to a two-stage treatment (UF/NF) to obtain fractions with increased activity. In this study 4 major fractions were obtained. Size chromatography was used to characterize the peptide composition in each fraction. Within some fractions it was possible to find a wide size distribution and not only the expected size (Bourseau et al., 2009). Other similar reports using fish protein as a substrate in a UF/NF fractionation system have been reported (Picot, 2010; Vandanjon et al., 2009).

Milk proteins have been widely studied as a source of bioactive peptides with a broad range of activities, for this reason Ting et al. (2007) proposed a NF treatment for the fractionation of peptides derived from whey protein. In this work, peptides were selectively separated, based on their charge and size. NF separation of peptides and amino acids is very dependent on physicochemical parameters such as pH and ionic strength. At pH 9, acidic peptides carry a net negative charge. Since both the peptides and the membrane are negatively charged at this pH, it is probable that electrostatic repulsion between the membrane and the acidic peptides prevented the peptides from permeating the membrane, meanwhile the basic peptides which were positively charged at this pH were present in the permeate.

Fractionation of a β-lactoglobulin (β-LG) peptide mixture by nanofiltration (NF) membranes was investigated by Lapointe et al., (2003). Peptide mixture was prepared by tryptic hydrolysis of commercial β-LG followed by ultrafiltration (UF) for enzyme removal. In this paper, important change in NF selectivity as affected by hydrodynamic conditions and recirculation time via possible peptide–peptide interactions occurring in the so-called weakly attached layer were reported. Roufik et al., (2007) showed that the hypotensive peptide β-lactoglobulin (β-Lg) f142-148, known as lactokinin, bind to bovine β-lactoglobulin variant A (β-Lg A) and this complex could delay the hydrolysis of this peptide by digestive enzymes.

The separation of peptides contained in rapeseed protein hydrolysate using UF/NF membranes was studied by Tessier et al. (2006). This process was also based on selective separation depending on the peptide and membrane charge. As in the previous example peptides with the same charge as the membrane (co-ions) were concentrated in the retentate whereas peptides with the opposite charge (counter-ions) were able to get through the membrane in the permeate. A pre-treatment step by acid precipitation to remove high molar mass substances was followed by the UF step (3kDa), in which the concentration of small peptides was allowed. Later the mixture of small peptides was desalted using a NF unit to assess the influence of ionic strength on the fractionation selectivity. The results showed that changes in pH and/or ionic strength modified the nature and the intensity of electrostatic interactions between co-ions, counter-ions and membrane. Thus the control of pH and ionic strength should be key parameter in the selective separation process using NF membranes. Butylina et al., (2006) described the fractionation and further isolation and characterisation of peptides and proteins present in sweet whey by means of ultrafiltration using a regenerated cellulose membrane with a nominal molar mass cut-off value of 10 kg/mol and nanofiltration through sulphonated polyether sulphone membrane with a cut-off of 1 kg/mol. The concentration of whey proteins was done below the critical flux. The sieving coefficients for the whey components (proteins, lactose and salts) were estimated. Whey proteins were completely rejected by the ultrafiltration membrane. Nanofiltration of whey permeates obtained after ultrafiltration was conducted at two pH values (9.5 and 3.0) that corresponded to the different charged states of the membrane and of the peptides. The transmission of peptides, amino acids and lactose was found to be mainly affected by the permeability of the fouling layer. The selectivity of the nanofiltration membranes toward peptides compared to lactose was calculated as 0.82 and 6.81 at pH 9.5 and 3.0, respectively.

3.5 Other applications

3.5.1 Fusel alcohols

Although fusel alcohols provide flavour in rice spirits, a high concentration of these compounds, can lead to off-flavors, cloudy appearance and cause headaches and dizziness. Based on this, NF was proposed to remove fusel alcohols from rice spirits. The best removal results were observed with a 150-300 Da cut-off and an operating pressure of 488.95 kPa. For these conditions the remaining fusel alcohols content decreased to 2.54 g/L. The alcohol content, soluble solids content and pH (organic acids) varied slightly in the rice spirits processed with NF, however the sensory evaluation results showed that the NF treatment, improved the taste and clarity of rice spirits. Thus NF treatment effectively improved the wine quality (Hsieh et al., 2010).

3.5.2 Alternative sweeteners

A new alternative non-caloric sweetener can be produce from *Stevia rebaudiana*. The production of this sweetener could involve the use of organic solvents and methanol and ethanol. For this reason a membrane based process has been proposed to manufacture this sweetener. It is possible to use a NF membrane process as a post-treatment or polishing stage after applying a previous concentration treatment. This is the case of a recent study, in which an integrated process with a microfiltration, ultrafiltration and nanofiltration membrane stage was used for the purification of this sweetener (Vanneste et al., 2011).

3.5.3 Soy sauce

Soy sauce with a reduced salt content was obtained after a NF treatment (150 Da) for removal of NaCl and recovery of nutritional components such as amino acid and fragrance from raw soy sauce. The process consisted in a dilution step followed by concentration and then diafiltration. This mode of operation showed the least processing time and least water consumption, along with a high salt removal and high nutrient retention. The permeate produced in soy sauce desalination could be either reused as processing water or as feed to produce light color soy sauce, all the materials including NaCl, amino acids and water could be fully utilized. Consequently, this process was proposed as an alternative to produce soy sauce with low sodium content (Luo et al., 2009).

4. Conclusion

In consideration of the growing interest in promoting health through food, there is a continuous development of food-grade delivery systems to encapsulate, protect and deliver bioactive compounds (i.e. antioxidants, vitamins, bioactive peptides, antimicrobials).

Thus, there is now the possibility to produce food for a healthier population and improve its health and wellness through nanoscale technology; because of this, the "nano-food" market is expected to grow very positively within the next few years.

Materials such as proteins, polysaccharides and lipids have shown to exert good results as nanomaterials and would contribute to create less fear among consumers given the fact that these polymers are part of the human diet.

As part of the nanotechnology involved in the food processing, operations such as nanofiltration represent a useful alternative for the separation, concentration, fractionation and purification of food components with added value since many of them can exert a beneficial effect upon health.

Nanofiltration in combination with controlled hydrolysis also provides an interesting alternative for production of bioactive compounds derived from food ingredients with some advantages such as the use of mild temperatures, the reuse of the enzymes, the final product free of enzyme or other compounds, making this operation economically attractive and highly likely to be used at an industrial scale.

5. Acknowledgment

This work was supported by Consejo Nacional de Ciencia y Tecnología (CONACyT) as part of the scholarship to pursue T.V. Nieto-Nieto Doctoral work.

6. References

Acosta, E. (2009). Bioavailability of nanoparticles in nutrient and nutraceutical delivery. *Current Opinion in Colloid & Interface Science*, 14(1), 3-15.

Augustin, M. A., & Hemar, Y. (2009). Nano- and micro-structured assemblies for encapsulation of food ingredients. *Chemical Society Reviews*, 38(4), 902-912.

Augustin, M. A., & Sanguansri, P. (2009). Nanostructured materials in the food industry. *In Advances in food and nutrition research*, Taylor, S.L. (Ed.), (pp. 183-213) Academic Press.

Barile, D., Tao, N., Lebrilla, C. B., Coisson, J., Arlorio, M., & German, J. B. (2009). Permeate from cheese whey ultrafiltration is a source of milk oligosaccharides. *International Dairy Journal*, 19(9), 524-530.

Benedetti, S., Prudêncio, E. S., Mandarino, J. M. G., Rezzadori, K., & Petrus, J. C. C. (2011). Concentration of soybean isoflavones by nanofiltration and the effects of thermal treatments on the concentrate. *Food Research International*, In Press, Corrected Proof

Bengoechea, C., Peinado, I., & McClements, D. J. (2011). Formation of protein nanoparticles by controlled heat treatment of lactoferrin: Factors affecting particle characteristics. *Food Hydrocolloids*, 25(5), 1354-1360.

Botelho-Cunha, V. A., Mateus, M., Petrus, J. C. C., & de Pinho, M. N. (2010). Tailoring the enzymatic synthesis and nanofiltration fractionation of galacto-oligosaccharides. *Biochemical Engineering Journal*, 50(1-2), 29-36.

Bourseau, P., Vandanjon, L., Jaouen, P., Chaplain-Derouiniot, M., Massé, A., Guérard, F., Chabeaud, A., Fouchereau-Péron, M., Le Gal, Y., Ravalleec-Plé, R., Bergé, J.-P., Picot, L., Piot, J.-M., Batista, I., Thorkelsson, G., Delannoy, C., Jakobsen, G., & Johansson, I. (2009). Fractionation of fish protein hydrolysates by ultrafiltration and nanofiltration: Impact on peptidic populations. *Desalination*, 244(1-3), 303-320.

Bunjes, H. (2010). Lipid nanoparticles for the delivery of poorly water-soluble drugs. *Journal of Pharmacy and Pharmacology*, 62(11), 1637-1645.

Butylina, S., Luque, S., & Nyström, M. (2006). Fractionation of whey-derived peptides using a combination of ultrafiltration and nanofiltration. Journal of Membrane Science, 280(1/2), 418-426.

Catarino, I., Minhalma, M., Beal, L. L., Mateus, M., & de Pinho, M. N. (2008). Assessment of saccharide fractionation by ultrafiltration and nanofiltration. *Journal of Membrane Science*, 312(1–2), 34–40.

Chen, L. Y., Remondetto, G. E., & Subirade, M. (2006) Food protein-based materials as nutraceutical delivery systems. *Trends in Food Science & Technology*, 17(5):272-83.

Conidi, C., Cassano, A., & Drioli, E. (2011). A membrane-based study for the recovery of polyphenols from bergamot juice. *Journal of Membrane Science*, 375(1-2), 182-190.

Coral, L. A., de, O. P., Bassetti, F. d. J., & Lapolli, F. R. (2011). Nanofiltration membranes applied to the removal of saxitoxin and congeners. *Desalination & Water Treatment*, 27(1-3), 8-17.

Couto, D. S., Dornier, M., Pallet, D., Reynes, M., Dijoux, D., Freitas, S. P., et al. (2011). Evaluation of nanofiltration membranes for the retention of anthocyanins of açai (euterpe oleracea mart.) juice. *Desalination & Water Treatment*, 27(1-3), 108-113.

Darnoko, D., & Cheryan, M. (2006). Carotenoids from red palm methyl esters by nanofiltration. *Journal of the American Oil Chemists' Society*, 83(4), 365-370.

Donsi, F., Senatore, B., Huang, Q. R., & Ferrari, G. (2010). Development of novel pea protein-based nanoemulsions for delivery of nutraceuticals. *Journal of Agricultural and Food Chemistry*, 58(19):10653-10660.

Feng, Y. M., Chang, X. L., Wang, W. H., & Ma, R. Y. (2009). Separation of galacto-oligosaccharides mixture by nanofiltration. *Journal of the Taiwan Institute of Chemical Engineers*, 40(3), 326-332.

Garcia, M., Forbe, T., & Gonzalez, E. (2010). Potential applications of nanotechnology in the agro-food sector. *Ciencia e Tecnologia De Alimentos*, 30(3), 573-581.

García-Martín, N., Perez-Magariño, S., Ortega-Heras, M., González-Huerta, C., Mihnea, M., González-Sanjosé, M. L., Palacio, L., Prádanos, P., & Hernández, A. (2010). Sugar reduction in musts with nanofiltration membranes to obtain low alcohol-content wines. *Separation and Purification Technology*, 76(2), 158-170.

Giorno, L., & Drioli, E. (2000). Biocatalytic membrane reactors: Applications and perspectives. *Trends in biotechnology*, 18(8):339-349.

Goncharuk, V. V., Kavitskaya, A. A., & Skil'skaya, M. D. (2011). Nanofiltration in drinking water supply. *Journal of Water Chemistry and Technology*, 33(1), 37-54.

Gosling, A., Stevens, G., Barber, A., Kentish, S., & Gras, S. (2010). Recent advances refining galactooligosaccharide production from lactose. *Food Chemistry*, 121(2), 307-318.

Goulas, A. K., Grandison, A. S., & Rastall, R. A. (2003). Fractionation of oligosaccharides by nanofiltration. *Journal of the Science of Food and Agriculture*, 83(7), 675–680.

Goulas, A. K., Kapasakalidis, P. G., Sinclair, H. R., Rastall, R. A., & Grandison, A. S. (2002). Purification of oligosaccharides by nanofiltration. *Journal of Membrane Science*, 209(1), 321–335.

Graveland-Bikker, J. F., Schaap, I. A. T., Schmidt, C. F., & De Kruif, C. G. (2006). Structural and mechanical study of a self-assembling protein nanotube. *Nano Letters*, 6(4), 616-621.

Gullón, B., Gullón, P., Sanz, Y., Alonso, J. L., & Parajó, J. C. (2011). Prebiotic potential of a refined product containing pectic oligosaccharides. *LWT -Food Science and Technology*, 44(8), 1687-1696.

Gullón, P., González-Muñoz, M. J., Domínguez, H., & Parajó, J. C. (2008). Membrane processing of liquors from eucalyptus globulus autohydrolysis. *Journal of Food Engineering*, 87(2), 257-265.

Hartmann, R., & Meisel, H. (2007). Food-derived peptides with biological activity: From research to food applications. *Current Opinion in Biotechnology*, 18(2), 163-169.

Hentschel, A., Gramdorf, S., Muller, R. H., & Kurz, T. (2008). Beta-carotene-loaded nanostructured lipid carriers. *Journal of Food Science*, 73(2), N1-N6.

Hsieh, C., Huang, Y., Lai, C., Ho, W., & Ko, W. (2010). Develop a novel method for removing fusel alcohols from rice spirits using Nanofiltration. *Journal of Food Science*, 75(2), N25-N29.

Jones, O. G., & McClements, D. J. (2010). Functional biopolymer particles: Design, fabrication, and applications. *Comprehensive Reviews in food Science and Food Safety*, 9(4), 374-397.

Jones, O. G., Decker, E. A., & McClements, D. J. (2010). Comparison of protein-polysaccharide nanoparticle fabrication methods: Impact of biopolymer complexation before or after particle formation. *Journal of Colloid and Interface Science*, 344(1), 21-29.

Kuhn, R. C., Maugeri Filho, F., Silva, V., Palacio, L., Hernández, A., & Prádanos, P. (2010). Mass transfer and transport during purification of fructooligosaccharides by nanofiltration. *Journal of Membrane Science*, 365(1-2), 356-365.

Kuo, P. (2010). The application of nanotechnology to functional foods and nutraceuticals to enhance their bioactivities. In *Biotechnology in functional foods and nutraceuticals*, B. Debasis, L. Francis & K. G. Dilip (Eds.), (pp. 447-462) CRC Press.

Lapointe, J.-F., Gauthier, S. F., Pouliot, Y., & Bouchard, C. R. (2003). Effect of hydrodynamic conditions on fractionation of b-lactoglobulin tryptic peptides using nanofiltration membranes. *Journal of Membrane Science*, 212(1-2), 55–67.

Liu, Z., Jiao, Y., Wang, Y., Zhou, C., & Zhang, Z. (2008). Polysaccharides-based nanoparticles as drug delivery systems. *Advanced Drug Delivery Reviews*, 60(15), 1650-1662.

Livney, Y. D. (2010). Milk proteins as vehicles for bioactives. *Current Opinion in Colloid & Interface Science*, 15(1-2), 73-83.

Luo, J., Ding, L., Chen, X., & Wan, Y. (2009). Desalination of soy sauce by nanofiltration. *Separation and Purification Technology*, 66(3), 429-437.

Martinez-Ferez, A., Guadix, A., Zapata-Montoya, J. E., & Guadix, E. M. (2008). Influence of transmembrane pressure on the separation of caprine milk oligosaccharides from protein by cross-flow ultrafiltration. *International Journal of Dairy Technology*, 61(4), 333-339.

Martinez-Ferez, A., Rudloff, S., Guadix, A., Henkel, C., Pohlentz, G., Boza, J., Guadix, E., & Kunz, C. (2006) Goats' milk as a natural source of lactose-derived oligosaccharides: Isolation by membrane technology. *International Dairy Journal*, 16(2), 173-181

Matalanis, A., Jones, O. G., & McClements, D. J. (2011). Structured biopolymer-based delivery systems for encapsulation, protection, and release of lipophilic compounds. *Food Hydrocolloids*, 25(8),1865-1880.

Moraru, C. I., Panchapakesan, C. P., Huang, Q., Takhistov, P., Liu, S., & Kokini, J. L. (2003). Nanotechnology: A new frontier in food science. *Food Technology*, 57(12), 24-29.

Mozafari, M. R. (2010). Nanoliposomes: Preparation and analysis. *Methods in Molecular Biology (Clifton, N.J.)*, 605, 29-50.

Muchow, M., Schmitz, E. I., Despatova, N., Maincent, P., & Müller, ,R.H. (2009). Omega-3 fatty acids-loaded lipid nanoparticles for patient-convenient oral bioavailability enhancement. *Die Pharmazie*, 64(8), 499-504.

Muthukumarappan, K., & Marella, C. (2010). Membrane processing. In: *Mathematical modeling of food processing*, Farid, M.M. (Ed.), (pp. 735-758) CRC Press.

Negrão Murakami, A.N., de Mello Castanho Amboni, R.D., Prudêncio, E.S., Amante, E. R., de Moraes Zanotta, L., Maraschin, M., Cunha Petrus, J.C., & Teófilo, R.F. (2011). Concentration of phenolic compounds in aqueous mate (*Ilex paraguariensis* A. St. Hil) extract through nanofiltration. *LWT - Food Science and Technology*, 44(10), 2211-2216.

Pan, K., Song, Q., Wang, L., & Cao, B. (2011). A study of demineralization of whey by nanofiltration membrane. *Desalination*, 267(2-3), 217-221.

Peters, R., Dam, G. t., Bouwmeester, H., Helsper, H., Allmaier, G., Kammer, F. v., Ramsch, R. Solans, C., Tomaniová, M., Hajslova, J., & Weigel, S. (2011). Identification and characterization of organic nanoparticles in food. *Trends in Analytical Chemistry*, 30(1), 100-112.

Picot, L., Ravallec, R., Martine, F. - P., Vandanjon, L., Jaouen, P., Chaplain-Derouiniot, M., Guérard. F., Chabeaud, A., LeGal, Y., Martinez Alvarez, O., Bergé J. -P., Piot, J. -M., Batista, I., Pires, C., Thorkelsson, G., Delannoy, C., Jakobsen, G., Johansson, I. & Bourseau, P. (2010). Impact of ultrafiltration and nanofiltration of an industrial fish protein hydrolysate on its bioactive properties. *Journal of the Science of Food and Agriculture*, 90(11), 1819-1826.

Pisal, D. S., Kosloski, M. P., & Balu-Iyer, S. V. (2010). Delivery of therapeutic proteins. Journal *of Pharmaceutical Sciences*, 99(6), 2557-2575.

Podaralla, S., & Perumal, O. (2010). Preparation of zein nanoparticles by pH controlled nanoprecipitation. *Journal of Biomedical Nanotechnology*, 6(4), 312-317.

Pouliot, Y. (2008). Membrane processes in dairy technology—From a simple idea to worldwide panacea. *International Dairy Journal*, 18(7), 735-740.

Rahimnejad, M., Mokhtarian, N., & Ghasemi, M. (2009). Production of protein nanoparticles for food and drug delivery systems. *African Journal of Biotechnology*, 8(19), 4738-4743.

Rios, G. M., Belleville, M. P., Paolucci, D., & Sanchez, J. (2004). Progress in enzymatic membrane reactors - a review. *Journal of Membrane Science*, 242(1), 189-196.

Ron, N., Zimet, P., Bargarum, J., & Livney, Y. D. (2010). Beta-lactoglobulin-polysaccharide complexes as nanovehicles for hydrophobic nutraceuticals in non-fat foods and clear beverages. *International Dairy Journal*, 20(10), 686-693.

Roufik, S., Gauthier, S. F., & Turgeon, S. L. (2007). Physicochemical characterization and in vitro digestibility of β-lactoglobulin/β-Lg f142-148 complexes. *International Dairy Journal*, 17(5), 471-480.

Sabaté, J., Labanda, J., & Llorens, J. (2008). Nanofiltration of biogenic amines in acidic conditions: Influence of operation variables and modeling. *Journal of Membrane Science*, 310(1-2), 594-601.

Sanguansri, P., & Augustin, M. A. (2006). Nanoscale materials development - a food industry perspective. *Trends in Food Science & Technology*, 17(10), 547-556.

Saxena, A., Tripathi, B. P., Kumar, M., & Shahi, V. K. (2009). Membrane-based techniques for the separation and purification of proteins: An overview. *Advances in Colloid and Interface Science*, 145(1-2), 1-22.

Sereewatthanawut, I., Baptista, I. I. R., Boam, A. T., Hodgson, A., & Livingston, A. G. (2011). Nanofiltration process for the nutritional enrichment and refining of rice bran oil. *Journal of Food Engineering*, 102(1), 16-24.

Shidhaye, S. S., Vaidya, R., Sutar, S., Patwardhan, A., & Kadam, V. J. (2008). Solid lipid nanoparticles and nanostructured lipid carriers -- innovative generations of solid lipid carriers. *Current Drug Delivery*, 5(4), 324-331.

Shukat, R., & Relkin, P. (2011). Lipid nanoparticles as vitamin matrix carriers in liquid food systems: On the role of high-pressure homogenisation, droplet size and adsorbed materials. *Colloids & Surfaces B: Biointerfaces*, 86(1), 119-124.

Souto, E. B., & Müller, R. H. (2006). Applications of lipid nanoparticles (SLN and NLC) in food industry. *Journal of Food Technology*, 4(1), 90-95.

Suárez, E., Lobo, A., Alvarez, S., Riera, F. A., & Álvarez, R. (2009). Demineralization of whey and milk ultrafiltration permeate by means of nanofiltration. *Desalination*, 241(1-3), 272-280.

Swanson, J. E. (2003). *Encyclopedia of food and culture; bioactive food components*, Charles Scribner's Sons, New York.

Takacs, L., Korany, K., & Vatai, G. (2010). Process modelling in the production of low alcohol content wines by direct concentration and diafiltration using nanofiltration membranes. *Acta Alimentaria*, 39(4),397-412.

Tessier, B., Harscoat-Schiavo, C., & Marc, I. (2006). Selective separation of peptides contained in a rapeseed (*Brassica campestris* L.) protein hydrolysate using UF/NF membranes. *Journal of Agricultural and Food Chemistry*, 54(10), 3578-3584.

Thassu, D., Pathak, Y., & Deleers, M. (2007). Nanoparticulate drug-delivery systems: an overview. In: *Nanoparticulate Drug Delivery Systems*, Thassu, D. (Ed.), (pp. 1-31) Informa Heaalthcare.

Ting, B. P. C. P., Gauthier, S. F., & Pouliot, Y. (2007). Fractionation of β-lactoglobulin tryptic peptides using spiral wound nanofiltration membranes. *Separation Science and Technology*, 42(11), 2419-2433.

Tylkowski, B., Tsibranska, I., Kochanov, R., Peev, G., & Giamberini, M. Concentration of biologically active compounds extracted from Sideritis ssp. L. by nanofiltration. *Food and Bioproducts Processing*, In Press, Corrected Proof.

Vandanjon, L., Grignon, M., Courois, E., Bourseau, P., & Jaouen, P. (2009). Fractionating white fish fillet hydrolysates by ultrafiltration and nanofiltration. *Journal of Food Engineering*, 95(1), 36-44.

Vandanjon, L., Johannsson, R., Derouiniot, M., Bourseau, P., & Jaouen, P. (2007). Concentration and purification of blue whiting peptide hydrolysates by membrane processes. *Journal of Food Engineering*, 83(4), 581-589.

Vanneste, J., Sotto, A., Courtin, C. M., Van Craeyveld, V., Bernaerts, K., Van Impe, J., Vandeur, J., Taes, S., & Van der Bruggen, B. (2011). Application of tailor-made membranes in a multi-stage process for the purification of sweeteners from *Stevia rebaudiana*. *Journal of Food Engineering*, 103(3), 285-293.

Vegas, R., Moure, A., Domínguez, H., Parajó, J. C., Alvarez, J. R., & Luque, S. (2008). Evaluation of ultra- and nanofiltration for refining soluble products from rice husk xylan. *Bioresource Technology*, 99(13), 5341-5351.

Walha, K., Ben Amar, R., Massé, A., Bourseau, P., Cardinal, M., Cornet, J., Prost, C., & Jaouen, P. (2011). Aromas potentiality of tuna cooking juice concentrated by nanofiltration. *LWT - Food Science and Technology*, 44(1), 153-157.

Wang, L., Shao, C., & Wang, H. (2009). Nanofiltration continuous process control: Recovery of oligosaccharides from streamed soybean waste water. *Membrane Science and Technology -Lanzhou-*, 29(1), 79-82.

Zhang, Z., Yang, R., Zhang, S., Zhao, H., & Hua, X. (2011). Purification of lactulose syrup by using nanofiltration in a diafiltration mode. *Journal of Food Engineering*, 105(1), 112-118.

8

Diversity of Microorganisms Hosted by the Albanian Medicinal Plants and the Antimicrobial Effect of the Chemical Compounds Extracted by Them

E. Troja[1], N. Dalanaj[2], R. Ceci[2], R. Troja[2], V. Toska[1], A. Mele[2], A. Como[2], A. Petre[2] and D. Prifti[2]
[1]*Faculty of Medicine, Department of Pharmacy,*
[2]*Faculty of Natural Sciences, Department of Chemistry & Department of Industrial Chemistry,*
Albania

1. Introduction

Aromatic medicinal plants are a very important part of the Albanian biological diversity. They are distributed in natural habitats, with different characteristics of relief and climate and are considered also as an important national richness.

Medicinal plants are evaluated for:

- their interest related with the production of the essential oils;
- their medical and industrial importance;
- their efficiency in different therapies;
- the economical values.

Herbal medicinal plants and the traditional medicine derived from them, have been used by the Albanian ancients and the Mediterranean neighbours, to treat the digestive disorders, dermatitis, toothaches, headaches, infections of the urinary tract, to reduce the high levels of the cholesterol and other different uses.

Recently, the application of the modern techniques and technologies with the final objective the preparation of herbal teas and drugs, derived from the aromatic medicinal plants and their extracts, have increased the interest about their use as bio-products with a lot of benefits from their curative properties. The production of dietary supplements derived from the plants is realized in recent days, through the cooperation between pharmacologists, microbiologists, botanists and chemists. Their use by children, adults, elderly and groups with specific conditions, is clearly expressed by the correlation between the tradition and new approaches.

The research study here presented, is designed by taking into consideration the huge number of the medicinal plants cultivated in Albanian habitats and their positive role in human health. It is realized by a team of the specialists in Industrial Microbiology, Food Technology, Pharmacognosy and Physic Chemistry.

The study is focused mainly in the evaluation of the microorganisms hosted by the selected fresh and dried medicinal plants and the role, the organic chemical compounds of these plants play in the growth of the isolated and identified microbial populations. In the experimental plan there was involved the study of the impact of some aromatic additives in food products, to increase their safety and the shelf life. The obtained results are the important dates to continue in the same research scientific field.

2. General considerations

Aromatic medicinal plants grow in the whole area of Albania, on the sea level, hills and mountains.

The recent studies are focused in their curative role-herbal teas produced in the country & imported and in the production of the essential oils. The dates about their endemic micro flora and added microorganisms are limited and actually their study is a very important issue of the experimental work. The presence of the microbial strains in these plants depends not only on the growing areas and their climate, but also by the chemical compounds in the anatomic parts of the plants and the relationships *plant microorganisms* and *microorganisms between each other* (hosted in the same plant). Research teams are cooperating on the above issues to profit by the common experiences. They have as target, the use of the benefits of the microbial world for the development of modern biotechnologies and for the enrichment of the microbiological Collections with important strains, in order to use them for the industrial enzymatic processes and for the production of bio molecules. The achievements through the common projects in the taxonomic studies of the yeasts and moulds, isolated from natural habitats, bear witness to the raw materials, products and wastes can serve as favourable Carbon sources, with low cost, for the development of the microbiological strains, as total micro flora and/or for the screening of the special strains. The importance of the microbiological control of the aromatic medicinal plants is related today with the consume also of the herbal teas (mentioned above) and the use of the aromatic plants as natural antimicrobials and food additives, in order to improve the quality and safety of the different food products to improve their odour and test.

Albanian natural habitats are full of Mediterranean plants, which serve also as raw materials for several technological treatments. Microbiological control of the medicinal plants is still unknown completely, comparing with the full studies in chemical and medicinal aspects. The experience in the microbiological control of these plants shows the presence of mould resistant strains, even the plants are dried or heat treated (termotolerant strains or the resistant strains by the low values of water activity - aw 0.3-0.4, corresponding the monomolecular stratum). These strains are taking part in the micro flora of herbal teas, dealing in the market dried, shredded, milled and packaged. Their total microbial charge can not be "*endemic*", because of the added microbial populations derived from the "packaging system" and those from the surrounding environment, when the opened packets are used for a long time.

The experimental work here presented is a modest contribution in:

- the evaluation of the total microbial charge of the selected aromatic and medicinal plants of the Albania (fresh and dried) as *Matricaria camomilla, Laurus nobilis, Folium sennae, Thymus vulgaricus, Thymus serpilium, Thymus longicaulis,* etc.;
- the exploration of the special microbial strains-yeasts and moulds with a very important role in the human health and biotechnology;
- the use of the taxonomic methods in order to make the identification of the isolated and purified strains derived from the selected plants, having as target the creation of a Collection of these microorganisms;
- the study of the relationship between plants and microorganisms;
- making into evidence the chemical compounds presented in some selected plants, used to obtain essential oils by effective techniques of the extraction;
- the evaluation of the role of some aromatic herbs, to reduce the total microbial charges in different food products, animal proteins and other products, to increase their safety and to improve the quality (some *Ready to Eat* food products);
- the study, after the microbiological control, of the morphological and physiological characteristics of the isolated strains, in order to evaluate the presence and the effect of the selected strains in the safety of the above products.

The experimental work was realised in a five years period, in the framework of the projects R&D, financed by the Albanian Government and other projects of the bilateral cooperations with foreign Institutions. The scientific research was done in the Department of Industrial Chemistry, Section of Food Technology and Microbiology, Faculty of Natural Sciences in the University of Tirana, with the cooperation of the Departments of Chemistry and Pharmacy of the same University. The natural selected habitats were selected according to a detailed experimental plan, determining also the specific herbal plants, characteristic for the areas under the examination. These plants offer more possibilities for the isolation and identification of the typical microbiological strains. In the experimental plan were involved also products originated by the above plants as herbal teas prepared in Albania and some food products of the Albanian market, used to observe the effect of the herbal additives on the reduction of their total microbial charge.

3. Materials and methods

3.1 Sampling

Based on the objectives of the experimental studies, the sampling involved:

- Fresh endemic plants, grown in fields and in a highness about 700-900 m above the sea level, as *Laurus nobilis, Rosa canina, Thymus vulgaris, Thymus longicaulis, Saturea montana, Matricaria camomila, Folium senae, etc.* and some herbal teas originated by them . These plants were known and used by the Albanians, even when the technologies of the medicinal plants did not existed.
- Food products with a high distribution in the Albanian market, consumed by the Albanian populations. There were selected products with a high percentage of the animal proteins and a low percentage of triglycerides (anatomic parts of chickens of the Albanian farms). There were selected also the vegetable conserves (ketch up, other simple & composed tomato sauces and green sauces), used to improve the quality of

food products-commodities, produced in a short time, with a low cost and preferred by the Albanian consumers.

The selection was based in analogue studies. According to them, chemical compounds (aromatic and medicinal compounds) of the above plants have a natural antimicrobial behaviour towards epiphyte and added microorganisms. They reduce and/or eliminate the total charges or specific strains in the same time with their effect to improve the taste, the odour and the quality of the products where they are used as food additives.

From the animal proteinic foods were selected the anatomic parts of the chickens - *white meat* and the *skin*. The white meat was selected, because there is a fast increasing of its consume on the global level. It is agreeable by the consumers and offers low energetic values. The skin was selected, because, for all the anatomic parts of the poultry, being in the market without special treatments, as packaging in vacuum, or in the controlled atmospheres, the skin is the host of the added microorganisms, which are not characteristic flora, but a contaminated micro flora for the above product. Thus, it has a potential negative impact in the safety of the whole food product and in the consumers protection. The skin is also the part with a direct contact with the surrounding environment and with the selected natural antimicrobials, also. There were selected from the vegetable foods, principally red sauces, ketch up and green sauces. Red tomato sauces are part of the Albanian daily food. They are prepared by the Albanian companies with classic and modern technologies. The Albanians find them in the market together with analogue imported products from Italy, Greece and other countries. The consumers prefer simple and composed sauces. Their daily use is related with the modern tendencies of the nutrition and their use as additives in pastes and macaroni, other global foods, or as additives in the portions of the fast food network (for example the global consume of the ketch up). Green sauces were known later by the Albanian consumers, except the traditional mixers of herbs, individually prepared for special food portions. Actually they are used by the consumers and liked by the market, because of the new tendencies of the nutrition, transferred from the neighbours (Italy is known all over the world for the preparation and the consume of the green sauces, example " *pesto alla Genovese*").

Microbiological studies of the research team were focused last years in principal food products, as cereals and their products, milk and dairy products, alcoholic & non-alcoholic beverages and others.

Sauces were not part of these studies because they were only added in principal food products; they were termically treated and most of them were prepared with substances used for their preservation, as ascorbic acid. Actually they are more distributed in the food portions, more used than before and most of them are prepared without chemical substances for preservation, labeling as "bio"products. The importance of their microbiological control is related also with a long period of their use, (they often are preserved for a long period of time, not hermetically closed, in home refrigerators or in cold storages of fast food network). So, actually it should not be neglected their impact in the safety of the prepared foods and also the role of the herbal additives (herbal medicinal plants) in the composed sauces, to reduce their total microbial charge and to increase the confidence of the consumers towards their use as simple food, or as part of the other food products.

A detailed description about the three important groups of the selected samples, mentioned above, is as follows:

The characteristics of some of the selected aromatic medicinal plants:

- *Matricaria chamomilla*- an annual plant distributed and grown near populated areas in Europe and Northwest Asia; it is considered as native plant of the above areas. It has a right stalk with ramifications on the upper part of it, with characteristic white-yellowish flowers. Pharmacological effect is related with the presence of chemical compounds as bisabolol, camasulen, spathulenol, flavonoides, hydroxicoumarins, rhamnogalacturonanes, etc. These substances are responsible for the antibacterial effect of the plant in open wounds, the antimicrobial effect upon *Staphylococcus* and *Candida*, that caused skin diseases. One of the most important chemical compounds -camasulen, is well-known for its antirritant and antioxidant effect. Alfa-bisabolol is responsible for the gastrointestinal effect, while flavonoides are known for a very positive effect against cough and bronchial diseases.
- *Cassia angustifolia (senna)*- a plant of tropical and subtropical areas, native in Africa, America & naturalized and more uses in Europe. It is a brush wood with double leafs, always green and with yellow flowers. It is commonly used as herbal laxative tea with an effect originated by the derivatives of anthracene –A, A1,B,C,D sennosides (anthraquinone glycosides), present in the plant together with naphthacene derivatives. It is used in the treatment of constipation, stimulated the intestinal peristalsis.
- *Satureja montana* – endemic plant of Europe and other geographic areas. It is grown in Albania as a perennial sub shrub with yellow flowers. It is the origin of the essential oils composed by p-cimene, linalool, carvacrol, thymol, myrcene, limonene, pinene, etc. The chemical composition is responsible not only for the aromatic effect of the respective essencial oil, but for the strong antibacterial effect, overall against G+ bacteria, as *Staphylococcus aureus, Streptococcus pyogenes and Sarcina sp..*

3.1.1 Albanian aromatic medicinal plants as row materials for the production of the essential oils

A part of the whole experimental work, here presented, is the study of the extraction of the essential oils from *Thymus vulgaris L. (thyme), Thymus Longicaulis C Preisl.(creeping thyme)* and *Rosmarinus officinalis L (rosemary):*

- *Thymus vulgaris* is a culinary and a medicinal herb, used in Mediterranean areas, in Albania also. It is used both fresh and dried to flavour meats, sauces and other food products, together with other vegetable ingredients. The essential oil of thymus contains a high percentage of the antiseptic substance, called *thymol.*
- *Thymus Longicaulis C Preisl* is an evergreen perennial plant with pink flowers in summer and small green pine scented leaves. Its hydrodistillated essential oil are considered with an antioxidant activity in beta-carotene/linoleic acid system.
- *Rosmarinus officinalis L* is a woody, perennial herb with fragrant, evergreen leaves and part of the mint family. Mediterranean areas are familiar with it, using the leaves for the culinary purposes. Rosemary is rich with biologically active compounds, including antioxidants, such as carnosic acid and rosmarinic acid.

The selected samples were taken from the Department of Pharmacy (Pharmacognosy Scientific Laboratory) and certified wholesalers, responsable for the medicinal plants export activities.

3.1.2 The characteristics of the selected animal proteinic food products and respective antimicrobial additives

During the experiments, there are used also *clove essence* and *oregano* in order to observe their effect towards epiphyte and added micro organisms in selected food products, overall to evaluate the antimicrobial effect of the herbal plants, more used in food industries and in the preparation of food portions. Oregano is considered as a herb with double effect - to improve the sensorial characteristics and to reduce the microbial charge. The use of both above substances was limited in the study of their effect regard to the microbial populations in food with animal proteins. For the future it is planed a detailed study related with their micro flora and the relationship *endemic micro flora/chemical compounds*. Antimicrobial effect of them was tested in selected poultry products, simply packaged, with a high distribution in the Albanian market- poultry farms of *Korca, Patosi* and *Gjirokaster.*

3.1.3 The characteristics of the vegetable composed conserves, selected for the microbiological tests

Sampling includes also red tomato sauces, ketch up and green sauces of the Albanian market. Microbiological study of these samples was realised referred the presence of *laurel,* thyme, oregano and basil as important ingredients of composed sauces. The selection of these conserves was made having as target the illustration of the antimicrobial effect of the above ingredients in composed food products. There are used basil sauces, Bolognese sauces-B, with thyme as ingredient; ketch-up's nominated with initials S and K, with the natural herbs: laurel, basil and thyme, as ingredients. There were selected also green basil sauces – "Pesto" G and T. The obtained results of the composed sauces were compared with the results of the simple sauces. (the above initials are used for the confidentiality of the trade-marks).

General characteristics of the above herb ingredients, studied for the moment only as additives, are as follows:

- Cloves- *Syzygium aromaticum,* are aromatic flower buds of a special tree, grown first in Indonesia and other Asiatic countries. Dried cloves are used recently in the cuisines of all the world. They are used, for their medicinally properties, principally in dentistry and to improve the peristalsis also. The active compound, *eugenol,* comprises high percentage of the essential oil, extracted from the cloves (about 90%).
- Oregano- *Origanum vulgare* is a simple specie of *Origanum* from the mint family. The plant is originated by Asia, Mediterranean areas and Balkan Peninsula. It is grown as a perennial herb in average temperatures, in a pH range between 6.0 and 8.0. Chemical compounds of the essence are carvacrol, thymol, limonene, pinene, p-cimene, cis-ocimene, linalool and others. Oregano has a powerful antiseptic effect and a high antioxidant effect because of the high content of the phenol acids and flavonoides. The antimicrobial effect is well-known against the microorganisms, contaminants of food products as *Listeria monocytogenes.*
- Basil- *Ocimum basilicum,* is a medicinal plant with sedative, antispastic, antimicrobial and anti irritant effects. Antiseptic properties are related principally with the chemical compounds of the essential oil, prepared by the basil leaves, which is considered as the main active compound of the plant. These chemical compounds are eugenol, methyleugenol, linalool and others.

- Laurel-*Laurus nobilis,* is a shrub with green and glossy leaves. Its leaves are aromatic and are used fresh or dried, usually in Mediterranean food products. They are used also in aromatherapy and for curative purposes. The essential oil of laurel is very important and full of chemical compounds as *cineole* or *eucalyptol, terpenes, sesquiterpenes, methyleugenol,* and others.
- Thyme-*Thymus vulgaris* is described in details on the above paragraphs.

The structure of the experimental work, including the selected samples, was designed to point out the effects on a wide range of the herbal medicinal plants: as additives or ingredients in food industries, as safety and quality herbal teas and as producers of essential oils with very positive effects for the human being.

3.2 Methods

The applied methods of analysis are chosen, based on the structure of the experimental plan. They include both microbiological and analytical methods.

Microbiological control of samples was realised using the followed steps and methodology:

- Preparation of the samples applying all the rules for their preservation as *time, temperature, packaging,* etc..
- Application of the *Method of the Limited Dilutions,* in order to determine exactly the total charge of the micro organisms.
- The use of the selective media to determine the presence of bacteria (MPA-Meat Peptone Agar - growth medium), yeasts (MA-Malt Agar, YEPD-Yeast Extract Peptone Dextrose, YED-Yeast Extract Dextrose- growth media), moulds (CAPEK and YEPD - growth media). Streptomycin was added in the selective media of the moulds, in order to prevent the development of the bacteria strains.
- The determination of the number of the present micro organisms using the numeration methods - *Petri dishes divided in sectors* and/or numerations in whole *Petri dishes* using a *digit all colonies counter.*
- The use of the classic methods for the isolation *(screening)* and the identification of the special strains (the case of *Rhodotorula sp.* grown in *Malt Agar* and *Rose Bengalis*-media).
- The use of the *taxonomic tests* (morphological, physiological and technological tests) for the specific isolated strains of yeasts and moulds (*Barnet* and *Lodder&Kreger van Rij*).
- The classification of the moulds were made referring the cultural characteristics as the colour, the structure of the colonies and their knitting behind the Petri dishes.

3.2.1 Methods for the extraction of the essential oils

The obtaining of the essential oils from the selected medicinal plants was done using the new effective methods of the extraction. One of them, a very effective extraction method, is the use of fluids in critical conditions, for example CO_2, based on the properties of pressurized effluents and in the specific characteristics of the above chemical compound. CO_2 is a clean effluent, cheaper than others, nontoxic and very effective to obtain good percentages of extracts, eliminating or reducing the environmental pollution also.

There are used in the experimental work, here presented, classic methods (hydrodistillation) of extraction & subcritical fluid extraction with CO_2.

There are realised the comparative studies between classic and modern methods of extraction, identifying also with chromatographic methods (*Gas chromatography–mass spectrometry, GC-MS*), the organic compounds with industrial and scientific importance, presented in the extracted essential oils.

The quantitative determination of essential oils was performed using a *Clevenger* apparatus, following methodologies from Italian Pharmacopeia and WHO (*World Health Organization*) manuals. Distillation apparatus was composed from the following elements : 1) heat system, able to offer and maintain a controlled optimal temperature; 2) glass flask; 3) distillation apparatus Clevenger type; 4) vertical refrigerator. All these elements were provided from the Departments of Pharmacy and Chemistry, University of Tirana. There were used 100g. plant materials for each distillation procedure.

The chosen round-bottom flask had a volume of 3000 ml, in which were added 1000 ml water. Distillation time duration was planned to be 5 hours. It was added 1 ml of organic solvent (hexane), in order to collect the essential oil. The solvent was chosen based on GC-MS analyses technique. The average distillation rate was monitored to be around 2-3 ml/min.

The percentage of the obtained essential oil was determined using the formula :

$$\% \text{ of essential oil } (\text{ml} / \text{gram}) = \frac{100\ (V_2 - V_1)}{a}$$

V_2 - volume of the obtained essential oil + hexane (ml)

V_1 – volume of the quantity of the added hexane (ml)

a - quantity of plant material distilled (g.)

During the experimental work were used also analytical methods based on the official methods of analysis to determine the physicochemical characteristics of the selected food products.

4. Results and the discussion

4.1 Microbiological analysis of the fresh aromatic medicinal plants

Microbiological analysis, to determine the total number of the microorganisms presented in the different anatomic parts of *Rosa cannina, Satureja montana and Thymus longicaulis Presl.* (roots, stalks, flowers and leaves) were realised, using selective media for each group of microorganisms. In general, there were observed an infinitive number of microorganisms in the first dilution. So, there were analysed the second, the third and the fourth dilutions, in order to see the dominant strains. There were observed a presence of the pigmented yeasts and different characteristic moulds. There were isolated, purified and identified the yeast strains of *Rhodotorula sp.*. There were observed also the strains of *Aureobasidium* and others, which are in the process of the preservation in selective media, having as target, making them part of the Collection of the isolated microorganisms. The identified moulds were principally *Ascomycetes* as *Aspergillus niger, Aspergillus candidum, Aspergillus terreus, Penicillium rubrum,* etc.; *Phycomycetes* as *Mucor hiemalis* and *Rhizopus nigricans; Fungi imperfecti* as *Cladosporium sp., Alternaria sp., Trichoderma viride, Trichotecium roseum,*

Helminthosporium sp. and others. The same strains were observed during the repeated annual microbiological analysis of the fresh plants.

4.2 Microbiological analysis of the herbal teas

The determination of the presence of the micro organisms, bacteria, yeasts and moulds in herbal teas: *laurel, chamomile, savory* and *senna* is presented in the histogram of the Fig. 1. The graphic is the function of the logarithm of the total numbers of the micro organisms by the type of the herbal tea. The results presented in the Fig.1 are summarised from the parallel microbial tests for the determination of the micro organisms in the second dilutions of the herbal selected teas. Bacteria were dominant, while yeasts and moulds were numerable. The determination of the total microbial charge was made not as a target of the experiments, but as an indispensability to observe the microbiological differences (epiphyte and added micro organisms) between herbs with different active components. According to this, the interpretation of each result is focused on the peculiarities of the herbal teas.

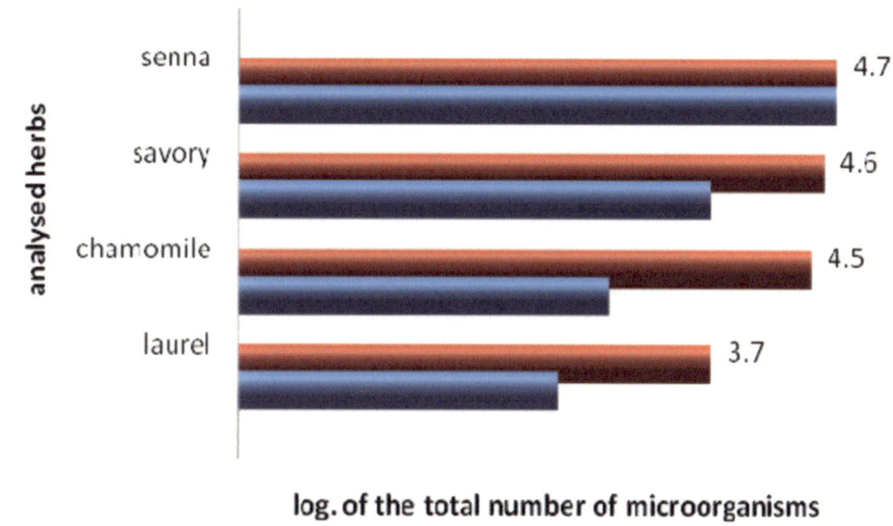

Fig. 1. The evaluation of the total charge of microorganisms, present in the selected herbal teas

Laurel- laurel is a very good host for a big number of bacteria. Based on the experience of analogue studies, the total charge of bacteria is not related with epiphyte strains but with added strains in the dried laurel leaves with low values of the activity of water (dried leaves staying for a long time in open packets are hosts for different bacterial strains). The number of moulds in the selective media is high, but lower that total mould charge in the fresh leaves. This is related with the active components of the plant (moulds are grown also in a wide interval of values of the water activity). The number of the yeasts is practically 0 and it is related with the presence of alkaloids, which (together with the low values of the activity of water) reduce drastically the number of the yeasts. Characteristic identified strains of the moulds are *Aspergillus flavus , Aspergillus niger, Aspergillus tereus, Penicillium islandicum, Penicillium rubrum* and *Penicillium cyclopium.*

Chamomile - chamomile is the host of an infinite number of bacteria. There were identified using taxonomic methods, the added types of *Pseudomonas sp.* and *Bacillus megatherium.* There were not observed *Staphylococcus and Candida.* This is related with the repressive action of the plant on the above strains. The prevention of their growth is related with physic chemical conditions of the plant also. There was observed an average charge of moulds, which was not an indicator of the pollution, caused by added strains. *Phycomycetes-Mucor sp.*and *Rhisopus nigricans* were dominant.

Savory-making a distinction from other herbal teas, savory was host of the pigmented yeast, *Rhodotorula sp.* From the former experimental work related with the microbiology of the fresh plants, the same psychrotolerant strain was isolated in the fresh plant grown in low temperatures in mountains. The above specie is a mesophilic strain having as attribute the antioxidant properties, known and evaluated in the similar isolated strains, (Fig.3).

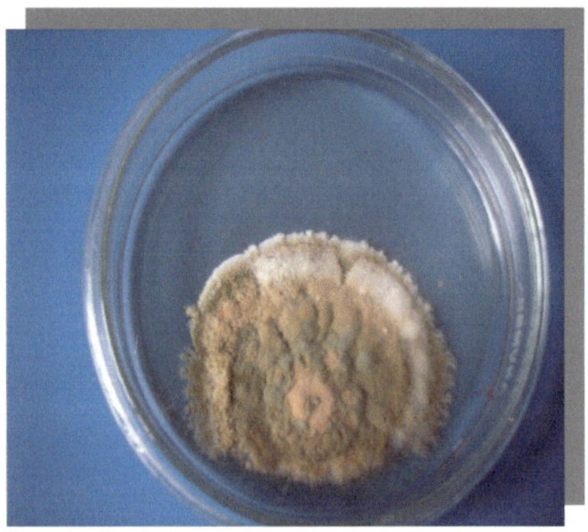

Fig. 2. *Penicillium cyclopium* in dried leaves of the laurel

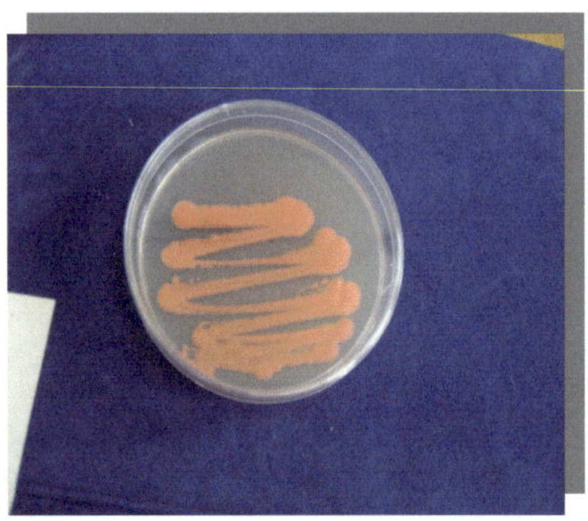

Fig. 3. *Rhodotorula sp.* isolated during the experimental work

Senna - there were observed some other specific mould strains hosted in senna dried leaves, *Cladosporium sp.* and *Aureobasidium sp.*. Based on the experience of the former experimental work with fresh plants as samples, *Aurobasidium* is attendant of the micro flora of the plants and other strains with antioxidant effect; while *Cladosporium* is epiphyte mesophilic strain of the areas with Mediterranean vegetation and is observed as a part of the micro flora of the respective earths.

4.3 The extraction of the essential oils by the selected herbal plants and the study of their composition and properties

The determination of essential oil content using fluid extraction in critical conditions (subcritical CO_2)

Plant materials from including thyme, creeping thyme and rosemary were extracted by liquid CO_2, under liquid-vapour conditions near its critical conditions. The used apparatus is shown in Fig.4. The liquid CO_2 extracts were compared with hydro distillates using GC-MS. For each plant was observed and compared the presence of important chemical compounds in hydro distillates and CO_2 extracts, analysing selectivity, differences and future objectives. The quantitatively determined yields of the CO_2 extractions were 2.5 to 4.5 times larger than the yields of hydro distillations. For the extraction of the plant materials approximately 3-4 hours were needed and for the hydro distillation 5 hours. The yield of the extractions as function of time could be described with a simple equation.

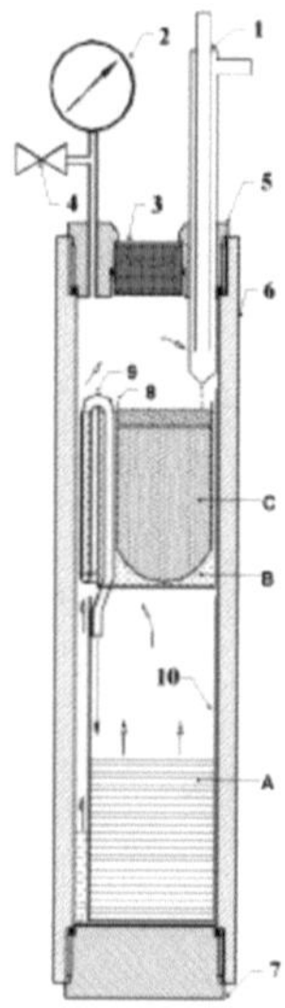

1) Cooling finger; 2) Pressure gauge; 3) Sapphire Window; 4) Valve; 5) Upper cover; 6) Steel Cylinder; 7) Bottom cover; 8) Extracting thimble; 9) Siphon; 10) Product dropping glass; A) Product; B) Extract solution; C) Plant material

Fig. 4. The equipment for the high pressure extraction with liquid CO_2 under liquid - vapor equilibrium conditions.

The determination of the essential oil chemical composition with Gas chromatography - Mass spectrometry (GC-MS)

The extracts obtained from hydro distillation and CO_2 subcritical methods were analyzed using Gas chromatography–Mass spectrometry (Fig.5), based on standardized techniques, described in details as follows:

Essential oil extracts were analyzed using a GC- 2010 (Shimadzu, Japan), equipped with a split/split less injector, electronic pressure control, AOC-20i auto injector, GCMS-QP2010 Plus mass spectrometer detector, and a GCMS Solution software.

The column used was a ZB-5 (Zebron) capillary column, 30m×0.25mmI.D. and 0.25_mphase thickness. Helium, 99.996% was used as a carrier gas at a flow of 1mL/min. Oven temperature programming was 60 ◦C isothermal for 4min then increased to 106 ◦Cat 2.5 ◦C/min and from 106 ◦C to 130 ◦Cat1 ◦C/min and finally from 130 ◦C to 250 ◦C at 20◦C/min, this temperature was kept constant for 10.2 min. Sample injections (1_L) were performed in split mode (1:20).

The inlet pressure of the carrier gas was 57.5 KPa. Injector temperature was of 250 ◦C and MS ion source and interface temperatures were 230 ◦C and 280 ◦C, respectively. The mass spectrometer was used in TIC mode, and samples were scanned from 40 to 500 amu. Thymol, borneol, camphor, eucalyptol and linalool were identified by the comparison with standard mass spectra, obtained in the same conditions and compared with the mass spectra from library Wiley 229. Rests of the compounds were identified by the comparison with the mass spectra from Wiley and Nils library.

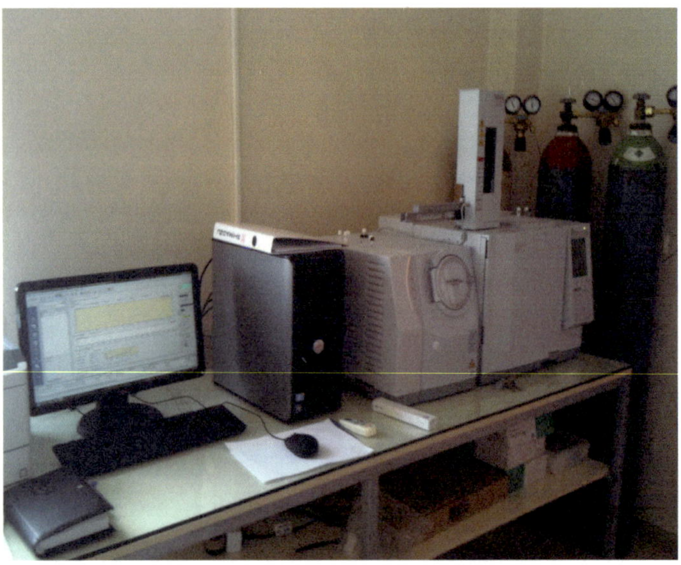

Fig. 5. GC - MS apparatus, used for the chemical composition analyses

The first step was the exact identification of the chemical compounds presented in *thymus, creeping thymus* and *rosemary* liquid essences. The above essences have been studied before but the methods, equipments and identification libraries used in this experimental work are modern and contemporary giving a high certainty for the chemical components and the whole essence.

There was confirmed by the obtained results, the predominant character of the phenolic structures in the liquid essences. The flavors and the distinguished antiseptic properties are related directly with the above components. Some of the principal chemical substances of

Thymus vulgaris, Thymus longicaulis and Rosmarinus officinalis are: α- pinene, p-cimene, camphor, 1,4 - terpineol, thymol, carvacrol, linalool, 1-8 cineol, limonene, camfene, β-myrcene, verbenone, cariophyllene and cariophyllene oxides. For the first time, during a long experimental work, was made possible the chemical analysis of the extracts taken by the use of subcritical CO_2, in the liquid state with unique particularities. The information about the yield was enriched with the dates about the chemical content. The obtained results were very interesting. There was confirmed the idea that the CO_2

extracts have another characteristics compared with those of hydro distillation. They are rich with new components, including terpene hydrocarbons and/or resins. Analyzing the extracts after the 21-st cycles, there is confirmed that the principal components, identified in the hydro distillation essence, are taken with very small differences using the CO_2 subcritical method for the three aromatic selected plants. Observing the obtained chromatograms, analyzing also especially, the obtained fractions after a number of cycles (3, 9, 15, 21), it was confirmed a very interesting particularity related with two principal chemical substances, thymol and carvacrol. The extraction method with CO_2 shows a preference and a specific affinity related with both components. They were extracted first (there were clearly observed their picks after the first three cycles); their extraction continued after 9 and 15 cycles, arriving in maximum values after the 21-st cycle. Their highness of the picks was considerable, with a clear difference from other components as phenols or terpene hydrocarbons. The above conclusion is very interesting, because of the confirmation of the specificities presented by the CO_2 method of the extraction related with both most important chemical compounds of the studied herbal plants. These two components show their selective antimicrobial effectivity during the microbiological tests, including in the experimental work.

The above observation is fundamental and serves to stimulate the scientific research activities in the same issues. The short term objective is the work with pure reference standards of thymol, carvacrol and other chemical compounds, in order to make the quantitative evaluation of them, profited at the end of the work cycles, including also the final comparison with hydro distillation.

High total yilds of the extracts taken from the CO_2 method and the specificity of it toward thymol and carvacrol, reinforce the idea that their quantity values may be higher that them on the liquid.

4.4 Aromatic medicinal plants and their essential oils as antimicrobials in animal proteinic food products

Medicinal plants, fresh and dried, together with the respective essential oils, produced by them in modern technological processes play a very important role to prevent the growth of the microbiological strains in food products. In the meantime they make the lyses of the existing cells, acting as the reducers of the presented strains. So, there is an antagonist relationship between bacteria, molds and yeasts and the active compounds of the plants or their essential oils, which explains the repressive effect of the specific active compounds toward different microbial strains.

As it was mentioned above, there were selected the samples of *white meat* and *poultry skin* in order to explicate and evaluate the antagonist effects of some herbs and essential oils in animal products.

White meat and the skin of poultry products, selected directly on the market, were treated with o*regano* and with the *essential oil of the cloves* also, in order to improve the taste and to reduce the microbiological charges. The results are summarized in Fig. 6 and Fig.7. There are presented in the graphic of the Fig.6 the total microbial charges of the three untreated skin samples from the Albanian farms (lp1, lp2, lp3) and the same treated samples with oregano (lr1, lr2, lr3) & with the essential oil of the cloves (le1, le2, le3). Referring to the corresponding results, there was observed a big number of bacteria comparing with the existence of the moulds and yeasts. This was an expected result, related with the amount of the triglycerides in the poultry skin, as well as with the environmental microbial contaminants which penetrate the simple packaging (without vacuum, without controlled or modified atmospheres). The reduction of the bacterial charges is evident and the effectivity of the action of eugenol was clear and really expected.

There are presented in the graphic of the Fig.7 the total microbial charges of the three untreated white meat samples from the Albanian farms (mp1, mp2, mp3) and the same treated samples with oregano (mr1, mr2, mr3) & with the essential oil of the cloves (me1, me2, me3).

The number of the bacterial strains in white meat samples was lower because of the lower content of the lipids; the tendencies of the microbial charge reductions under the action of oregano and eugenol were the same. Knowing the good practices of the production of the meat, the simple use of oregano is familiar principally for the red meat. There is planned for the future experimental work, the use of oregano essential , as a natural antimicrobial of red and white meat. The research team is working now to extract it from the fresh plant, using new techniques of the extraction.

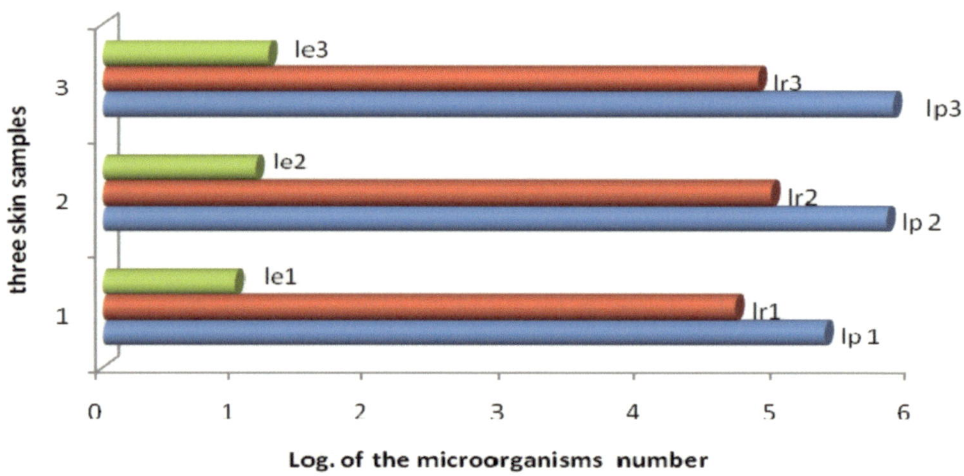

Fig. 6. The reduction of the number of microorganisms in poultry skins treated with oregano and the essential oil of the cloves

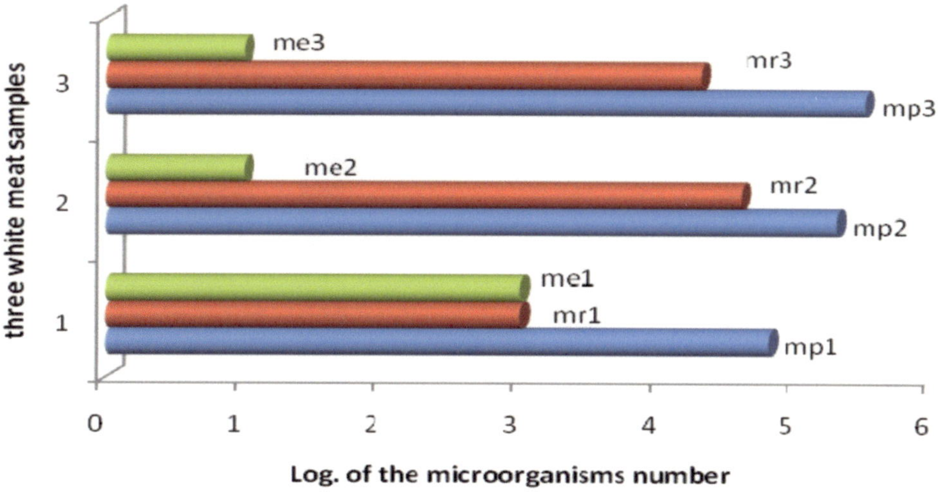

Fig. 7. The reduction of the number of microorganisms in poultry white meat samples treated with oregano and the essential oil of the cloves

4.5 Aromatic plants as additives in different types of sauces

Microbiological study of the some selected samples of sauces, was realised using two selected media MPA and YEPD-agar, to confirm the presence of the bacterial and mould charges. The summarized results are presented in the graphs below, Fig. 8 and Fig.9. The hermetically closed samples (S1, S2, S3, S4-red sauces; S5, S6- ketch up-s and S7, S8- green sauces of trade marks G and T) were first analysed in order to veryfie the microbiological safety of the products and after, during the preservation period.

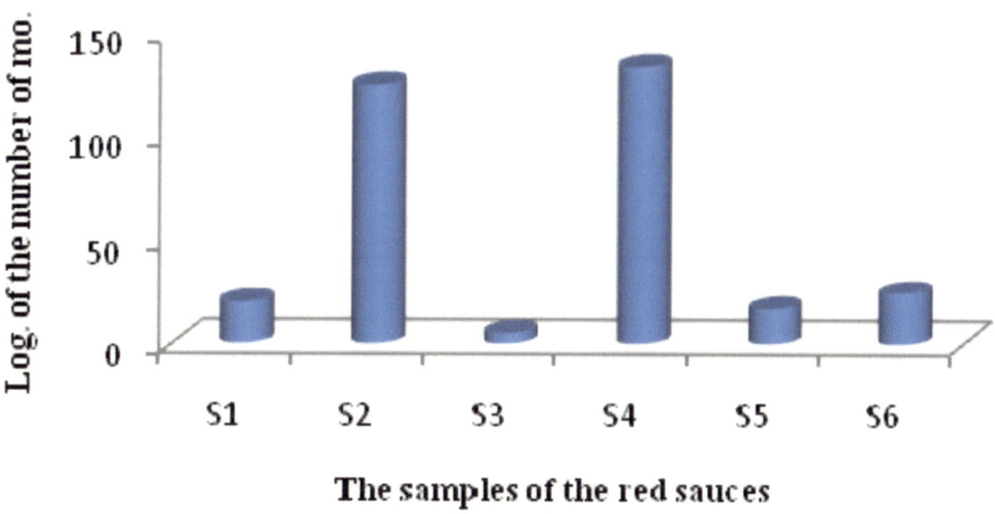

Fig. 8. The total microbial charge of the sauces samples

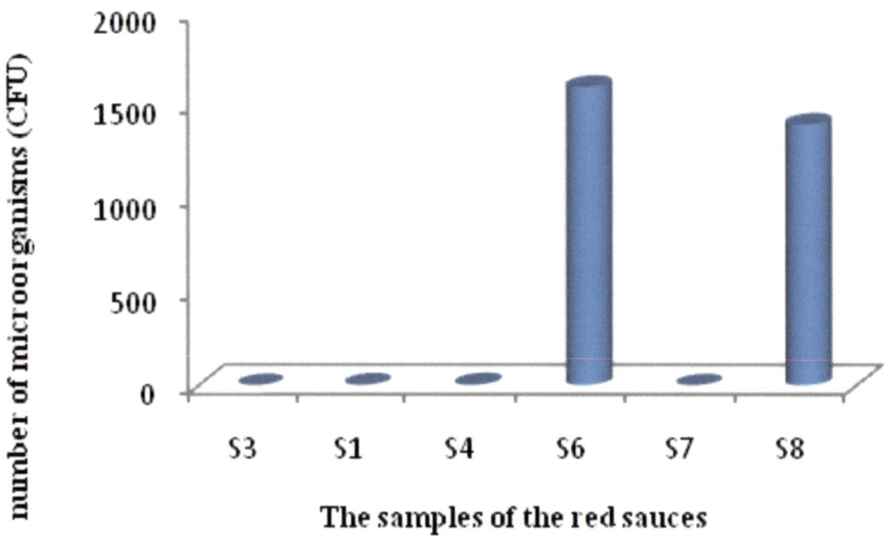

Fig. 9. The total number of the moulds in some of the sauces samples

The samples were identified by the initials and the detailed results were saved in order to respect the privacy of the trade marks. By the arithmetic mean results, taken by the parallel tests, in order to reduce the uncertainty of the measurements, there were observed the growth of some specific mould strains, presented in the not hermetically closed cans, even there were preserved in refrigerators. There is a quite difference between simple and composed sauces, with different herbs as ingredients. The above conclusion was achieved after many repeated microbiological tests of the same and analogue simple and composed samples.

S1 and S3 products were simple sauces with a relative low total charge, but with a high possibility to be increased, just after the opening. Sample S2 is a sauce with a high percentage of carbohydrates. This C source is in favour of the growth of the added strains in the opened samples and usually a principal component of the selected liquid and solid nutrients A specific case is that of S4, a spicy one, where the pepper as ingredient, must be a powerful microbial charge reducer. Its effect really is decelerated by the presence of the animal proteins, part of the composed product. This proteins may be an appropriate medium for the growth of microorganisms. The antimicrobial effect of the active herbs, as ingredients, is obvious in S5 and S6 samples. S5 is a composed product, with a large amount of laurel and thyme, while S6 is prepared with basil and savory ingredients. The laurel of S6 acts against added microorganisms because of tannic substances and a small quantity of the eugenol as a chemical compound of its essential oil. Thus, the sample S6 resulted a safety product regarding to the total charge of microorganisms, preserved for a long time in low temperatures. The presence of the moulds were evident in the samples S6 and S8, while the added yeasts charge was practically 0. The mould growth was related with a huge interval of the water activity (0.6-0.9), but also with the presence of basil as an added component in S6, S7 and S8. Concerning the bacterial strains in red sauces, the role of the chemical compounds of the aromatic plants to decrease the microbial charges was clear. Green concentrated sauces, rich on basil herb and other ingredients, were an appropriate substrate to stimulate the growth of the moulds. S7 sample was rich of a large amount of lactic acid and anacardic acid (6-pentadecylsalicylic acid), originated from some dried and woody fruits. That is why the green concentrated sauces offer a high resistance toward bacterial cells (example S7 with pine sticks and green herbs).

There are presented on the images of the Fig. 10 some specific identified microbial strains of the analysed samples.

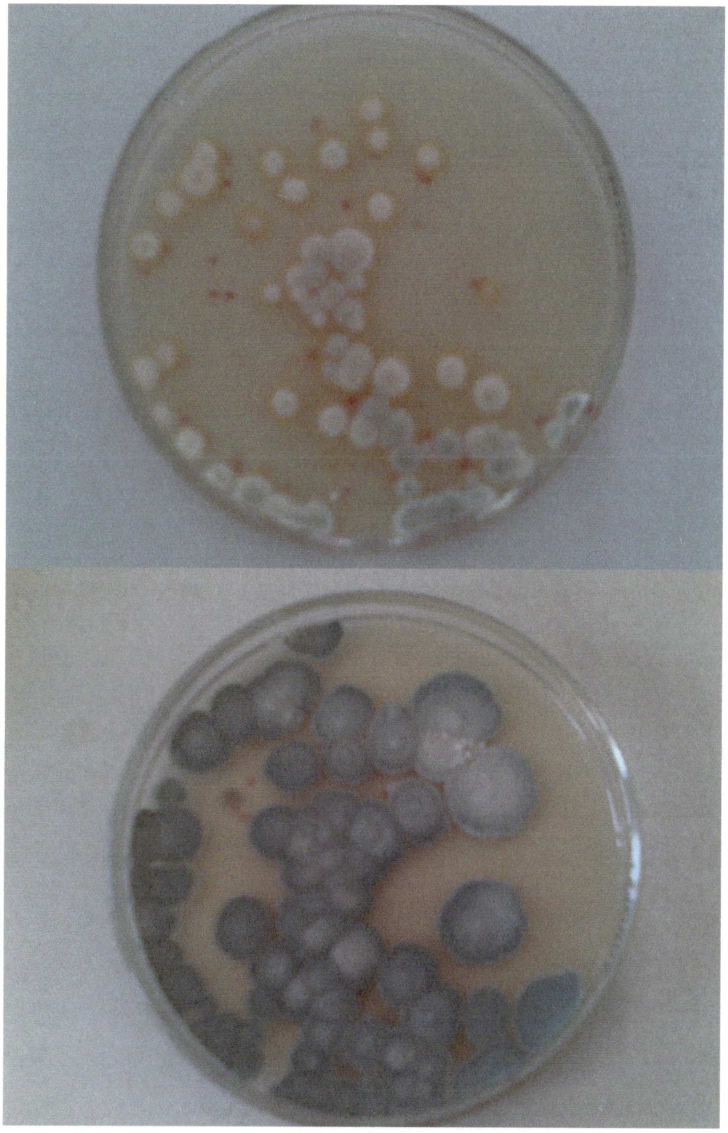

Fig. 10. Indentified moulds of the genus *Penicillium ciclopum* together with the bacterial pigmented strains, isolated by the composed sauces

5. Conclusions

The anatomic parts, the structure of the cells and the chemical compounds of the aromatic medicinal plants can play a very important role in the growth and the development of the different microbial populations, which are hosted in the plants, food and pharmaceutical products derived from them, as native or added microflora.

The relationships between chemical substances and the microbial cells are antagonists, when most of these chemical components offer a powerful repressive action toward microbial cells, decelerating the growth and the reproducibility of the cells, or causing their lyses. Typical "poison" chemical substances are principally in the meantime "the active components" of the plants and their essential oils, tannins, alkaloids and other groups, specified on the text.

The development of the different genera of the plant microorganisms is related not only with the plant characteristics, but with the geographic vegetative areas where they are grown, the climate, the different plant varieties and with also, all the technological processing's of their transformation into final products, originated by herbal plants. The change of the above conditions is accompanied with big differences related to their presence and their morphology. Some of the microbial strains modify their behaviour, adopting their growth in new substrates, sporulating when they are sporeformings, reflecting obviously their new situation and their coexistence with other new strains.

The presence and the growth of the microorganisms are conditioned by the *activity of water*, *temperature* and the *preservation time* of the substrates.

Aromatic medicinal plants have shown specific properties related to their medical applications. Their use in obtaining of pure active ingredients or starting compounds (modified during chemical synthesis), in order to be part of effective new drugs, are some examples of their importance. Also, from another point of view, these natural plants can be considered as substrates for the development of the special microbial strains having extra and intracellular biomolecules as antibacterial and antioxidants, (the presence of *β-carotenes* in *Rhodotorula* genius and *Aureobazidium* genius, identified during the experimental work).

Essential oils, produced by the Albanian endemic and naturalised plants, act with their specific and significant active compounds, reducing the total charges of microorganisms, eliminating or preventing in the meantime the other strains and species.

Essential oils extracted from herbal plants, having also as target the optimization of their effects as antimicrobials in some food products, will be a future objective for advanced scientific researches on the same field.

The use of modern methods of extraction (critical fluid extraction) and new techniques for their chemical analysis (GCMS equipment), has offered the first results, related to the chemical compound of the extracted essential oils, making possible the study of the relationship between chemical compounds and identified strains.

The controlled tests related to the action of essential oils in food products and also the preventive action of fresh parts of the herbal plants, promise a future use of them in a wide range as antiseptic and antioxidant ingredients or additives of many food products, overall commodities. Their antimicrobial effect is accompanied together with the improvement of the sensorial properties of the food products. Referring to the experimental work, there were evaluated some tests with a very high reduction of the microbial survival, about 10-30% reduction.

In the future are planned new experiments, to optimize the effect of the extracted essential oils, towards composed, analogue food products.

The determination of the total number of the microorganisms in red and green sauces confirmed the predominance of mould strains and served also to enlarge and orient the experimental plan toward taxonomic studies for the identification purposes.

The comparison of the results between simple and composed sauces, rich with herbs, reflects an obvious reduction of the microbial pollution particularly in the presence of laurel, thymus and oregano. A high reduction is evident, compared also with simple sauces produced with chemical conservants. All the detailed interpretations of these results are given in specific paragraphs of the publication.

6. References

Asllani, U. (1992). " Chemical Nature of the Albanian Essential Oils . Thier polychimism and thir quality improvement", Durres, Albania.

Barnet, H. L. & Hunter, B. (1972). „Ilustrated genera of Imperfect Fungi",Burgess Pub. Co , 3rd. edition, ISBN-10: 0808702661, Burgess Pub. Co.

Buzzini, P. (2000). " The rising power of yeasts in science and industry", Teenth International Symposium on Yeasts. 27 august-1 september 2000.

Buzzini, P. (1999). " Current Genetics" (Eukaryotes with emphasion on: Yeasts, Fungi, Protists, Cell Organellas.) XIX International Conference on Yeasts Genetics and Molecular Biology, 25-30 May, Rimini, Italy.

Campbell, I.& Duffus, J.H.(1988). "Yeast a practical approach" ; ISBN-13: 978-0947946807 Oxford University Press, USA.

Capelli, P.& Vannuci, V. (2005). "Chimica degli alimenti. Conservazione e Transfomazione" 3rd edition, ISBN 8808075893, Italy.

Demiri, M. (1979). "Plant Determinant" SHBLU , Tirana, Albania.

Frashëri, M.; Prifti, D. (1997) " Practicum of Technical Microbiology", Tirana, Albania.

Kurtzman, C.P. (2011)."The yeasts: a taxonomic study" 5 Edition, Publisher Elsevier, ISBN0123847087, 9780123847089

Mónica, R. Gonzalo, V. Guillermo, R. Fornari, T. (2011) "Fractionation of thyme (Thymus) vulgaris L.) by supercritical fluid extraction and Chromatography ", Journal of Supercritical Fluids 55, 949–954.

Hemalatha, S. Vivekananda, M. Yogesh, M. (2007)"Microwave Assisted Extraction - An Innovative and Promising Extraction Tool for Medicinal Plant Research", Pharmacognosy Reviews, Jan - May 2007.

Sarikurkcu, C. Sabih, O.M. Tepe, B. Can, S. Mete, E. (2010)"Essential oil composition and antioxidant activity of Thymus longicaulis C Preisl, subsp. Longicaulis, var. longicaulis". Food. Chem. Toxicol. 48(7), Elsevier, April 2010.

Sima, Z. (1994). "Pharmacognosy", SHBLU, Tirana, Albania.

Tortora, G.J.; Funke B. R.; Case Ch. L.; (2010)"Microbiology- An introduction", the Benjamin/Cummings Publishing Company, ISBN 0321550072.

WHO, (1999)"Monographs on selected medicinal plants", Volume 2, Geneve .

Part 4

Food Components and Food Additives as Bioengineering Materials

9

Poultry Products with Increased Content of CoQ_{10} Prepared from Chickens Fed with Supplemental CoQ_{10}

Petra Jazbec Križman[1], Mirko Prošek[1], Andrej Šmidovnik[1], Alenka Golc Wondra[1], Roman Glaser[2], Brigita Vindiš-Zelenko[2] and Marko Volk[3]
[1]*National Institute of Chemistry, Ljubljana,*
[2]*Perutnina d.d., Ptuj,*
[3]*Faculty of Agriculture and Life Science, Hoče,*
Slovenia

1. Introduction

Nutritional science has become more and more focused on foods and food components that have the potential to optimize the physical and mental state of the consumer, as well as to reduce the risk of disease. In our laboratory, we aim to design food products for a healthy diet and to decrease the risk of suffering from chronic diseases, especially those that become more prevalent with advancing age.

Our emphasis is on food or food additives and feed with added coenzyme Q_{10} (CoQ_{10}), or ubiquinone. CoQ_{10} is a key component in the inner mitochondrial membrane, where it plays important role in oxidative phosphorylation (Lenaz et al., 2007). It is also present in other subcellular fractions and in plasma lipoproteins, where it acts as an important antioxidant (Bentinger et al., 2007). CoQ_{10} has also been shown to have an effect on gene expression (Littarru & Tiano, 2007). These three functions are important for its use in clinical practice and in food supplementation.

CoQ_{10} can be found in various foods in different concentrations. The richest nutritional sources of CoQ_{10} are meat, fish, nuts, and some oils. CoQ_{10} can also be found in vegetables, fruits, cereals, and dairy products but at much lower levels (Mattila & Kumpulainen, 2001). The average contribution from food sources is 3-5 mg of CoQ_{10} a day for a healthy individual according to Weber et al. (1997). This amount can be easily consumed with normal food. In cases of deficiency, the contribution from food intake would have to be higher than 100 mg per day. This amount cannot be obtained only through food and therefore supplemental doses of CoQ_{10} must be given to the patient. It is known that the absorption of exogenously administering CoQ_{10} is slow and limited due to its lipophilic nature and relatively high molecular weight (M_r = 863), which is consequently reflected in its relatively poor bioavailability. However, increasing its solubility in aqueous medium should, in most cases, increase its bioavailability, which is why, we developed a soluble form of CoQ_{10} that was later patented in the form of complex with β-cyclodextrin (β-CD) (Prošek et al., 2005). This new form of CoQ_{10} exhibits increased water solubility and, consequently, better bioavailability in comparison to powder and oil-based CoQ_{10} preparations (Prošek et al., 2008; Žmitek et al., 2008). This new form of CoQ_{10} has already been commercialized as a functional food additive for human nutrition.

Our first selection for the practical use of this product was a group of dairy products (Stražišar et al., 2005). Their blend of fat and water was the ideal base for application of a functional food additive with an increased amount of CoQ_{10}. Later, we found different meats (fish, poultry, beef and pork) and different kinds of pates suitable for fortification with water soluble CoQ_{10}.

Our first attempt to evaluate the effect of CoQ_{10} on living organisms was on free living non-laboratory broiler chickens. That study covered the accumulation of the CoQ_{10} in different parts of the chicken's body (blood, heart, liver, and different muscle tissue) after different periods of administration (up to 40 days). For this purpose, 200 chickens, provenance ROSS 308, were used. Tests were carried out under optimal breeding and health conditions. Two different types of fodders (BRO-G and BRO-F2) were prepared in order to cover the needs of 40-day period. An adequate amount of water-soluble substance in the form of a paste with 7.5 % CoQ_{10} was weighed and mixed in consecutive steps until a final concentration was obtained. The daily dose was set at 5 mg CoQ_{10} per animal per day. For a better understanding of CoQ_{10} accumulation, its distribution on the cellular level was studies by analyzing several subcellular fractions of breast muscle tissue (Jazbec et al., 2009).

We also tested the performance of poultry meat in industrial production with an increased quantity of CoQ_{10} in an industrial scale experiment, where 37000 chickens were fed fodder enriched with CoQ_{10} for the last 20 days before slaughtering. Fortified fodder was prepared as shown later in the text. In such a manner, we prepared natural functional meat products, which are more resistant to the potentially harmful effects of free radicals.

2. Ubiquinone

2.1 Properties of ubiquinone

Coenzyme Q_{10} (CoQ_{10}) is a lipid soluble molecule composed of a quinoid head and a hydrophobic tail, which contains 10 isoprenoid units (Fig. 1). It is an essential player in oxidative phosphorylation in the mitochondria and has an important role in the formation of ATP (Ernster & Dallner, 1995). It also maintains the fluidity of cellular and mitochondrial membranes and acts as an important antioxidant, which efficiently protects phospholipids, mitochondrial DNA, and membrane proteins from free radicals (Ernster & Dallner, 1995; Littarru, 1994; Crane, 2001; Bentinger et al., 2007)

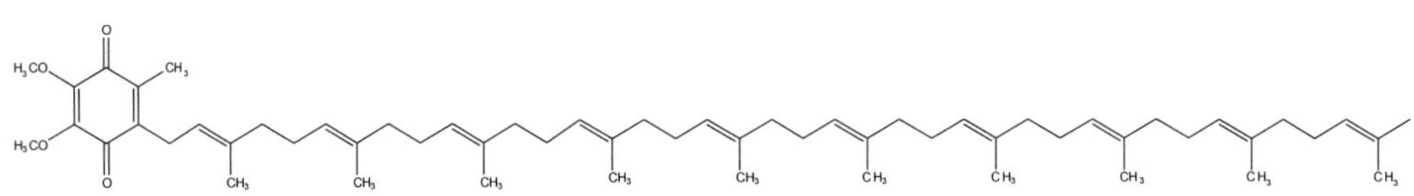

Fig. 1. Structure of CoQ_{10} (in the form of ubiquinol)

2.2 Biosynthesis of ubiquinone

CoQ_{10} is endogenously synthesized in all human and animal cells (Olson & Rudney, 1983; Elmberger et al., 1987). It can also be administered via dietary uptake through food and dietary supplements. The biosynthesis of CoQ_{10} includes two pathways. The biosynthesis of the polyprenyl side chain runs through the mevalonate pathway. The reaction starts from acetyl-coenzyme A and ends up with farnesyl pyrophosphate (FPP). FPP is also the substrate for the biosynthesis of isoprenylated proteins, dolichol and cholesterol. The quinone head is synthesized from tyrosine and in some cases from phenylalanine (Turunen et al., 2004).

2.3 Intake of CoQ_{10} with food and their absorption

CoQ_{10} can be administered with plant and animal food. The richest sources of dietary CoQ_{10} are meat and fishes, due to their relatively high levels of mitochondria. Dairy products and vegetables are much poorer in CoQ_{10}, when compared to animal tissues. The average daily intake of CoQ_{10} is estimated to be under 10 mg (Weber et al., 1997; Mattila & Kumpulainen, 2001).

Due to the lipophilic nature of CoQ_{10} its absorption follows the same process as that of fat-soluble nutrients in the gastrointestinal tract (Bhagavan & Chopra, 2006). Following absorption, CoQ_{10} is first incorporated into chylomicrons and is taken up rapidly by the liver where CoQ_{10} is repackaged into very low density protein/low density protein particles (VLDL/LDL) and released into circulation (Kaikkonen et al., 2002).

3. Complex of CoQ_{10} with β-cyclodextrine

CoQ_{10} is classified as a lipophilic compound and is practically insoluble in aqueous solutions. Due to its high molecular weight and poor water solubility, it is absorbed from the gastrointestinal tract poorly and slowly. Mainly soft and hard gel capsules filled with powder or CoQ_{10} dispersed in sesame or soybean oil are available as a nutritional supplements on the market. Therefore, it has been a challenge to develop a CoQ_{10} formulation for oral administration with better water-solubility and therefore better bioavailability.

Different methods have been used to improve the solubility of CoQ_{10}. Some approaches have included the preparation of nanoparticles incorporating CoQ_{10} (Hsu et al., 2003; Ankola et al., 2007), solubilization in a blend of sorbitan monooleate, polysorbate 80, medium chain triglycerides, propylene glycol, α-tocopherol and poly vinyl pyrrolidone (Chopra et al., 1998), preparation of redispersible dry emulsion (Takeuchi et al., 1992), solid dispersion of CoQ_{10} with Eudragit® (Nazzal et al., 2002), and fine oil-in-water emulsion via the development of a self-emulsifying drug delivery system (Kommuru et al., 2001). The most recent preparation of CoQ_{10} is All-Q® 10% CWS/S, which is based on food grade starch (Ullmann et al., 2005). Some of the most useful enzyme-modified starch derivatives are cyclodextrins, which are well known as inclusion-complexing agents for small and large molecules (Szejtli, 1998).

The most common cyclodextrin (CD) is β-CD, a ring molecule consisting of seven glucopyranose units. Having a hydrophilic outer surface and a hydrophobic inner cavity gives it a unique ability to form inclusion complexes with lipophilic compounds and increase their water-solubility, stability and/or bioavailability (Doorne, 1993). Lutka and Pawlaczyk (1995) tried to synthesize inclusion complexes of CoQ_{10} with various CDs. They prepared inclusion complexes of CoQ_{10} with γ-cyclodextrin (γ-CD) and substituted γ-CDs using "kneading" and "heating" methods, but they could not confirm a complex with either non-substituted β-CD or with β-CD.

In the Laboratory for Food Chemistry at the National Institute of Chemistry Slovenia, we participated in the development of a water-soluble form of CoQ_{10}, which could be used in the preparation of functional foods. For this purpose, we prepared complexes of CoQ_{10} with β-CD by a co-precipitation method in aqueous solution (Fig. 2). The complex of CoQ_{10} with β-CD was patented (Prošek et al., 2005). The physicochemical characteristics of the resulting complexes, such as solubility in relation to temperature and pH, and the influence of temperature and ultra-violet (UV) light on its stability were also examined (Fir Milivojević et al., 2009 a, 2009 b).

The prepared complex of CoQ_{10} with β-CD was characterized and quantified using chromatographic and spectroscopic techniques; thin layer chromatography (TLC), high performance liquid chomatography (HPLC), high performance liquid chromatography coupled with mass spectrometry (HPLC/MS), infra red spectrometry (IR), nuclear magnetic resonance (NMR). Among these sophisticated techniques, the relatively simple and inexpensive TLC gave us some very important and informative results. Identification and quantification of CoQ_{10}/β-CD were done with three different TLC procedures (Prošek et al., 2002), one and two dimensional TLC (Prošek et al., 2004).

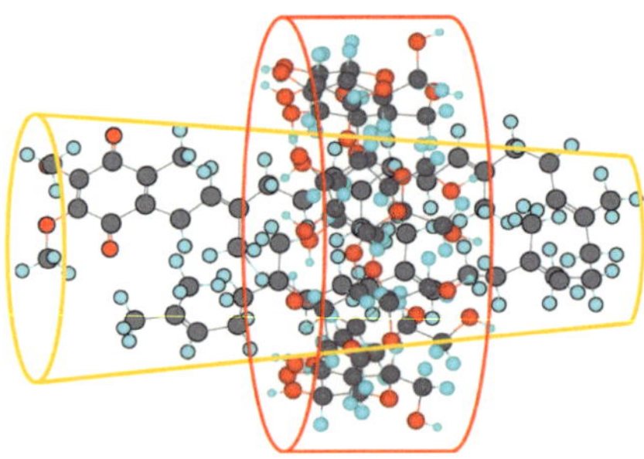

Fig. 2. Expected 3D structure of the inclusion complex of β-CD and CoQ_{10} with folded isoprene chain

To prove the enhanced bioavailability of the CoQ_{10}-βCD complex, we performed two bioequivalence studies. In the first study, the relative bioavailability was investigated for two forms: our water-soluble CoQ_{10} formulation and a commercially available oil-based CoQ_{10} in the form of soft-gel capsules. The bioavailability was determined by measuring the plasma CoQ_{10} levels periodically after administration to a group of beagle dogs. The mean value for the baseline plasma concentrations, maximum plasma concentrations (C_{max}), time of maximum plasma concentration (T_{max}), area under the plasma concentration curve $AUC_{(0-48h)}$, and the elimination half life ($t_{1/2}$), were determined for both formulations. The results of the experiment show the advantage of water-soluble CoQ_{10} over commercially available soft-gel capsules (Fig. 3). This is shown by the nearly three times higher $AUC_{(0-48h)}$, nearly two times higher C_{max}, and the shortened T_{max} from 6 to 4h, where $AUC_{(0-48h)}$ represents the area under the plasma concentration curve, C_{max} the maximum plasma concentration, and T_{max} the elimination half life (Prošek et al., 2008).

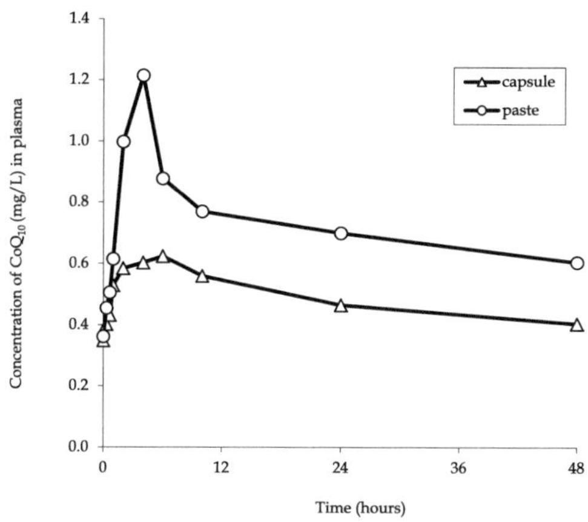

Fig. 3. Comparison of CoQ_{10} absorption in oil capsules and in complexed CoQ_{10}-βCD in plasma (Prošek et al., 2008)

A second bioequivalence study was also performed for two formulations, a novel CoQ_{10} paste with increased water-solubility and to soft-gel capsules with CoQ_{10} in soybean oil, but this time on human subjects. This single-dose bioequivalence study once again revealed, that the oral absorption and bioavailability of CoQ_{10} can be significantly affected by increasing the water solubility with the formation of CoQ_{10}-βCD complex, because it demonstrates superior bioavailability over the soft-gel capsules (Žmitek et al., 2008).

3.1 Food with added complex CoQ_{10}-βCD

The experiments showed that complex CoQ_{10}-βCD can be easily and uniformly mixed with many food products. We found that the most appropriate foods for this purpose are dairy and meat products. These kinds of products can be easily used to increase the daily intake of CoQ_{10}, especially for persons having problems with digesting fats and vegetable oils, which are normally used as a matrix in non water soluble preparation of CoQ_{10}. Additionally, when inclusion complex CoQ_{10}-βCD is placed into an environment with a pH below 3, it is disintegrated and CoQ_{10} is released from the β-CD carrier in its natural form. This is the reason why a water-soluble substance in the form of a paste with 7-10% of CoQ_{10} is used as a very efficient dietary supplement. It is also suitable for animals and humans who dislike or cannot swallow relatively big capsules.

3.1.1 Dairy products fortified with CoQ_{10}

To establish which kind of food products are most suitable carriers for our new additive we investigated different kinds of dairy products. Nearly 100 different products from retail stores were selected and analyzed, as we needed to know the approximate amount of CoQ_{10} in different milk products. CoQ_{10} was extracted with a combination of ethanol - n-hexane extraction and analyzed using HPLC-MS. In Table 1 shows the concentration of CoQ_{10} in milk produced in different Slovenian regions, with different amounts of fat and production procedures. The results were measured in the oxidized state as ubiquinone and represent the total amount of CoQ_{10} in the samples. Unfortunately, few studies with reliable results have been published with which we could compare our results. The analyzed dairy products were divided into 6 groups: milk, yogurts, sour milk products, curds, creams and soybean "milk" products. From the obtained measurements it is evident that milk with higher amounts of fat has a higher amount of CoQ_{10} and that sterilized milk prepared from concentrates has a lower amount of CoQ_{10}, regardless of the amount of fat in the milk.

Food products		**Fat (%) declared**	**CoQ_{10} (mg/kg)**
Milk	Fresh cow milk from local farm	3.6	1.90
	Cow milk from the alpine region	3.5	1.57
	Cow milk from the alpine region	1.6	0.66
	Ultra-heat-treated homogenized milk	3.5	1.70
	Ultra-heat-treated homogenized milk	1.6	1.16
	Ultra-heat-treated homogenized milk	0.5	0.46
Yogurt product	Yogurt from goat and sheep milk	6.0	0.32
	Yogurt	3.2	1.13
	Yogurt	1.3	0.70
	Yogurt with fruits	3.2	0.72
	Yogurt	2.8	0.78
	Liquid yogurt	1.6	0.76
	Yogurt bioactive	1.5	1.36
	Yogurt bioactive with inulin	1.5	0.86
	Light yogurt with Ca	0.1	0.24
	Yogurt without fat	0.0	0.06
Sour milk products and probiotics	Sour milk	0.1	0.0
	Sour milk	3.2	0.51
	Sour milk	1.6	0.50
	Liquid sour milk	3.6	0.55
	Acidofil drink	3.2	0.91
	Probiotic drink	2.0	0.60
	Kefir	3.5	0.94
	Kefir	1.6	0.68
Curd and cream products	Cream	35.0	0.92
	Sour cream	21.5	0.90
	Cream for coffee	10.0	0.11
	Curd	34.0	0.68
	Curd pressed	13.4	0.68
Soy milk and soybean product	Soy drink	-	0.03
	Soy yogurt	-	0.04
	Soy yogurt with fruit	-	0.03

Table 1. Concentrations of CoQ_{10} in some milk and milk products, as well as soy milk products

From acquired results is evident that milk with higher concentration of fat has also higher amount of CoQ_{10} and that sterilized milk prepared from concentrates has a lower amount of CoQ_{10}, regardless of the fat concentration.

The second group shows the concentrations of CoQ_{10} in different kinds of yogurts are. The concentrations of CoQ_{10} vary from 1 mg/kg in yogurt from natural cow's milk with 3.2% of fat to practical zero in special light versions with no declared fat (0%). The sample preparation and extractions were not very effective, due to different types of additives in some yogurts, especially of plant origin, such as inulin and other sugar polymers. Our results show that yogurts produced from natural cow s milk with standardized amounts of fat (3.2%) are the right origin for dietary CoQ_{10}. In the sour milk group, the kefir and acidophilous drinks' amount of CoQ_{10} vary from sample to sample, but it is evident that the amount of fat correlates with the amounts of CoQ_{10}. The next group shows CoQ_{10} concentrations in creams and curds. The measured concentrations are not very high, especially if compared with the concentrations of fat in the samples. These types of milk products are not very suitable sources of natural CoQ_{10}, as the consumer would have to eat too much fats. On the other hand, this type of food, mostly used as a desert, is very good for enrichment with herbs and CoQ_{10}. Several soybean products were also analyzed, but we found only small amounts of CoQ_{10} in purchased soy bean drinks and yogurts.

The measured values together with consumers' nutritional habits show that the milk-based dairy products are very convenient candidates for supplementation. Milk products with a suitable combination of fat and water are an ideal basise for functional food products containing increased amounts of CoQ_{10}. Ubiquinone is a fat soluble substance and a certain amount of fat in the carrier food is advantageous.

3.1.2 Pates fortified with CoQ_{10}

The second set of food products which can easily be enriched with water soluble CoQ_{10} are different meats (fish, poultry, beef and pork) and liver pates. A certain group of such products were purchased from local stores and analyzed. In addition to the concentrations of CoQ_{10}, the total amount of fat, fatty acid profiles, and cholesterol were also determined.

The analyzed samples may be sorted in five groups. Quantitative results are shown in Table 2. In the first group, typical pork pates are shown; these usually contains: 10-25% pork liver, 5-30% pork meat, up to 10% pork fat, as well as ham, bacon, intestines, proteins, and carbohydrates. Results show a very high amount of CoQ_{10}. The mean value of 9 different samples is 12.3 mg/kg, and the maximum value 27.4 mg/kg of CoQ_{10}. There is a high level of extracted fat, with a mean value of nearly 24%. In the extracted fat, 40.3% is saturated fatty acids and only 17.7% is unsaturated, but nearly 40% is C18:1, which makes these products quite resistant to fat oxidation. On the downside, these types of pates do have high amount of cholesterol, with an average value of 0.46 g/kg.

The second group contains pates prepared from poultry meat and liver, with additives in the form of milk and soy proteins, vegetable fats, and hydrolyzed carbohydrates. This group showed the highest amount of CoQ_{10} from among the entire group of analyzed samples. The mean value of CoQ_{10} in the 9 different pates is 17.9 mg/kg. The level of total fat is not high but is very variable: the mean value is 19.9% and the maximal value is 30.8%. The variability is probably the result of different concentrations of vegetable fat added to the final products.

The amount of saturated fatty acids in fat is only 26.4%. The combination of C18:1 and other unsaturated fatty acids means there are up to 72.4% unsaturated fatty acids in these poultry products. But once again, as in the first group, the main drawback is the very high level of cholesterol, with an average value of 0.73 g/kg. This high value is probably the result of the high amount of fatty poultry skin in the product.

Pate	**Pork pate**	**Poultry pate**	**Pork pate with vegetables fats**	**Beef pate**	**Fish pate**
Components	Pork liver, fat, meat	Poultry meat, liver	Pork meat, vegetable fats & additives	Beef meat	Fish meat
Saturated fatty acids (%)	40.3	26.4	39.6	33.1	22.5
Unsaturated fatty acids (%)	17.7	39.7	16.0	27.6	53.0
C18:1 (%)	39.7	32.7	42.2	36.9	24.2
Unsaturated fatty acids (sum total) (%)	57.5	72.4	58.2	64.5	77.2
Fat g/100 g	23.9	19.9	25.5	21.2	17.0
CoQ_{10} g/100 g	1.23	1.79	0.64	0.98	0.63
Cholesterol g/100 g	0.046	0.073	0.031	0.072	0.050

Table 2. Quantitative results of fat, fatty acids, CoQ_{10}, and cholesterol in some meat and liver pates

In the third group, pates produced from pork meat and added vegetable fat and carbohydrates are presented. In this group the total amount of fat is high and its concentration is stable. The main reason is the controlled addition of vegetable fat, which also influences the amount of C18:1 and not the high amount of cholesterol, where the average value is only 0.3 g/kg. These positive effects are reduced by the relatively low amount of CoQ_{10}, averaging only 6.4 mg/kg.

The fourth group consists of samples prepared from beef meat. These pates have small amount of fats, only 21.2%. The level of unsaturated fatty acids is relatively high, the level of CoQ_{10} is acceptable, with an average value of 9.8 mg/kg, but the concentration of cholesterol is very high, with an average value of 0.7 g/kg.

In the last group are pates prepared from fish meat. Although only a small number of products were analyzed we can conclude from the obtained results that this group has a low amount of CoQ_{10}: the mean value of 4 different pates is 6.3 mg/kg and the maximum value is 8.4 mg/kg. The level of total fat is very low, only 17.0%. The amount of saturated fatty acids is small (22.5%) and the amount of unsaturated fatty acids is extremely high (53.0%), if we include C18:1 it can reach up to 77.2%. The level of cholesterol is low, with an average value of 0.50 g/kg.

Results show that meat and liver pates can be a very good dietary source of CoQ_{10}. The concentration is higher than in any other food products, such as in raw milk where the concentration of CoQ_{10} is less than 3 mg/L. The only problem is the high amount of cholesterol, which is correlated with amount of CoQ_{10}. In a product with a high amount of ubiquinone there is also a high amount of cholesterol present. The relation between CoQ_{10} and cholesterol in analyzed samples is shown in Fig. 4.

A strong correlation between CoQ_{10} and cholesterol was found; however, there was no correlation between total fat and CoQ_{10} and total fat and cholesterol. This is logical due to the different product formulation of meat pates even in cases where they have similar labels.

The obtained results show that CoQ_{10} in meat pates is a reasonable supplement. It can change the ratio of CoQ_{10} to cholesterol present in the meat products and makes them more suitable for the consumer.

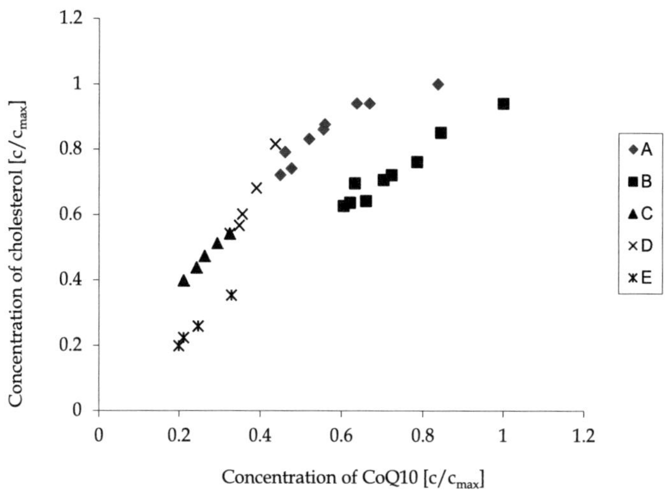

Fig. 4. Relation between concentrations of cholesterol and concentrations of CoQ_{10} in different pates, expressed as the ratio between measured value and the max value on each coordinate ($c_{max}(CoQ_{10})$ = 2.56 mg/100 g, c_{max}(cholesterol) = 201 mg/100 g). Group A consists of pork liver, fat and meat; B of poultry meat and liver; C of pork meat, vegetable fats and additives; D of beef meat; E of fish meat

4. Poultry products with increased content of CoQ_{10}

The distribution of exogenous CoQ_{10} in different tissues has been described mostly in rats and mice (Kwong et al., 2002; Kamzalov et al., 2003). In both rats and mice, the predominant endogenous form of CoQ is CoQ_9, due to their relatively short life span. In humans and animals with longer life spans, including chicken, the major homologue is CoQ_{10}. There is a lack of studies on feeding long-lived animals with dietary CoQ_{10} and its distribution in tissue, which may explain the accumulation of CoQ_{10} in animals whose predominant form of CoQ_{10}. Until now, there was also not a lot of information available to describe the effect of long term controlled supplementation with CoQ_{10}.

4.1 Influence of added CoQ_{10} in chicken feed on breeding and its accumulation in chicken tissue

In this study, we present the influence of CoQ_{10} used as a food additive on the health and physical condition of chickens during the feeding period. The amounts of CoQ_{10} and cholesterol in chickens' blood before slaughtering and in different tissues after slaughtering are determined.

4.1.1 Animals and experimental design

Two hundred 1-day-old ROSS 308 male chicken broilers were provided from a local hatchery (Perutnina Ptuj d.d., Slovenia). The study was carried out under optimal health and growing conditions, according to the prepared protocol. During the 40-day production period all animals were treated under identical controlled environmental and growing conditions with deep litter technology, except for the feed. One hundred chickens were treated as the control group (G0) and the other hundred (the group in the study) were distributed into four subgroups (G1-G4). The control group of animals was fed plain feed for the whole period, while the study group received feed fortified with water-soluble CoQ_{10} paste with 7.5% of CoQ_{10} in the form of inclusion complex with β-cyclodextrin. The active substance was synthesized in our laboratory (Laboratory for Food Chemistry, National Institute of Chemistry, Ljubljana, Slovenia) and mixed with the fodder in two steps in order to give a final concentration of 0.0042% CoQ_{10}. The concentration of CoQ_{10} in feed was calculated so that each animal received an average of approximately 5 mg of CoQ_{10} per day. The subgroup G4 was fed with fortified feed for the whole period of 40 days. The subgroups G1, G2, and G3 were fed with fortified feed for the last 10, 20 and 30 days of the 40-day production period, respectively (Table 3).

Group of chicken	G0	G1	G2	G3	G4
Start of feeding (on day of age)	/	30th	20th	10th	0th
Time of feeding with CoQ_{10} (number of days)	/	10	20	30	40
Content of CoQ_{10} in feed (mg/day)	/	5	5	5	5
Number of chicken	109	25	25	25	25

Table 3. Experimental design of feeding chicken with added CoQ_{10}

At the end of day 40, before slaughtering, blood samples from 20 randomly chosen birds were collected in commercially available heparinized tubes, centrifuged at 3000 g for 15 min and immediately frozen at -80 °C. After slaughtering, 6 birds were randomly chosen from each of the subgroups and several chicken parts (heart, liver, breast, leg, wing and body fat) were separately sampled, labelled, packed, and frozen at -20°C. After about 24 hours the collected frozen samples were transported to the laboratory facility for long-term storage at -80°C and kept so until needed for analysis.

4.1.2 Growth of chicken

During the 40 day growth period, chickens were monitored by measuring their body weight and by feed intake weighing. Significant changes were noticed in the physical condition of chickens over the growth period of 40 days. In Fig. 5, the increase in the chicken weight is shown. The net increase in chickens weight in each group is compared to the average value of the control group at the corresponding measurement time. In this way the influences of higher starting weight of the control groups and non-uniform distribution of groups formation (10, 20, 30, 40 days) and weighting times (10, 21, 29, 36, and 40 days), are eliminated. Our study showed that a major increase in chicken body mass was seen in group

G4, followed by group G2. From the obtained results, we can conclude that the most economical results would be obtained with non-stop twenty days foddering.

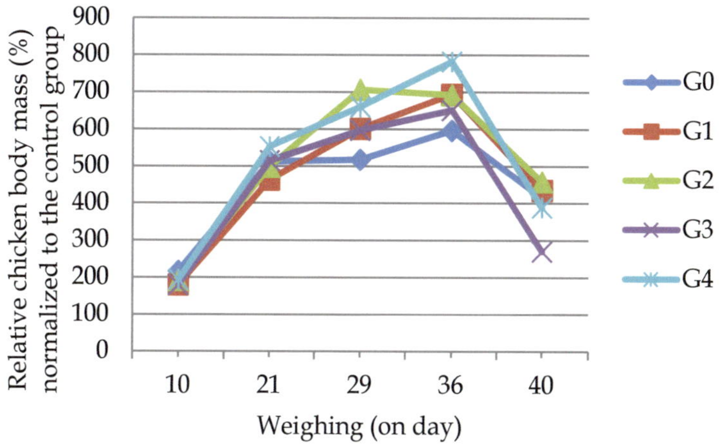

Fig. 5. Influence of exogenous CoQ_{10} on chicken body mass (g) in groups (G0, G1, G2, G3 in G4) on day 10 (1st weighing), on day 21 (2nd weighing), on day 29 (3rd weighing), on day 36 (4th weighing) and on day 40 (5th weighing) of chicken age

We also observed that the addition of CoQ_{10} had a strong influence health during breeding. Due to their fast growth, chickens are vulnerable to a number of diseases. Ascites is the most common metabolic disease in chicken and occur worldwide, especially at high altitudes. The disease has a complex aetiology and is predisposed to by reduced ventilation, and respiratory disease (Currie, 1999). The literature data has demonstrated the positive effect of exogenous CoQ_{10} in reducing ascites mortality in broilers (Geng et al., 2004).

4.1.3 Accumulation of CoQ_{10} in plasma and various tissues

Four different groups of broiler chicken were administered CoQ_{10} for the last 10, 20, 30, and 40 days prior to slaughter, after which the amounts of CoQ_{10} and cholesterol were measured in plasma, liver, heart, breast, wings, and legs. The results for CoQ_{10} (Fig. 6) and cholesterol concentrations (Fig. 7) in different samples from chickens were obtained by HPLC-MS and TLC methods, respectively.

After ingestion of feed enriched with CoQ_{10}, molecules of CoQ_{10} were transferred with other lipids via chylomicrons to liver cells. The results showed that the concentration of CoQ_{10} in the liver after feeding with feed fortified with CoQ_{10} in the various test groups is not statistically significantly changed in comparison to the control group. The liver, probably through unknown mechanisms, regulated the concentration of endogenously synthesized CoQ_{10}. In liver tissue cholesterol concentration increased.

In the liver, CoQ_{10} was incorporated into lipoproteins, mostly into VLDL/LDL particles and released into the circulation. The major function of CoQ_{10} in the blood is as an antioxidant. CoQ_{10} protects LDL from lipid peroxidation by scavenging peroxyl radicals (Alleva et al., 1995, 1997). The antioxidant protection is important for broiler chickens because of their rapid growth and consequent higher feed intake per time-unit and higher metabolic rate.

The concentration of CoQ_{10} in plasma increased in all test groups by approximately 1.6-fold over that in the control group. CoQ_{10} levels in blood also exhibited a noticeable maximum increase of about 80% in the case of group G3. However, the concentrations declined with prolonged, 40-day supplementation (group G4) to 140% of the levels found in the control group (G0). In the case of blood levels, a variety of values within each group were found, leading to large differences in the relative standard deviations of the measurements. The highest relative standard deviations values were found in the blood samples of groups G0 and G4, where they reached up to 20% and 28%, respectively, while in other groups the values were in the range of 13-16%. The variability in blood levels, on the other hand, reflects the differences among individual animals in response to CoQ_{10} supplementation. Moreover, blood levels are more time-dependent over a short timescale compared to tissues, as blood is the transport medium of an organism. It was necessary to average the results for blood CoQ_{10} levels to obtain a meaningful value as the levels varied so much among individual animals.

The concentration of cholesterol decreased in blood in all groups. The biggest changes were noted in G4 group, after 40 days of feedings chickens with CoQ_{10}. This kind of results are expecting according to the data from literature (Honda et al., 2010).

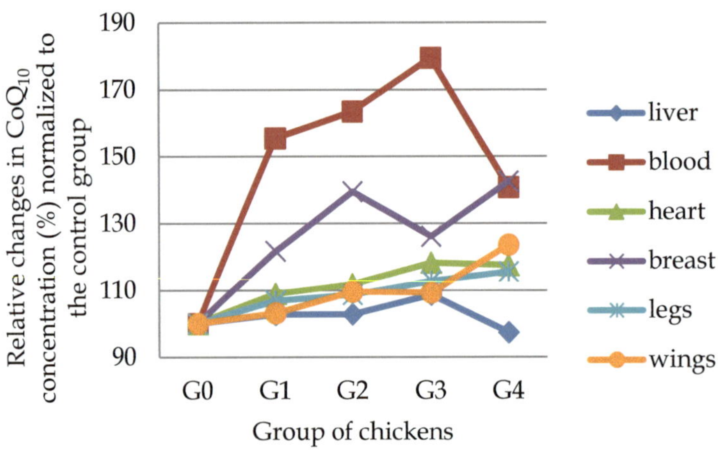

Fig. 6. Relative changes in CoQ_{10} concentration during the 40 day feeding period

CoQ_{10} is transferred from the blood into various tissue and organs. In this study the concentration of CoQ10 was examined in heart and muscle tissues (legs, breast and wings). CoQ_{10} is present in all tissue and cells but, on a weight basis, in variable amounts. The highest content of CoQ_{10} was founded in the most active organs like the heart, kidney, and liver (Ernsten & Dallner, 1995).

Statistically, the content of CoQ_{10} increased in the hearts of chickens with longer periods of feeding with supplemental CoQ_{10}, but only in the range of 7.3% - 11.3%, which is low compared with muscle tissue. Since the supplemented chickens are young organisms, we therefore assumed that the heart does not require additional CoQ_{10}. In heart tissue, concentration of cholesterol was nearly 10 % lower than in the control group.

In the analysis of the muscle tissue, we found that most of the CoQ_{10} was found in legs (cca 22 mg/kg), with significantly less in breast (cca 9 mg/kg) and wings (cca 7 mg/kg). The results of our study showed that the added CoQ_{10} increased CoQ_{10} concentration in all muscle tissue (legs, wings and breast). After 40 days of feeding with feed fortified with CoQ_{10} the content of CoQ_{10} increased most in breast (45%), followed by wings (25%) and legs (16%). At the same time, concentration of cholesterol in breast, wings and legs are not significantly changed.

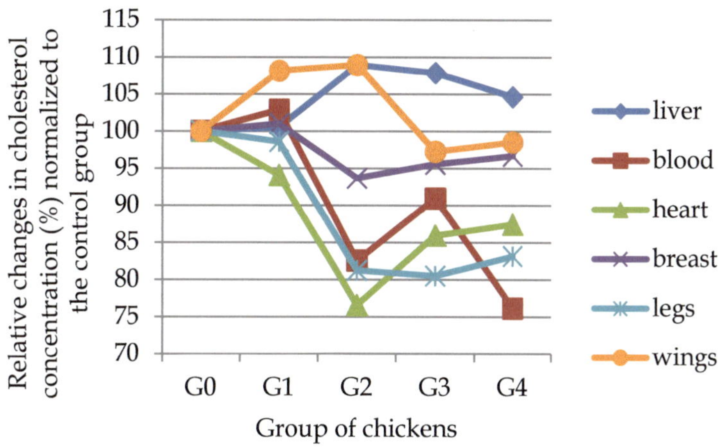

Fig. 7. Relative changes in cholesterol concentration during the 40 day feeding period

The influence of added CoQ_{10} on cholesterol concentration was not statistically significant. In the analyzed chicken meat we calculated the QCI (CoQ_{10}/cholesterol index) (1), which represents a measure of the improvement in the quality of the meat (Fig.8).

$$QCI = \frac{concentration\, of\ CoQ10\left(\frac{mg}{kg}\right)*1000}{concentration\, of\ cholesterol\left(\frac{mg}{kg}\right)} \tag{1}$$

The calculated QCI values obtained from the analyses of individual samples from individual animals proved to be a better means of studying CoQ_{10} profiles, than the use of absolute CoQ_{10} and cholesterol concentrations. The QCI was used, since according to our previous experiments, it is a very reliable indicator of the oxidative status and possible oxidative stresses activated by the food ingredients. We hope that this index will soon receive adequate attention as an informative factor in the evaluation of possible harmful effects of oxidants in foods. The QCI increased in all parts of the chicken, except in the wings where after 10 days of supplemental CoQ_{10} feeding, it fell by 9% and after 20 days of supplemental CoQ_{10} feeding the value by comparison with the control group hardly changed. The biggest improvement was achieved in the breast, where, after 20 days of feeding, the QCI relative change index increased by almost 50%.

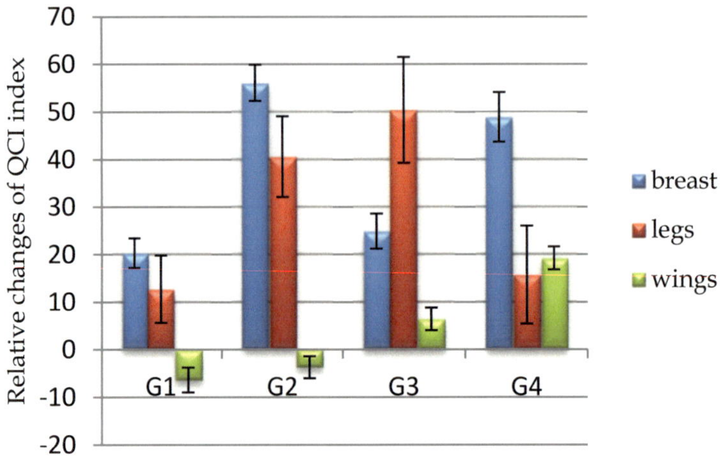

Fig. 8. Relative changes in the QCI (QCI = CoQ_{10} (mg/kg)/cholesterol (mg/kg)*1000 index) in different chicken muscle tissue in test groups with different periods of feeding with CoQ_{10}

The optimal time of feeding with CoQ_{10} in comparison with the content of supplemental CoQ_{10} (mg) during chicken raising was also of interest to this study (Fig. 9). Chicken under prolonged feeding with exogenous CoQ_{10} received different amount of CoQ_{10}. Chicken in group G0, G1, G2, G3 and G4 received 0, 50, 100, 150 and 200 mg of CoQ_{10}, respectively. The results indicated that the highest increase in the concentration of CoQ_{10} was after 20 days of feeding, when chicken received 100 mg of CoQ_{10}.

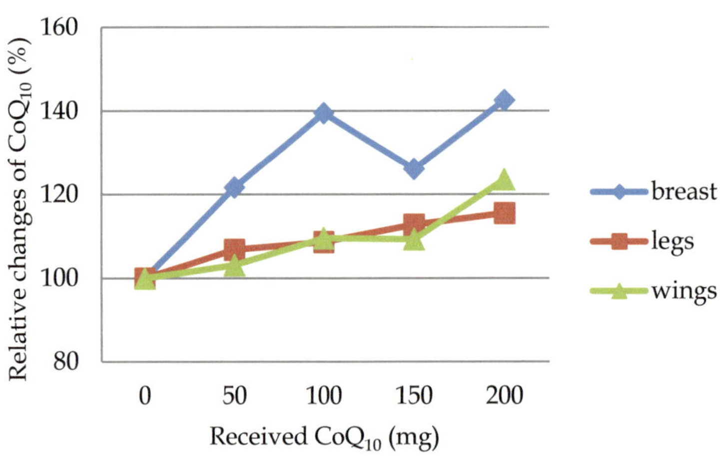

Fig. 9. Relative changes in CoQ_{10} (%) in chicken meat (legs, wings and breast) in comparison with the content of supplemental CoQ_{10} (mg) during chicken raising

4.2 Transfer of dietary CoQ_{10} into different parts of chicken breast cells

The biggest difference between the concentration of CoQ_{10} in the test group G0 and control group G4 was found in the breast tissue (Fig.10). This is the reason why this material was selected for further processing. The selected breast tissue was fractionated into four fractions (Fig. 11) essentially following the procedure from the literature (Casado et al., 1992). Fraction P1 contain mainly nucleus and cell debris, P2 included mitochondria, P3 consist of smaller cell organelles and the remaining S3 fraction was composed of cytosol and small section of disintegrated cell and organelle membranes.

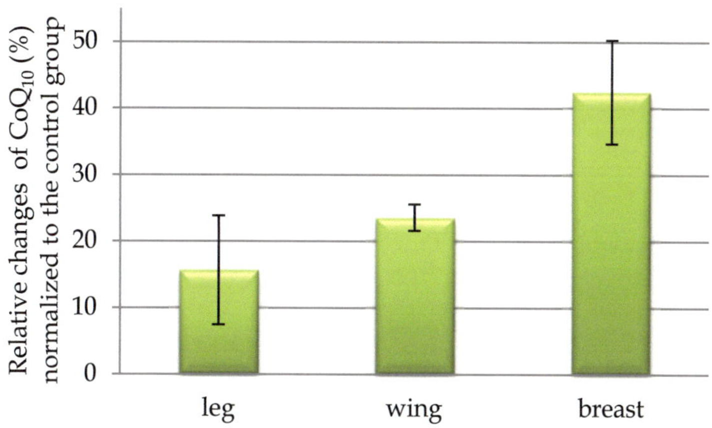

Fig. 10. Relative changes in the concentration of CoQ_{10} in several chicken tissue (legs, breast and wings) after 40 days of chicken feeding (G4 group) normalized to the control group (G0 group)

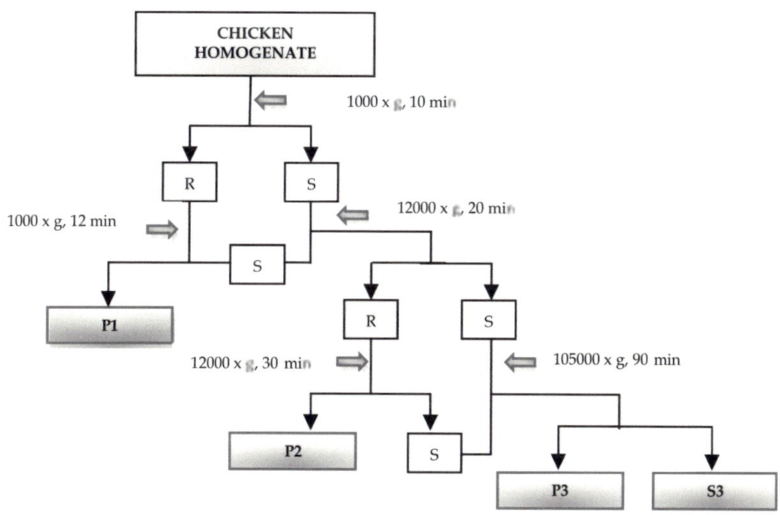

Fig. 11. Fractionation scheme of chicken homogenate (P1 is the nuclear fraction, P2 is the mitochondrial fraction, P3 is the microsomal fraction and S3 is the remaining supernatant)

The concentration of cholesterol differed between the subcellular fractions within each group (P1-P3, S3), but showed practically no difference between groups (G0, G4). At the same time, there were significant differences in the CoQ_{10} concentrations with in groups and between fractions in the control and test groups (Table 2). With regard to these results, we decided to normalize the concentrations of CoQ_{10} to cholesterol concentrations and not to the protein concentrations that are normally used. The QCI was calculated for all fractions in both groups. The results in Fig. 12 show that the CoQ_{10} concentration in the P1 fraction of the G4 group increased by approximately 3 times compared to the control group, meanwhile the increase in the P3 and P4 fraction was merely approximately 30%, but in P2 the amount of CoQ_{10} remained unchanged.

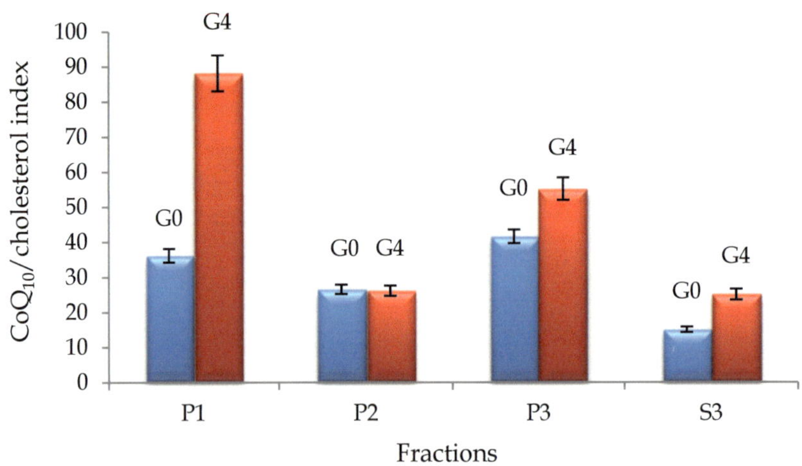

Fig. 12. Values of the QCI analysed by HPLC in different fractions and in the control group G0 and in the G4 test group that received CoQ_{10} (5mg/day) for 40 days

The obtained results showed that most of the administered CoQ_{10} was located inside the cell membranes. This is seen from the increase in CoQ_{10} concentrations in fraction F1, which consisted mostly of the nucleus, partially disintegrated cells, and large cell debris, and cytosol fraction F4 in which small parts of completely disintegrated cell membranes were found. A small increase was also noticed in fraction F3, in which smaller organelles were accumulated. It is very interesting that there is no change in the concentrations of CoQ_{10} and cholesterol in the crude mitochondrion fraction (F2). The observed variations in CoQ_{10} distribution indicate that supplementary ubiquinone was mainly built into the cell membranes and not into the mitochondria. Theoretically, the maximum increase was expected for the mitochondrial fraction, due to the essential role of CoQ_{10} in energy conversion. However, the observed accumulation pattern of CoQ_{10} in the nuclear and cytosolic fractions is likely due to other less known or even unknown metabolic functions. In addition, our results are in significant accordance with the concept of the antioxidant action of exogenous supplied CoQ_{10}, which is now promoted by Littarru (1994).

In our study, young and healthy animals, who likely have sufficient amounts of endogenous CoQ_{10}, were used and this could be the main reason why the exogenous addition was built into the cell membranes and not into the mitochondrion inner membranes.

4.3 Poultry products with an increased content of CoQ_{10}

The majority of consumers refuse to eat meat with higher levels of fat, due largely to the possible association between high levels of cholesterol and heart disease. Poultry meat and poultry products are widely consumed due to their lower content of cholesterol than other meat, faster digestibility, and price accessibility. Meat is a rich source of essential amino acids, minerals, some vitamins, and long chain polyunsaturated fatty acids. Thus, a moderate intake of meat is important part of balanced diet.

Supplementary dietary CoQ_{10} present in food or taken as a dietary supplement is widely consumed because of its beneficial effect on human health. Supplementation with CoQ_{10} has been shown to improve the resistance of LDL particles to oxidation and to prevent atherogenicity (Hanaki et al., 1993; Witting et al., 2000).

The aim of our work was to prepare food with an increased content of CoQ_{10} from chicken fed with supplemental CoQ_{10}. This procedure of preparing food with supplemental CoQ_{10} by direct feeding provided also health benefits to the chicken.

4.3.1 Animals and experimental design

Two groups of 1-day-old ROSS 308 male chickens were obtained from the regular hatching process in a poultry hatchery. The control group of animals (36600 chickens) was fed plain feed for the whole period, while the study group (37000 chickens) received the last 20 day feed fortified with water-soluble CoQ_{10}-βCD. The shortened feeding period was selected because the laboratory experiment showed that a major increase in the concentration of CoQ_{10} in meat tissue was provoked after the twenty-day administration period. The concentration of CoQ_{10} in feed was calculated so that each animal received an average of approximately 5 mg of CoQ_{10} per day. A set of poultry food products (breaded chicken wings, breaded chicken drumsticks, breaded chicken fillet, chicken nuggets, extra chicken sausages, and chicken liver pate) was prepared from the meat and organs (liver, heart) of the test animals according to usual industrial procedures. Concentrations of CoQ_{10} and cholesterol were evaluated using two reliable analytical procedures, quantitative HPLC-MS and semi-quantitative TLC.

4.3.2 QCI in poultry products

The content of CoQ_{10} increased as expected in all chicken products, while the concentration of cholesterol were only slightly reduced. The content of CoQ_{10} concentration in different chicken meat and their products are shown in Fig. 13. The results show that the poultry pates contain the highest value of CoQ_{10}. The concentration of CoQ_{10} increased in all poultry products according to the increased value of CoQ_{10} in the fortified chicken meat.

From the obtained results we calculated the QCI to determine the improvement in meat quality. The relative changes in the QCI concentration normalized to the control group are presented in Fig. 14. The results show the improvement for the fortified poultry products. The highest increase in the QCI was obtained in fortified chicken nuggets (220%) and in breaded chicken wings (206%).

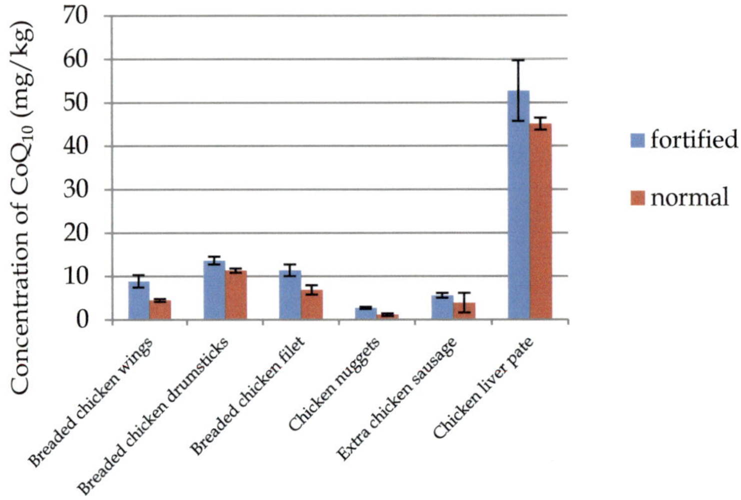

Fig. 13. CoQ_{10} concentration (mg/kg) in different meat (breast, wings and legs) and in different poultry products (breaded chicken wings, breaded chicken drumsticks, breaded chicken fillet, chicken nuggets, extra chicken sausages, and chicken liver pate)

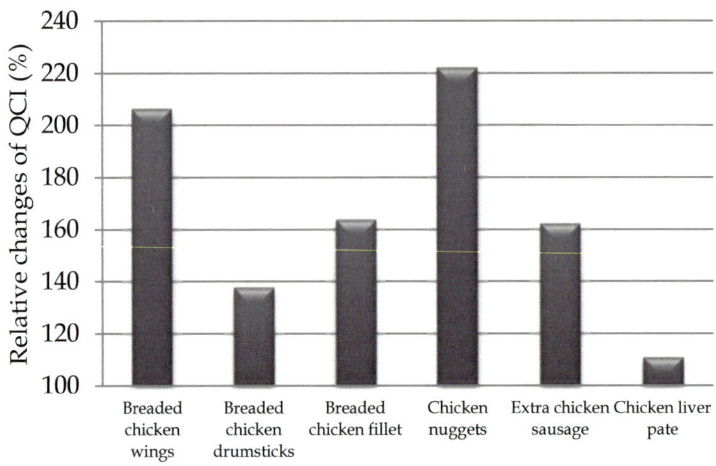

Fig. 14. Relative changes in the QCI (%) in different poultry products normalized to control groups

The CoQ_{10} content in people's diets was determined to be 3-6 mg per day, primarily derived from meat. The fortified chicken liver pate contains an average of 52.7 mg CoQ_{10}/kg sample, meaning that if the consumer ingests 100 g of the product, he/she gets approximately 5.3 mg of CoQ_{10}. Breaded chicken wings, breaded chicken drumsticks and breaded chicken fillets contain large amounts of CoQ_{10}, but these are products that require heating prior to ingestion. Weber et al. (1997) studied the loss of CoQ_{10} during food preparation. The results showed that frying reduces CoQ_{10} levels by 14-32%, while boiling does not reduce CoQ_{10} levels.

Food that contains CoQ_{10}, is a more complex matrix than capsules. The absorption of the various components of food from the gastrointestinal tract is one of the major determinants of bioavailability. Literature data indicate that intestinal absorption of food accelerates CoQ_{10} absorption (Ochiai et al., 2007). It is known that foods containing fat promote the excretion of bile acids, which form micelles water-insoluble components and thus increase the possibility of absorbtion through the gastrointestinal tract. Fear that the enriched products will exceed the current recommended daily dose, which should amount to 30 mg/day, is not grounded. In fact, these fortified products would increase the intake of CoQ_{10}, which numerous studies indicate to be positive.

5. Conclusions

Coenzyme Q_{10} is a key component in the inner mitochondrial membrane, where it plays important role in oxidative phosphorylation. It is also present in other subcellular fractions and in plasma lipoproteins, where it has antioxidant properties. CoQ_{10} has also been recognized to have an effect on gene expression. These three functions are important for its use in clinical practice and as a food supplement. A large number of clinical studies indicate that dietary CoQ_{10} administration has beneficial effects, particularly in cardiomyopathies, as well as degenerative muscle and neurodegenerative diseases.

The bioavailability of CoQ_{10} is relatively low due to its lipophilic nature and relatively high molecular mass. Increasing the availability of CoQ_{10} in aqueous medium could consequently also increase its bioavailability, which can be achieved by complexing CoQ_{10} with β-CD.

The aim of our work was to study the effect of added CoQ_{10} in water soluble form in chicken feed on broiler chicken and to prepare functional chicken products with biologically incorporated CoQ_{10}. The weighing of chicken showed that the greatest growth was achieved in chicken groups where the chickens were fed with added CoQ_{10} for 40 days.

The results of our study showed that added CoQ_{10} increased CoQ_{10} concentration in plasma, heart, and in chicken meat (breast, legs, wings). The added CoQ_{10} decreased cholesterol concentration in the heart and in plasma. The highest increase in CoQ_{10} concentration was observed in chicken breast cells after 40 days of feeding with CoQ_{10} added in feed. Fractionation of chicken breast cells showed that the added CoQ_{10} was mainly incorporated into cell membranes and not into mitochondria. This confirmed our hypothesis that CoQ_{10} added in chicken tissues acts mostly as an antioxidant.

In the industrial experiment, chickens were fed with CoQ_{10} added in feed for the last 20 days before slaughtering and the functional products were prepared from meat fortified with CoQ_{10}. As excepted, the content of CoQ_{10} increased in all chicken products especially in breaded chicken fillet, while nearly in all meat tissues concentration of cholesterol was slightly reduced which increased the QCI. We presume that, QCI can be used as useful indicator in meat nutritional quality which also indicates better growing conditions for chickens regarding to oxidative stress. Fortified chicken products can have a positive effect on consumer's health. Accumulated CoQ_{10} together with simultaneously absorbed fats can protect non-saturated fats from uncontrolled oxidation and consequently reduce the risk of different diseases.

In the next research period, we will deal with the importance of antioxidant network in preventing of oxidative stress. We also expect that the obtained results will help us to get enough information to design new functional food and feed additives and to provide better breeding production of chickens.

6. References

Albers, G.A.A. & Groot, I.A. (1998). Future trends in poultry breeding. *World Poultry*, Vol.14, No.8, (1998), pp. 42-43

Alleva, R., Tomasetti, M., Battino, M., Curatola, G., Littarru, G.P. & Folkers, K. (1995). The roles of coenzyme Q and vitamin E on the peroxidation of human low density lipoprotein subfractions. *Proceedings of the National Academy of Sciences of the United States of America*, Vol.92, No.20, (September 1995), pp. 9388-9391

Alleva, R., Tomasetti, M., Bompadre, S. & Littaru, P. (1997). Oxidation of LDL and their subfractions: kinetic ascpects and CoQ10 content. *Molecular Apects of Medicine*, Vol.18, Suppl., (1997), pp. S105-S112

Ankola, D.D., Viswanad, B., Bhardwaj, V., Ramarao, P. & Kumar, M.N. (2007). Development of potent oral nanoparticulate formulation of coenzyme Q10 for treatment of hypertension: Can the simple nutritional supplements be used as first line therapeutic agents fot prophylaxis/therapy? *European Journal of Pharmaceutics and Biopharmaceutics*, Vol.67, No.2, (September 2007), pp. 361-369

Bhagavan, H.N. & Chopra, R.K. (2006). Coenzyme Q10: absorption, tissue uptake, metabolism and pharmacokinetics. *Free Radical Research*, Vol.40, No.5, (2006), pp. 445-453

Bentinger, M., Brismar, K. & Dallner, G. (2007). The antioxidant role of coenzyme Q, *Mitochondrion*, Vol.7, Suppl., (June 2007), pp. S41-S50

Casado, V., Luis, C., Canela, E., Franco, R. & Mallol, J. (1992). The distribution of A1 adenosine receptor and 5'-nucleotidase in pig brain cortex subcellular fractions. *Neurochemical Research*, Vol.17, No.2, (1992), pp. 129-139

Chopra, R.K., Goldman, R., Sinatra, S.T. & Bhagavan, H.N. (1998). Relative bioavailability of coenzyme Q10 formulations in human subjects. *International Journal for Vitamin and Nutrition Research*, Vol.68, No.2, (1998), pp. 109-113

Crane, F.L. (2001). Biochemical Functions of Coenzyme Q_{10}. *Journal of the American College and Nutrition*, Vol.20, No.6, (December 2001), pp. 591-598

Currie, R.J.W. (1999). Ascites in poultry: Recent investigations. *Avian Pathology*, Vol.28, No.4, (1999), pp. 313-326

Elmberger, P.G., Kalen, A., Appelkvist, E.L. & Dallner, G. (1987). In vitro and in vivo synthesis of dolichol and other main mevalonate products in various organs of the rat. *European Journal of Biochemistry*, Vol.168, No.1, (October, 1987), pp. 1-11

Ernster, L. & Dallner, G. (1995). Biochemical, physiological and medical aspects of ubiquinone function. *Biochimica et Biophysica Acta*, Vol.24, No.1, (May 1995), pp. 195-204

Fir Milivojević, M., Šmidovnik, A, Milivojević, L., Žmitek, J. & Prošek, M. (2009). Studies of CoQ10 and cyclodextrine complexes: solubility, thermo- and photo stability. *Journal of Inclusion Phenomena and Macrocyclic Chemistry*, Vol.64, No.3-4, (2009), pp. 225-232

Fir Milivojević, M., Milivojević, L., Prošek, M. & Šmidovnik, A. (2009). Property studies of coenzyme (Q10)-cyclodextrins complexes. *Acta Chimica Slovenica*, Vol.56, No.4, (2009), pp. 885-891, ISSN 1318-0207

Geng, A.L., Guo, Y.M. & Yang, Y. (2004). Reduction of ascites mortality in broilers by coenzyme Q10. *Poultry Science*, Vol.83, No.9, (December 2004), pp. 1587-1593

Hanaki, Y., Sugiyama, S., Ozawa T. & Ohno M. (1993). Coenzyme Q10 and coronary disease. *Jounal of Clinical Investigation*, Vol.71, Suppl.8, (1993), pp. S112-S155

Honda, K., Kamisoyama, H., Motoori, T., Saneyasu, T. & Hasegawa, S. (2010). Effect of dietary coenzyme Q10 on cholesterol metabolism in growing chickens. *Japan Poultry Science Association*, Vol.47, No.1, (2010), pp. 41-47

Hsu, C.H., Cui, Z., Mumper, R.J. & Jay, M. (2003). Preparation and characterization of novel coenzyme Q10 nanoparticles engineered from microemulsion precursors. *Official Journal of the American Association of Pharmaceutical Scientists*, Vol.3, No.3, (2003), pp. 24-35

Jazbec, P., Šmidovnik, A., Puklavec, M., Križman, M., Šribar, J., Milivojević, L. & Prošek, M. (2009). HPTLC and HPLC-MS quatification of coenzyme Q10 and cholesterol in fractionated chicken-breast tissue. *Journal of Planar Chromatography*, Vol.22, No.6, (2009), pp. 395-398

Kaikkonen, J., Tuomainen, T.P., Nyyssonen, K. & Salonen, J.T. (2002). Coenzyme Q10: absorption, antioxidative properties, determinants, and plasma levels. *Free Radical Research*, Vol.36, No.4, (April 2002), pp. 389-397

Kazmalov, S., Sumien, N., Forster, M.J. & Sohal, R.S. (2003). Coenzyme Q intake elevates the mitochondrial and tissue levels of Coenzyme Q and alpha-tocopherol in young mice. *Journal of Nutrition*, Vol.133, No.10, (October 2003), pp. 3175-3180

Kommuru, T.R., Gurley, B., Khan, M.A. & Reddy, I.K. (2001). Self-emulsifying drug delivery systems (SEDDS) of coenzyme Q10: formulation development and bioavailability assessment. *International Journal of Pharmaceutics*, Vol.212, No.2, (January 2001), pp. 233-246

Kwong, L.K., Kazmalov, S., Rebrin, I., Bayne, A.C., Jana, C.K., Morris, P., Forster, M.J. & Sohal, R.S. (2002). Effects of coenzyme Q(10) administration on its tissue concentrations, mitochondrial oxidant generation, and oxidateive stress in the rat. *Free Radical & Biology Medicine*, Vol.33, No.5, (September 2002), pp. 627-638

Lenaz, G., Fato, R., Formiggini, G. & Genova, M.L. (2007). The role of coenzyme Q in mitochondrial electron transport, *Mitochondrion*, Vol.7, Suppl. 1, (2007), pp. S8-S33

Littaru, G.P. (1994). *Energy and defense: facts and perspectives on coenzyme Q10 in biology and medicine.* Casa Editrice Scientifica Internazionale, Roma

Littaru, G.P. & Tiano L. (2007). Bioenergetic and antioxidant properties of CoQ10: recent developments. *Molecular Biotecnology*, Vol.37, No.1, (December 2007), pp. 31-37

Lutka, A. & Pawlaczyk, J. (1995). Inclusion complexation of coenzyme Q_{10} with cyclodextrins. *Acta Poloniae Pharmaceutica*, Vol.52, No.2, (1995), pp. 379-386

Mattila, P. & Kumpulainen, J. (2001). Coenzymes Q9 and Q10: contents in foods and dietary intake. *Journal of Food Compositinon and Analysis*, Vol.14, No.4, (December 2001), pp. 409-417

Nazzal, S., Guven, N., Reddy, I.K. & Khan, M.A. (2002). Preparation and characterization of coenzyme Q10-Eudragit® solid dispersion. *Drug Development and Indrial Pharmacy*, Vol.28, No.1, (January 2002), pp. 49-57

Ochiai, A., Itagaki, S., Kurokawa, T., Kobayashi, M., Hirano, T. & Isekia, K. (2007). Improvement in intestinal coenzyme Q10 absorption by food intake. *Journal of the Pharmaceutical Society of Japan*, Vol.127, No.8, (June 2007), pp. 1251-1254

Olson, R.E. & Rudney, H. (1983). Biosynthesis of ubiquinone. *Vitamin & Hormones*, Vol.40, pp.1-43, ISBN 0-12-709840-2

Prošek, M., Butinar, J., Lukanc, M., Milivojević, Fir M., Milivojević, L., Križman, M. & Šmidovnik, A. (2008). Bioavalibility of water soluble CoQ10 in beagle dogs. *Journal of Pharmaceutical and Biomedical Analysis*, Vol.47, No.4-5, (August 2008), pp. 918-922

Prošek, M., Šmidovnik, A., Fir, M., Stražišar, M., Golc Wondra, A., Andrenšek, S. & Žmitek, J. (2005). Water soluble form of coenzyme Q10 in a form of an inclusion complex with beta-cyclodextrine, process of preparing, and use thefore. International publication no. WO 2005/111224, (May, 2005), International patent application no. PCT/SI2005/000013

Prošek, M., Šmidovnik, A., Fir, M. & Stražišar, M. (2004). TLC identification and quantification of coenzyme Q10-[beta]-cyclodextrin complex. *Journal of planar chromatography and modern TLC*, Vol.17, No.3, (May-June 2004), pp. 181-185.

Prošek, M., Golc-Wondra, A. & Vovk, I. (2002). Uncertainty in quantitative thin-layer chromatography. *Journal of Chromatographic Science*, Vol.40, No.10, (November-December 2002), pp. 598-602

Stražišar, M., Fir, M., Golc Wondra, A., Milivojević, L., Prošek, M. & Abram, V. (2005). Quantitative determination of coenyzme Q10 by liquid chromatography and liquid chromatography/mass spectrometry in dairy products. *Journal of AOAC International*, Vol.88, No.4, (July-August 2005), pp.1020-1027

Szejtli, J. (1998). Intoduction and general overview of cyclodextrin chemistry. *Chemical Reviews*, Vol.98, No.5, (July 1998), pp. 1743-1753

Takeuchi, H., Sasaki, H., Niwa, T., Hino, T., Kawashima, Y., Uesugi, K. & Ozawa, H. (1992). Improvement of photostability of ubidecarenone in the formulation of a novel powdered dosage form termed redispersible dry emulsion. *International Journal of Pharmaceutics*, Vol.86, No.1, (October 1992), pp. 25-33

Turunen, M., Olsson, J. & Dallner, G. (2004). Metabolism and function of coenzyme Q, *Biochimica et Biophysica acta*, Vol.1660, No.1-2, (January 2004), pp. 171-199

Ullmann, U., Metzner, J., Schulz, C., Perkins, J. & Leuenberger, B. (2005). A New Coenzyme Q10 Tablet-Grade Formulation (all-Q®) Is Bioequivalent to Q-Gel® and Both Have Better Bioavailability Properties than Q-SorB®. *Journal of Medicinal Food*, Vol.8, No.3, (2005), pp. 397-399

Van Doorne, H. (1993). Interactions between Cyclodextrins and Opthalmic Drugs. *European Journal of Pharmaceutical and Biopharmceutics*, Vol.39, No.4, (1993), pp. 133-139

Witting, P.K., Peterson, K., Letters, J. & Stocker R. (2000)., Antiatherogenic effect of coenzyme Q10 in apolipoprotein E gene knockout mice, *Free Radical Biology and Medicine*, Vol.28, No.3-4, (August 2000), pp. 295-305

Weber, C., Bysted, A. & Holmer, G. (1997). The coenzyme Q10 content of the average Danish diet. *International Journal for Vitamin and Nutrition Research*, Vol.67, No.2, (1997), pp.123-129

Žmitek, J., Šmidovnik, A., Fir, M., Prošek, M., Žmitek, K., Walczak, J. & Pravst, I. (2008). Relative bioavalibility of two forms a novel soluble CoQ10. *Annals of Nutrition and Metabolism*, Vol.52, No.4., (2008), pp. 281-287

10

Aflatoxin Contamination in Foods and Feeds: A Special Focus on Africa

Makun Hussaini Anthony[1], Dutton Michael Francis[2], Njobeh Patrick Berka[2], Gbodi Timothy Ayinla[3] and Ogbadu Godwin Haruna[4]
[1]Department of Biochemistry, Federal University of Technology, Minna,
[2]Food, Environment and Health Research Group, University of Johannesburg,
[3]Ibrahim Badamasi Babangida University, Lapai,
[4]Sheda Science and Technology Complex,
Federal Ministry of Science and Technology, Abuja,
[1,3,4]Nigeria
[2]South Africa

1. Introduction

A groan can almost be audible from scientists who work or have an interest in the field of mycotoxicology (fungi and mycotoxins) to the effect of "not another review publication on AF". It is true that since the discovery of AF in the early 1960s, a huge literature on AF has developed. In one way, this literature can be considered appropriate, as these mycotoxins set the scene for a new burst of activity in the contamination of feeds and foods with mycotoxins, those produced by filamentous fungi, which has shown that these substances are not merely academic novelties, but have important effects in questions of food quality and human and animal health. Further, the principal member of this mycotoxin group, AFB_1, one of the most carcinogenic natural products formed in nature (D'Mello, 2003), a major cause of hepatocellular carcinoma (liver cancer) in both animals and humans is rated as a Class IB carcinogen by the International Agency for Research on Cancer (IARC) (IARC, 1993) meaning that it is a proven cancer-inducing agent. It also occurs ubiquitously in the environment contaminating many different food and feed commodities. Rather interestingly, critics of the excessive focus on AF may perhaps, point out that one of the reasons it is intensely investigated is the ease with which it can be detected and measured due to its strong fluorescence and ultra violet (UV) light absorbing properties (Bhatnagar and Ehrlich, 2002) which skews attention away from other important mycotoxins not containing a chromophore such as the trichothecenes (TH) and fumonisins (FB). Another irony of the kudos given to AF is because of it discovery stemming mainly from the condition observed in Britain called Turkey X disease, where it was associated with deaths of tens of thousands of turkey poults (Blount, 1961) is that many of the symptoms of the disease fit those of cyclopiazonic acid (CPA) and later analysis of the groundnut used to produce feed for these birds was found to contain (CPA) in addition to AFB_1.

Whatever the merits and demerits of again reviewing literature on AF, this review has a specific purpose which has not been thoroughly explored in the past. This is to look at the ravages that this toxin has caused in Africa, not only from an animal and human health point of view, but also from the economic consequence of having agricultural commodities contaminated with these toxins. Although most countries of the world can be affected by AF, it is sub-Saharan Africa (SSA) that has suffered the most. It is difficult to defend this statement because of the lack of detailed information on the occurrence of AF in African crops. This is because, much of SSA agriculture occurs in impoverished rural areas and the lack of technical infrastructure in many African countries does not allow for routine quality control of even commercially produced commodities, never mind those produced by rural populations for their own consumption. Much of the data on the incidence of mycotoxins in SSA countries is generated by non-African agencies (e.g., IARC) (IARC 1993) or by academic institutions. This review attempts to bring together information from studies conducted over the years and attempts to discuss it in a way that will give some idea on the mycotoxin problem in the continent both from an agricultural, economic and health points of view. For reasons already highlighted, occurrence and effects of AF, principally AFB_1, is the best representative in doing this.

2. Factors enchancing the prevalence of aflatoxigenic fungi and aflatoxins in Africa

Prior to AF contamination, the food material must be infected with fungi that have the genetic capacity to synthesize and deposit the toxins on the foods and feeds before or after harvest. Only species of the genus *Aspergillus* are endowed with the 23 genes responsible for synthesis of AF. Members belonging to this genus are most abundant in the tropics and as such, are major food spoilage agents in warmer climates. The genus is metabolically versatile producing over twenty mycotoxins. Of the over 180 species of *Aspergillus*, only a few are aflatoxigenic. After the discovery of AF in the 1960s, *A. flavus* and *A. parasiticus* of the section *Flavi* were the only known AF producers producing the B and B/G types of AF, respectively (Blount, 1961). Other aflatoxigenic species that subsequently emerged are *A. nomius* (B and G types), *A. bombycis* (B and G AF), *A. ochraceoroseus,* and *A. pseudotamarii* (B type), but they occur less frequently (Peterson *et al.*, 2001, Ito *et al.* 2001)). *A. tamarii, A. parvisclerotigenus* (B types), *A. rambellii* and certain members of *Aspergillus* subgenus *Nidulantes* namely: *Emericella venezualensis* (Frisvad *et al.*, 2005) and *E. astellata* (Frisvad *et al.*, 2004), have now been included in the growing list of aflatoxigenic species. *A. arachidicola* sp. Nov. and *A. minisclerotigenes* sp. Nov that produce both forms of the toxin, are the latest emerging aflatoxigenic species (Pildain *et al.* 2008). The unexpected new comer is A. *niger,* an ochratoxin (OT) producer which was discovered over four decades ago but was never associated with AF synthesis. However, in a search for aflatoxigenic fungi in Romanian medicinal herbs, Mircea *et al.* (2008) showed the capacity of some strains of *A. niger* to produce AFB_1. All these known AF producing fungi particularly *A. flavus* are common and widespread in nature, and have been shown as fungal contaminants of African foods and feeds according to Atehnkeng *et al.* (2008), Essono *et al.* (2009), Njobeh *et al.* (2009) and Makun *et al.* (2011).

Despite the fact that a strain of mould has the genetic potential to produce a particular mycotoxin, the level of production would in part be influenced by the nutrients available.

Typically, moulds require a source of energy in the form of carbohydrates or vegetable oils in addition to a source of nitrogen either organic or inorganic, trace elements and available moisture for growth and toxin production. Cereals particularly oilseeds adequately provide all these nutrients and so are considered ideal substrates for growth of fungi and consequently, toxin synthesis (FAO, 1983). However, even amongst the cereals, mycotoxins contamination varies with size and integrity of seed the coat with the small, compact grains (wheat, rice, oat, sorghum) and those encapsulated in hard seed coats (beans and soybeans) being less susceptible to fungal infection and mycotoxin formation than larger grains such as maize (Stössel, 1986). In Africa, grains contribute about 46% of the total energy intake (FAOSTAT, 2010), this figure may even be higher in rural SSA, where cereals and tubers and roots are virtually the sources of nutrition. This over dependence on grains, an ideal substrate for AF synthesis in rural Africa is a major reason for the high AF load in the continent.

Probably the two most important environmental components favouring mould growth and AF production are hot and humid conditions. Although the optimum temperature and moisture content for growth and toxin production for the various aflatoxigenic fungi varies, many of them achieve best growth and toxin synthesis between 24°C and 28°C (Schindler *et al.*, 1977) and seed moisture content of at least 17.5% (Trenk and Hartman, 1970; Ominski *et al.*, 1994). These conditions approximate the ambient climatic conditions in most parts of Africa and hence also account for the high prevalence of the toxins in the continent. As to how climate change will alter the AF situation in the continent, is still a matter of public debate. However, in an attempt to predict how extreme climatic conditions associated with climate change can affect AF contamination, Paterson and Lima (2010) adduced that while anticipated warmer (33°C) and more humid conditions might increase AF prevalence in Europe, perhaps the reverse might be the case in tropical Africa as most aflatoxigenic fungi will not survive the expected 40°C. Even though this prediction is plausible, it is pertinent to state herein that if there will be a rise in temperature and a rainfall reduction in all parts of Africa except for East Africa where an increase in mean annual rainfall is expected (IPCC, 2007), it is only reasonable to assume that many regions of the continent will experience droughts while East Africa would be hotter and more humid and given that these conditions favour AF production (Mutegi *et al.* 2009)), climate change might exacerbate the AF crisis in Africa. In agreement with this deduction, Paterson and Lima (2010) also speculated that if temperature does not increase as envisaged, then droughts might be more frequent. Drought conditions actually constitute stress factors to plants rendering them vulnerable to *Aspergillus* infection (Robertson, 2005; Holbrook *et al.*, 2004) with ensuing increase in AF pollution. An indelible sign that droughts prop up AF contamination is the fact that these conditions preceded the fatal outbreak of acute human aflatoxicosis that occurred in Kenya in 2004 (Afla-guard, 2005; CDC, 2004).

Soil is another natural factor that exerts a powerful influence on the incidence of fungi. Crops grown in different soil types may have significantly different levels of AF contamination. For example, peanuts grown in light sandy soils support rapid growth of the fungi, particularly under dry conditions, while heavier soils result in less contamination of peanuts due to their high water holding capacity which helps the plant to prevent drought stress (Codex Alimentarius Commission, 2004). Produce harvested from land on which groundnut has been planted the previous year were more highly infested by *A. flavus* and contained more AF than crops grown on land previously planted with rye, oats, melon or potatoes (Martin and Gilman, 1976). Likewise, previously fungicide treated soil has been shown to reduce incidence of *A. flavus* in groundnuts to very low levels. Accordingly, it might also be useful to add recommendations on appropriate crop rotation programmes and soil treatments in order to reduce the hazards from mycotoxins.

The presence of other microorganisms either bacteria or other fungi may alter AF elaboration on food materials. When *A. parasiticus* was grown in the presence of some bacteria; *Streptococcus lactis* and *Lactobacillius casei,* AF production was reduced (Ominski *et al.*, 1994). Meanwhile, fungal metabolites such as rubratoxins from *Penicillium purpurogenum;* cerulenin from *Cephalosporium caerulens;* and *Acrocylindrium oryzae* enhance AF production even though they repress growth of AF producing fungi (Smith and Moss, 1985). This type of positive interaction between fungi in the same food matrix with regards to AF synthesis couple with multi-occurrence of mycotoxins from the different fungi which could at the least have additive if not synergistic health impact on the host (Speijer and Speijer, 2004) worsen the AF plight in Africa because such simultaneous co-occurrence of fungi and mycotoxins in African agricultural commodities is very common phenomenon as indicated by many workers including Makun *et al.* (2007), Makun *et al.* (2009) and Njobeh *et al.* (2009).

Much more than in other parts of the world, insects, termites, rodents and birds constitute a major problem to food safety and availability in Africa. The accessibility of these pests and predators to crops is made possible by the available deplorable storage facilities (bags, 'rhumbu') and traditional postharvest preservation techniques (drying grains on rocks and bare floor) applied in the region. Besides eating up parts of the human rations, they boost the susceptibility of crops to microbial infestation and infection with a correlating swell in AF occurrence (Hell *et al.* 2000). In the course of feeding on crops in the field or during storage, these animals physically wound kernels and tubers. Mechanical damage resulting from such actions of pests disrupts the seed coat and facilitates the penetration of fungal inoculum into the interior of the grains. In addition to the physical damage, these animals transmit spores from other plants and environmental surfaces to inoculate the already defective kernels and tubers and as such, help to distribute moulds widely throughout a bulk mass of grain or feed. The metabolic activities of pests especially insect larvae also produce metabolic water and heat (Sinha, 1984) that are beneficial for mould growth. The activities of these insects and birds occasioned by poor storage amenities (Hell et al., 2000; Udoh *et al.*, 2000) provoke fungal growth and toxin production (Agboola, 1992; Wagacha and Muthomi, 2008). According to the review of Bankole and Adebanjo (2003) on the mycotoxin situation in West Africa, the commonest insects that spread *A. flavus* in preharvest maize in the region are lepidopteran ear borer *Mussidia nigrivenella, Sitophilus zeamais* and *Carpophilus dimidiatus.*

Agricultural practices have profound impact on AF contamination of foods. Population drift from agrarian rural areas to urban settlements in search for employment has led to reduced workers on the farm and so 'off season' harvesting during early rains is increasingly common. Also because farmers have improved varieties of grains particularly maize, crops are now grown and harvested twice during the planting season and so harvesting of the first set of these crops is done in the wet months of July and August. Such 'off season' harvesting promotes growth of aflatoxigenic fungi and toxin synthesis (Kaaya and Warren, 2005). Harvesting methods that enhance seed breakage would also increase the degree of mycotoxin formation. *Aspergillius flavus* was observed to be more abundant in kernels from pods gathered by combined harvesting than from hand harvested pods and the respective AFB_1 content were 1780 and 140ppb (Martin and Gilman, 1976). This is in agreement with the suggestion that certain modern agricultural management practices may create unique ecological niches which select toxigenic fungi (Bilgrami et al., 1981). Meanwhile, as Africa is experiencing a boost in mechanized farming, there are no commiserate control measures put in place to reduce the negative impact of this agricultural revolution on mycotoxin pollution.

Fungi are generally aerobic organisms; therefore storage atmosphere deficient in oxygen would lead to reduced metabolism and consequently mycotoxin production. Javis (1971) and Agboola (1992) reported that a reduction in oxygen content of storage environment from 5% to 1% and increase in carbon dioxide content to above 20% dramatically reduced the growth of *A. flavus* and AF production. Commodities are better stored anaerobically with the addition of organic acids such as propionic acid as preservatives in storage systems which do not absorb moisture or enhance moisture migration (Ominski *et al.* 1994). Unfortunately traditional storage facilities in Africa are devoid of such standard storage environments (Hell *et al.*, 2000; Udoh *et al.*, 2000).

Unwholesome trade practices have become problematic in guaranteeing food safety in the developing world. Since fungal proliferation and mycotoxin formation increase with duration of storage (Hell *et al.* 2000) when favourable conditions prevail, the hoarding of commodities under poor storage circumstances from October to July/August for sale during period of scarcity in order to maximize profits which has become a common practice in rural Africa exaggerates AF contamination problem. Similarly, the tradition of mixing grains of different grades in order to improve the quality of contaminated grains especially when one contains a large number of fungi spores will provide inoculum for the good grade and probably contaminate the toxin-free grain with AF. Other compelling factors that worsen the AF burden in Africa are public ignorance of the existence of the toxins; complete absence or lack of enforcement of regulatory limits; and introduction of contaminated food into the food chain which has become inevitable due to shortage of food supply caused by drought, wars and other socioeconomic and political insecurity (Wagacha and Muthomi, 2008).

3. Occurrence of AF in foods and feeds

AF producing fungi particularly *A. flavus* are common and widespread in nature and most often found when certain grains are grown under stressful conditions such as drought. The moulds occurs in soil, decaying vegetation, hay and grains undergoing microbiological deterioration and invades all types of organic substrates when the conditions are favourable for growth particularly in hot and humid situations (Ominski *et al.*,1994). Under such a suitable environment, aflatoxigenic fungi contaminate foods and feeds directly or indirectly. In direct contamination, the product is infected with aflatoxigenic fungi with subsequent toxin production. Indirect contamination occurs when food or feed was previously contaminated with AF producing fungi and although the fungi has been removed or killed during processing, AF still remains in the final product. Such contamination of cereals and oilseeds is the main point of entry of many mycotoxins in the human and animal dietary systems particularly in Africa (Smith and Moss, 1985). Fungal infection of agricultural produce is inevitable but while in developed countries, they get removed from the food chain, in most parts of Africa, moulded foods are part of the daily diets. All foods and feedstuffs are vulnerable to fungal contamination but the nature and degree of aflatoxigenic fungal contamination will depend on the presence or absence of AF in the product. Whereas the identification of toxigenic fungal contaminants is an undoubted pointer to a potential risk, positive conclusions can only be made with certainty by quantifying the suspected toxins which is why this critical review of levels of AF in food commodities is an empirical assessment of Africa's health condition with regards to mycotoxicoses. The crops that frequently support growth of AF producing fungi and subsequent toxin production are thus principal sources of exposure to AF that include but not limited to cereals (maize, sorghum, pearl millet, rice, wheat), oilseeds (peanut, soybean, sunflower, cotton), spices (chile peppers, black pepper, coriander, turmeric, ginger), and tree nuts (almond, pistachio, walnut, coconut, brazil nuts) (FAO, 1983). The other possible sources of entry of mycotoxin into animal and human systems include fruits, vegetables, animal tissues and animal products and fermented products (Jarvis, 1976).

There are at least 14 natural occurring AF known to exist, however, only six have public health and agricultural significance and they include AFB_1, B_2, G_1 and G_2. The other two are AFM_1 and M_2 which are the hydroxylated forms of AFB_1 and B_2, respectively that are secreted in animal (including humans) tissues and fluids. There is no region of the world that is free from AF problem but the strict food and feed quality control programmes put in place in the developed countries greatly reduce the AF burden in those countries. This however, is not the case in Africa and to some extent, in Asia. Besides the lack of regulation of mycotoxins in the continent, the prevailing hot, humid climate and subsistence on foods suitable for AF contamination and other factors discussed in Section 2.0, make Africa the most AF vulnerable region in the world. Hence, a current overview of its natural occurrence in different raw and processed food commodities that are major sources of the toxins in the continent which will not only reflect the disparity in human exposure to AF worldwide, but also contributes to a better understanding of Africa's health afflictions is the primary objective of this section.

3.1 Raw agricultural products

3.1.1 Nuts and oilseeds

These products are the most investigated of all foods and feeds with regards to AF contamination. The reasons being that they are the most susceptible to AF in addition to their high protein content particularly in the case of groundnut which has made them priceless components of many animal and human diets. Data on AF incidence and concentrations in groundnut are consistently incriminating with several reports regularly accounting for extremely high levels of the toxins (Table 1) in all regions of the continent. Table 1 shows over 60% prevalence of AF across Africa at very critical levels. The incidence and levels reported vary with season. Highly contaminated samples were obtained during the rainy season (90% frequency at 12-937 μg/kg) than in the dry season (53.2% prevalence at 15-390 μg/kg) (Kamika and Takoy, 2011). The highest contaminations noted in Malawi (ICRISAT, 2010) were in districts that were prone to late session rains which enhance favourable conditions for post harvest contamination. Similarly, Mutegi *et al.* (2009)

demonstrated that wet and humid weathers are associated with severe AF contamination than drier locations. Serious contamination of Nigerian groundnut is a recurring problem since the 1960s when up to 8000 μg/kg were reported in 1976 (Opadokun, 1992) from the Northern region of the country. Even after heat treatment that should reduce AF levels in contaminated samples (Ogunsanwo *et al.* 2004), dry-roasted groundnuts from South Western part of the country contain as high as 165 μg/kg (Bankole *et al.*2005). Bankole and Adebanjo (2003) reviewed the AF contents in groundnut cake (range: 20 - 455 μg/kg) and snack (30 μg/kg) from Nigeria and concluded that the levels of contamination in these products were toxicologically unsafe. Homemade and unrefined groundnut oil contains AF at levels ranging between 20 and 2000 μg/kg in Nigeria (Obidoa and Gugnani, 1992). It is pertinent to state herein that in 1988, the deaths of some primary school children in Nigeria were associated with incriminating levels of AF in groundnut cake '*kulikuli*' (Fapohunda, 2011). The AF problem of peanuts persists even in South Africa where food safety standards are adhered to. This came to limelight in 2001 when the South African Primary School Nutrition Programme received substantial media coverage on unacceptably high AF content in sandwich containing peanut butter given to school children. The government then established a national monitoring programme to survey AF in groundnuts and peanut products. Between July, 2003 and March 2004, 1140 peanut and groundnut products samples were analyzed and accordingly, about 30% of analyzed samples did not comply with the legal limit of 10 μg/kg (5 μg/kg for AFB_1) and concentrations as high as 560μg/kg were obtained (MRC, 2006).

AF contamination of cottonseed has been a major concern worldwide as extremely high contents ranging between 200,000 to 300,000μg/kg were reported much earlier in samples exported to the European markets (Smith and Moss, 1985). A survey of cottonseed for AF during such periods (between 1976 and 1979) when Nigeria was a major exporter of the crop, revealed over 60% (17/28) incidence rate with mean value of 105μg/kg recorded (Opadokun, 1992). Gbodi (1986) subsequently showed moderate contamination of the cash crop grown in the semi temperate climate of Plateau State, Nigeria (Table 1). Mycotoxin research on cottonseed in Nigeria has since ceased to exist as the country no longer depends on agriculture as its main source of revenue. When AF contaminated cottonseed is processed into oil, the toxins are concentrated in the residual cottonseed meal and cake which are often used as feed components of livestock feeds. The cottonseed oil also contains residual amount of the toxins at certain levels but this usually depends on the extent of contamination of the seeds from which it was obtained.

Aside from cottonseed, its products are also very important sources of human and animal exposure to AF. Abalaka (1984) found 26.1μg/kg and 12.6 μg/kg of AF in groundnut and cottonseed oil samples from the Guinea Savannah region of Nigeria. All cottonseed meal samples investigated for AF in South Africa (Table 1) contained the toxins in which 42 of the 60 samples analyzed exceeded the maximum level for feeds. The AF menace in cottonseed is not limited to the sub-Saharan region of the continent as aflatoxigenic fungi and the two B forms of the toxin were found naturally contaminating cottonseed, cottonseed meal and cake from Egypt (Mazen *et al.* 1990). Melon seed another important oilseed in West Africa has been shown to be prone to fungal and AF contamination (Table 1) at largely unsafe levels (Bankole *et al.* 2006). Also Opadokun (1992) reported high incidence (73%) of AF in Nigerian melon seed at mean content of 19μg/kg.

Commodity	Country	Type of Aflatoxin	Incidence rate	Range (µg/kg)	Mean level ± SD (µg/kg)	Reference
Barley	Tunisia	AFB1	11/25	3.5-11.5	18.4 ±27.3	Ghali *et al.* (2008)
Cheese	Libya	AFM_1	15/20	0.11-0.52		Elgerbi *et al.* (2004)
Cow milk	Nigeria	AFM_1	3/22	-	≤ 2.04	Atanda *et al.* (2007)
	Sudan	AFM_1	42/44	0.22 – 6.90	2.07	Elzupir and Elhussein, (2010)
	South Africa	AFM_1	42/42	0.04 – 1.32	0.12	Dutton *et al.* (2010)
	Kenya	AFM_1	474/613	0.005-0.78	0.064	Kang'ethe and Lang'a (2009)
	Cameroon	AFM_1	10/63	0.006-0.527	-	Tchana *et al.* (2010)
Cowpea	Benin	AFB1	3/92		3.58	Houssou *et al.* (2009)
	Cameroon	AF	5/15	0.2-6.2	2.4	Njobeh *et al.* (2010)
Dried Beef	Nigeria	AFB	10/10	0.003		Oyero and Oyefolu (2010)
		AFG	10/10	0.004		
Dried Chilli	West Africa	AFB1	1/30	3.2	3.2	Hell *et al.* (2009)
Dried figs	Morocco	AFB1	1/20	0.28	0.28	Juan *et al.* (2008)
Dried okra	West Africa	AFB1	3/30		5.4	Hell *et al.* (2009)
		AFB2	1/30		0.6	
Dried raisins	Morocco	AFB1	4/20	3.2-13.9	10.7±2.3	Juan *et al.* (2008)
Egg	Cameroon	AF	28/62	0.002-7.68	0.82± 1.71	Tchana *et al.* (2010)
Fresh Beef	Nigeria	AFB	10/10	0.02		Oyero and Oyefolu (2010)
		AFG	10/10	0.03		
Groundnut	DR Congo	AFB1	43/60	1.5-937	229.07	Kamika and Takoy, (2011)
Groundnut	Malawi	AFB1	/1189	0-3871		ICRISAT (2010)
Groundnut	Kenya	AF	170/769	0-7525	< 4	Mutegi *et al.*(2009)
Local beer	Malawi	AF	5/5	8.8-34.5	22.3±4.93	Matumba *et al.* (2011)
Maize	Malawi	AFB		0.0-1335		ICRISAT (2010)
Maize	Nigeria	AFB1	55/55	0-1874	257.82	Atehnkeng *et al.* (2008)
		AFB2		0-608		
		AFG1		0-937		
		AFG2		0-286		
Maize	Ghana	AF	30/30	6.20-29.50	13.596	Akrobortu (2008)
Maize	Uganda	AF	22/49	1.00-1000		Kaaya and Warren, (2005)
Maize flour	Morocco	AFB1	16/20	0.23-11.2	1.57± 0.78	Zinedine et al. (2007) [b]
Melon seed	Nigeria	AFB1	37/137	2.3-47.7	14.2	Bankole *et al.*(2006)
Milk[c]	South Africa	AFM1	98/114	Max: 2.07	0.15	Lishia et al. (data unpublished)
Milk[d]	South Africa	AFM1	85/85	Max: 2.48	0.14	
Millet	Nigeria	AFB1	12/49	1370.28-3495	2587.47±78.23	Makun *et al.*(2007)
Mouldy Sorghum	Nigeria	AFB1	93/168	0-1164	199.51-259.90	Makun *et al.* (2009)
Pasteurized milk	Morocco	AFM_1	48/54	0.001-0.117	0.018	Zinedine *et al.* (2007)[a]
Powdered milk	Nigeria	AFM_1	19/100	0.02-0.41	0.136	Makun *et al.* (2010)
Powdered soymilk	Nigeria	AFB1	30/30	4.58-19.76	11.53	Adebayo-Tayo *et al.* (2009)
		AFB2		2.57-11.54	6.04	

Commodity	Country	Type of Aflatoxin	Incidence rate	Range (μg/kg)	Mean level ± SD (μg/kg)	Reference
Raw cow Milk	Egypt	AFM_1	19/50	0.023 - 0.073	0.049 ± 0.017	Amer and Ibrahim (2010).
Rice	Nigeria	AF	21/21	27.7-371.9	82.5±16.9	Makun *et al.* (2011)
Roasted Groundnut	Nigeria	AFB1	68/106	5-165	25.5	Bankole *et al.* (2005)
		AFB2	28/106	6-26	10.7	
		AFG1	12/106	5-20	7.2	
		AFG2	3/106	7-10	8.0	
Smoke-dried fish	Nigeria	AFB1	11/11	1.505-8.11	3.46	Adebayo-Tayo *et al.* (2008)
		AFG1	11/11	1.810-4.51	2.94	
Sorghum	Tunisia	AFB1	58/93	0.34-52.9	9.9±11.5	Ghali *et al.* (2009)
		AFB2	45/93	0.11-3.7		
		AFG1	3/93	0.45-0.7		
Sorghum	Malawi	AFB1	2/15	1.7-3.0	2.35±0.65	Matumba *et al.*(2011)
Sorghum	Ethiopia	AFB1			0-26	Ayalew *et al.* (2006)
Soybean	Cameroon	AF	2/5	0.2-3.9	2.1	Njobeh *et al.* (2010)
Wheat	Kenya	AFB1	23/50	0-7	1.93	Muthomi *et al.* (2008)
	Tunisia	AFs	15/51	4.0-12.9	6.7±2.4	Ghali *et al.* (2008)
	Nigeria	AFB1		17.10-20.53	19.00±1.67	Odoemelam and Osu, (2009)
Wheat & products	Algeria	AFB1	30/53	0.13-37.42	>5	Riba *et al.* (2010)
	South Africa	AFB1	13/238	0.5-2.0	>2	Mashinini and Dutton, (2006)

[c]Milk obtained from rural communities in South Africa. [d]Milk obtained from commercial farms

Table 1. Some reports of aflatoxin levels in some foods commodities from Africa

Low AF contamination of cowpea, soybean and their products is reported in many parts of the world. The data on these nuts from Cameroon and Benin (Table 1) showed only trace amount of AF in few samples. Similarly, only 3 positive samples were demonstrated in 268 cowpea samples analyzed between 1975 and 1983 from Nigeria (Opadokun, 1992). In the same review, it was shown that all two samples of soybean oil assayed did not contain AF while 41 of 55 palm kernels were contaminated with low amount of the toxin. High seed coat integrity, ensuring limited access and low moisture content are responsible for the low susceptibilities of these nuts to aflatoxigenic fungi (Stossel, 1986) with consequent rare occurrence of the toxins in them. In spite of the low presence of AF in soybean, intolerable levels of these carcinogens were found in branded and unbranded powdered soymilk, a processed product of soybean in Nigeria (Adebayo-Tayo *et al.* 2009). Soymilk has become a popular infant recipe in Africa because it is a cheap source of water soluble protein, carbohydrates and oil and so human exposure to AF can arise in countries using it for human consumption. Although producing fungi particularly *Aspergillus flavus* are considerably natural contaminants of cocoa bean (Sanchev-Hervas *et al.* 2008), AF are rarely detected in cocoa bean and at very low concentration. In a survey for mycotoxins in cocoa and cocoa products sponsored by the Foundation of German Cocoa and Chocolate Industry, Hamburg, Beucker *et al.* (2005) screened over 200 samples and found AFB_1 at concentrations below 2μg/kg. Literature search for data on high AF in cocoa from Africa was unsuccessful and that could be a clear indication that these toxins are not problematic in this cash crop. However, ochratoxin A is the mycotoxin of interest to the cocoa (Arogeun and Jayeola, 2005) and soybean (Smith and Moss, 1985) industries.

3.1.2 Cereals

Cereals are rich sources of minerals, vitamins, carbohydrates, oils and proteins but when refined majority of the nutrients are lost leaving mostly carbohydrates and are therefore grown mainly for energy. Grains as they are sometimes called provide more food energy worldwide than any other crop and thus are staples. They are staple for two third of the earth's population, providing 85% of the world's food energy and protein intake (FAOSTAT, 2006). Cereal consumption is moderate in developed countries however in Africa and Asia, it is a daily sustenance. In Africa, cereals contribute 46% of the total energy intake; however, this figure could be as high as 78% in some African countries (FAOSTAT, 2010). According to figures made available by the afore mentioned statistics division of FAO, the six most cultivated and hence consumed grains worldwide in order of decreasing production are maize, rice, wheat, barley, Sorghum and millet, and of these major grains maize, wheat and rice together account for 87% of all cereal production worldwide and 43% of all food calories. Because of their rich nutrient composition, cereals support fungal growth and mycotoxin production excellently on the farm, during storage and after processing into foods and feeds with the small grains (sorghum, rice, wheat, millet and rice) being less susceptible to fungal and toxin contaminations than the larger grains like maize. And since these ideal substrates for mycotoxin contamination are highly consumed globally, they constitute the most remarkable sources of mycotoxins (especially the most prevalent of mycotoxins; AF) to animals and human beings. While there are other cereals (oats, rye, triticale, fonio, buckwheat and quinoa) of global significance, this review will focus on the six major ones which are pertinent to Africa.

Maize is one of the important staple foods in Africa and is now widely grown for animal feeds. It is the third most cultivated food commodity in the continent after cassava and sugar cane but it ranks the first in term of food energy supply in Africa. AF are regularly detected in maize throughout the world and the recent serious contamination which was associated with drought led to fatal human aflatoxicosis in Kenya (Afla-guard, 2005; CDC, 2004) as previously stated in Section 2.0. The natural occurrence of AF in maize from some African countries has been reviewed (Table1). A regional effect on the incidence of the toxins has been demonstrated with lower contents observed in the drier North African countries (Morocco; 0.11-11.2 μg/kg) than amounts reported for such SSA countries as Uganda, Nigeria, Ghana and Malawi having up to 1874 4μg/kg. Monitoring for AF in maize samples from different regions of Africa showed disturbingly high levels of contamination above 1000 μg/kg with many of the samples containing AF contents exceeding the CODEX regulatory limit of 20 μg/kg (five times more lenient than the EU guideline of 4μg/kg). Over 77% of the tested samples from Malawi were contaminated with AF with 90% of the positive samples not qualified for the EU export market (ICRISAT, 2010). Atehnkeng *et al.* (2008) previously analyzed 55 samples of maize from 11 districts across three agro-ecological zones of Nigeria for AF and mean values ranging between 30.9 -507.9 μg/kg were recovered from the 10 districts, which were far beyond all known acceptable levels. Maxwell *et al.* (2000) also found AF at alarming concentrations of between 3,000-138,000 μg/kg in Nigerian pre-harvest maize samples. In a survey for AF in Ghana which spanned for a decade, the toxin had over 50% prevalence with more than half of the investigated samples having levels exceeding the EU maximum limit (Akrobortu, 2008). Similarly, stored maize from Uganda contain unsafe AF levels (Kaaya and Warren, 2005; Kaaya and Kyamuhangire, 2006). Very high AF level of 46,400μg/kg was found in samples from Kenyan local markets (CDC, 2004).

There is a general paucity of reports on mycotoxin in African indigenous rice. Opadokun (1992) reported AFB in only 13 of the 279 rice samples analyzed in Nigeria with only one having a level above 5μg/kg. One reason why mycotoxicologists are not attentive to local African rice relative to other crops such as maize and peanuts for example is due to the fact that it is entirely consumed within the continent and not available in the European markets. Makun *et al.* (2011) demonstrated a 100% prevalence of AF in Nigerian rice at unsafe levels (range: 28 - 372 μg/kg) (Table 1), and also showed critical contamination by ochratoxin A (OTA) and presence of deoxynivalenol (DON), fumonisins (FB) and zearalenone (ZEA) at trace levels. Although, AF were not found in Ugandan grown rice, however, they have been reported in rice from Cote D'Ivoire (1.5 – 10 μg/kg) and Kenya (294-1050 μg/kg) (Kaaya and Warren, 2005). Data from these few studies underscore the need for more mycotoxin surveys of indigenous rice in Africa.

Wheat does not thrive well in tropical climates so African wheat production is concentrated in the narrow strip along the Mediterranean coast from Morocco to Tunisia, in the Nile valley (Egypt), and in parts of South Africa, Kenya, Ethiopia, Zimbabwe and Sudan. Very little is grown in West Africa however, there is an increase in wheat production in Nigeria and the Democratic Republic of Congo in recent years. The grain is produced mainly for bread making but its bran is increasingly used as component of animal feeds. In spite of the limited production (18.6 million tonnes per year) of the cereal in Africa, 50.4 million tonnes is consumed yearly in the continent (FAOSTAT, 2010). Data on AF in local African wheat (Table 1) reveal moderate contamination across the continent with the highest level of contamination occurring in Nigeria. Over 57% incidence rate of AFB_1 in 53 Algerian pre-harvest and stored wheat samples was established by Riba *et al.* (2010), with only 5 analyzed samples having contents above the EU legal limit. The Tunisian grown wheat had safe levels of AFB_1 (Ghali *et al.* 2008). Of the 51 samples screened from the country, 37% were positive for AFB_1 at levels below 3.4μg/kg. Similarly, with regards to AFB_1, all South African wheat samples analyzed by Mashinini and Dutton (2006) qualified for the EU market, though it occurred simultaneously with DON, ZEA, OT and FB in some of the samples. Although AFB_1 contents in the Kenyan wheat were low (<7 μg/kg), most of the samples were also concurrently contaminated with low but significant levels of DON, T-2 toxin, ZEA and AFB_1 (Muthomi *et al.* 2008). Such co-occurrence of mycotoxins could result in additive or synergistic effects in the host animal (Miller, 1995) as will be discussed in-depth subsequently. Considering the EU and other international and national legal limits, the Nigerian wheat with AFB_1 levels above 17 μg/kg (Odoemelam and Osu, 2009) is of low quality and thus, unfit for human consumption. Makun *et al.* (2010) also found extremely high AFB_1 contaminations of wheat marketed in Minna, Nigeria at unacceptable levels (range: 40-275 μg/kg) in 27 of the 50 tested samples. The severe contamination levels of this crop in Nigeria making it a principal source of mycotoxins raise public health concerns and underscore the need for the regulation of mycotoxins in SSA.

Because it was replaced by maize as a staple food commodity in many rural settlements in Africa (Bandyopadhyay *et al.* 2007), sorghum (also known as guinea corn) is another cereal that has been neglected for some time. However, its rising industrial profile as a suitable raw material for beer brewing has seen to its re-emergence at the world market such that as of 2007, sorghum production in Africa increased significantly even to the detriment of rice and wheat production (FAOSTAT, 2010). The renewed focus on sorghum is also because it is one of the most drought tolerable crops and such high water-use efficient characteristics makes it the crop of choice to boast food security in drought stricken regions of Africa and for the future against the anticipated water scarcity in the world. Sorghum is a staple grain for over 750 million people in Africa, Asia and Latin America (CODEX, 2011) that is traditionally grown mainly in the semi-arid tropics for human consumption and production of local alcoholic drinks is now a component of animal feeds. Regardless of its inherent resistance to mould infestation due to its high composition of fungicidal principles; phenols and tannin (US Grain Council, 2008), fungal contamination constitutes a major biotic constraint to sorghum improvement and production worldwide. It is estimated that annual economic losses in Asia and Africa due to mould are in excess of US $130 million (Chandrashekar *et al.* 2000). AF is the most investigated mycotoxin in this crop being reported in nine countries at levels of up to 3,282 µg/kg in Brazil (CODEX, 2011). A few reports from some major sorghum producing countries (Table 1) generally reveal a moderate prevalence and low concentrations of AF in the northern part of the continent though many of the contaminated samples from Tunisia (62.50%) had higher contents than the EU limit (Ghali *et al.* 2009). Meanwhile all the samples from Malawi met the EU standard (Matumba *et al.* 2011). The samples analyzed by Makun *et al.* (2009) were mouldy which might explain the observed increased incidence and levels of contamination. The relevance of such biased data cannot be overemphasized as mouldy grains do normally enter the human and animal food chain in Africa. In another unbiased study using representative samples, Odoemelam and Osu (2009) supported the findings of Makun *et al.* (2009) by reporting unacceptable levels of AFB_1 (between 27 to 36 µg/kg with a mean value of 31±3.4) in Nigerian grown sorghum. Twenty six of 69 sorghum samples from Uganda contained the toxins of which 10 of the positive ones had levels over 100µg/kg (Kaaya and Warren, 2005). Besides the entry of AF into human foods via contaminated sorghum, the carryover of substantial amount of these toxins from contaminated sorghum into traditional beer and beverages (Okoye, 1987; Matumba *et al.* 2011) poses additional risk of AF exposure. This should attract much attention to toxic contaminants of the grain but the reality is that there is limited information on mycotoxins in sorghum from Africa which is not commensurate to the escalating economic value of the cereal. Thus, the decision taken by the Codex Committee on Contaminants in Foods at her 5th session held in The Hague, The Netherlands in March, 2011, to prepare a discussion paper on 'mycotoxin in sorghum grain' is a very laudable initiative.

Millet *(Pennisetum* spp.) is the other traditional cereal replaced by maize in Africa in the last three decades. It is resistant to drought and so has been extensively cultivated in arid and semi arid regions. Millet ranks the sixth most important grain in the world, sustaining a third of the world's population and is the fifth most cultivated crop in Africa after maize, sorghum and wheat (Obilana *et al.* 2002). The occurrence of AF in this grain has largely been reported in West Africa particularly in Nigeria, its principal producer in the continent. In a review on chemical safety of traditional grains, Brimer (2011) reported that the average AF content of freshly harvested millet from West Africa was 4.6µg/kg. Odoemelam and Osu (2009) found higher values (range: 34 – 40 µg/kg; mean content: 37.5±2.5 µg/kg) in samples from the forest region of Nigeria. Makun *et al.* (2007) obtained the highest value of 3,495 µg/kg (Table 1) in wet season samples of millet that have been stored for over a year in Niger State, Nigeria. A 1971 report on AF contamination of Ugandan millet as reviewed by Kaaya and Warren (2005) revealed a low incidence (9/55) and AF level of 9 µg/kg. The limited reports on mycotoxins in millet in Africa are understandable as it is one of the 'lost crops of Africa' and that it is not an export crop. However, the anticipated 4 and 8 times reduction in risks of AF related problems if sorghum and millet, respectively, replace maize as primary staples (Bandyopadhyay *et al.* 2007), should bring these two African traditional crops to the front burner of mycotoxin research worldwide.

There is generally low natural incidence of AF in barley and derived products around the world. The levels reported in Tunisian grain (Table 1) is not an exception to this observation. Very low levels (<3.9 µg/kg) were also observed in barley from South Africa (Maenetje and Dutton, 2007). Because barley is grown in Africa for foods and lager beer production, good agricultural and manufacturing practices along with strict quality control measures are usually put in place in order to gain acceptability in competitive markets and this therefore accounts for the low prevalence of toxic substances in the grain as evident in these few reports. This sort of monitoring should be extended to all other foods and feedstuffs.

3.2 Animal feeds

A majority of animals reared in Africa are free ranged. During the dry season, pastoralists and their flocks travel hundreds of kilometres in search of greener pastures. Invariably, most domesticated African animals are raised on low grade cereals, informal pasture and domestic wastes. Commercially produced animals on the other hand, are essentially fed compounded feeds that are composed primarily of cereals, oilseeds and their by-products including those of animal origin. Even though the main ingredients of the diets of both the roaming and farm animals have been shown in subsection 3.1.1 and 3.1.2 to be heavily loaded with AF in the continent, a review on the occurrence of AF in animals feeds is still appropriate as processing might alter AF levels in the final product. More so, commercial livestock farming is now a major industry in Africa and thus feeds have become a major source of exposure of humans to food borne toxins via consumption of food products obtained from animals fed AF contaminated feeds. In view of the anticipated disparity in toxin content between the feed ingredients used and the compound feed produced, Mngadi *et al.* (2008) analyzed South African animal feeds and the raw ingredients used in manufacturing them for AF, FB, ZEA and OTA. Accordingly, data revealed 17 of the 23 samples tested positive for AF (Table 2) and over 7 of these samples having levels equal or exceeded the legislated limit of 20µg/kg for animal feeds. The raw ingredients namely cottonseed cake, sunflower oil cakes, molasses meal and bagasse were between 4 to 8 times more contaminated than the animal feeds and this disparity was attributed to the heat and other physical and chemical treatments (Rahana and Basappa, 1990) employed during processing that might have eliminated some of the toxins. As shown in Table 1, there is a high prevalence (84.6%) of carcinogenic AF in animal feeds from Kenya with 455 of the tested 830 samples having levels exceeding 5µg/kg, the WHO/FAO limits for feeds destined for diary animals. Poor storage facilities, use of moulded maize for feed production and the absence of monitoring for AF during processing were the reasons for the high frequency and levels of AF in Kenyan feeds (Kang'ethe and Lang'a, 2009). Data on AF in animal feeds from Nigeria (Table 1) clearly points out that while retailing from open bags provokes AF contamination, manufacturing of feeds using materials less susceptible to fungi including wheat offal and palm kernel limits contamination (Adebayo-Tayo and Ettah, 2010). The 6 positive samples among the 13 feeds screened were mainly from retailed shops by the aforementioned workers had no AF presence in the wheat and palm kernel base feeds. In accordance with EU regulations, poultry feeds from Morocco were safe for animal consumption (Zinedine *et al.* 2007). In light of the significant prevalence of AF in African feeds and feeding stuffs, it is recommended that the use of rapid and sensitive mycotoxin test kits by farmers, manufacturers and consumers to monitor the quality of products will largely lessen the AF burden in the continent.

Commodity	Country	Type of Aflatoxin	Incidence	Range (μg/kg)	Mean level ± SD (μg/kg)	Reference
Animal feeds	South Africa	AF	17/23	0.8-156	38.9	Mngadi *et al.* (2008)
Animal feeds	Kenya	AFB1	703/830	0.9-595	8.9-46.0	Kang'ethe and Lang'a (2009)
Animal feeds	Sudan	AF	36/56	4.1-579.9	130.6	Elzupir *et al.* (2009)
Animal feeds[b]	South Africa	AF	99/108	3.2-950	112.54	Mwanza *et al.* (data unpublished)
Bush mango seeds	Nigeria	AFB1	20/20	0.2-4.0	1.5	Adebayo-Tayo *et al.* (2006)
		AFG1	20/20	0.3-4.2	1.5	
Cottonseed	Nigeria	AFB1	3/8	0.0-271		Gbodi (1986)
		AFB2	3/8	0.0-36.6		
		AFG1	2/8	0.0-183		
		AFG2	1/8	0.0-9.1		
Cottonseed meal[a]	South Africa	AFB1	60/60	13.4-75.7	24.90	Reiter *et al.* (2011)
		AFB2		1.0-5.4	1.84	
		AFG1		6.6-64.3	17.91	
		AFG2		0.3-3.3	0.90	
Pistachio	Morocco	AFB1	9/20	0.04-1430	158±6.3	Juan *et al.* (2008)
Pistachio	Tunisia	AF	21/40	0.24-122.4	21.8±38.0	Ghali *et al.* (2009)
Poultry feeds	Morocco	AFB1	14/21	0.05-5.38	1.26± 0.65	Zinedine *et al.* (2007)
		AFB2		0.03-0.58	0.18±0.18	
		AFG1		ND	ND	
		AFG2		ND	ND	
Poultry/livestock feeds	Nigeria	AFB1	6/13	0.0-67.9	15.5	Adebayo and Etta (2010)

[a]Cottonseed cake was imported from Benin. [b]Animal feeds from rural community of Limpopo province. ND: Not detected

Table 2. Some reports of aflatoxin levels in some animal feeds and fruits from Africa.

3.3 Vegetables

Reports on fungal and mycotoxin contaminations of vegetables are exceptionally uncommon. One of the most recent and novel works in this area is that of Hell *et al.* (2009) who studied 180 dried vegetable samples across three countries in West Africa including Benin, Mali and Togo. AF concentrations were determined in dried okra, hot chilli, tomatoes, melon seed, onion and baobab leaves. The results as seen in Table 1 is one of the very first few reports on AF in dried okra and hot chilli after that of Obidoa and Gugnani (1992). AFs were not detected in the other dried commodities but however, the toxins have been detected in fresh vegetables elsewhere. For instance, Muhammad *et al.* (2004) found the toxins present in fresh tomatoes marketed in Sokoto, Nigeria. Similarly, Sahar *et al.* (2009) also showed the presence of AF in fresh tomatoes, pumpkin, powder chillies and coriander (dry) grown in Pakistan. Other fruits and vegetables in which the toxins were found in that study include cucumber, persimmon, peanut and peach. Interestingly, Obidoa and Gugnani (1992) had earlier found AF in dried okra, onions, dry pepper and table foods sold at restaurants. The ready-to-eat dishes mainly "gari" (cassava products) and beans served with vegetable soups had AFB1 at levels ranging between 8 and 61 µg/kg with total AF values ranging from 31.21 to 268µg/kg. The data suggest that AFs are common contaminants of most African table foods. Another vegetable that abhor aflatoxigenic fungi and consequently contain AF is oyster mushroom (Jonathan and Esho, 2010). The severe paucity of information on mycotoxins in vegetables necessitates increased research in the area.

3.4 Fruits

In a review of mycotoxins in fruits and fruit-processed products, Fernandez-Cruz *et al.* (2010) revealed that the commonest mycotoxin contaminants of fruits worldwide are patulin (PAT), OTA, AF and *Alternaria* toxins, and that natural AF contamination has been reported in oranges, apple and apple juices, dried apricots, dates, prunes, musts, dried figs and raisins. According to the review, AFs are most frequently reported in dried figs and raisins worldwide at significant levels of up to 550 and 63µg/kg, respectively. Furthermore, their occurrence was reported in apple juice from Egypt and dried raisins and figs from Morocco. Other reviewed literature of interest in this area is that provided by Trucksess and Scott (2008) and Barkai-Golan and Paster (2008). Reports on mycotoxins in fruits are not many in Africa with one reported in pawpaw and pineapple being contaminated by *Aspergillus flavus* in Maiduguri, Nigeria (Akinmusire, 2011). Baiyewu *et al.* (2007) showed the presence of the toxins in pawpaw from South Western Nigeria. Some data on AF incidence in Africa (Table 2) presented low occurrence of the toxins in African fruits. Juan *et al.* (2008) found AFB_1 at levels exceeding the EU limit of 2µg/kg in 5 and 20% of 20 samples each of pistachio and dried raisin, respectively, with the highest value of 1430 µg/kg detected in pistachio. While all dried fig samples were found fit for the EU market with regards to AFB_1 contents, 15% of them had AF levels above the 4 µg/kg, a maximum recommended limit of EU. The use of traditional processing and preservation methods for fruits in rural Morocco that provide optimal conditions for mould growth and mycotoxin formation was adduced as the cause for the increased prevalence of these AF in fruits. AF analysis in Tunisian pistachio performed by Ghali *et al.* (2009) found unacceptable (according to EU standards) concentrations of AF in 17% of the samples. Bush mango (*Irvingia* spp) is a sweet tasting fruit. The pulverized form of the dried seeds is used as condiment and thickeners for soups and stew in West Africa. Investigation into its AF content in Nigeria by Adebayo-Tayo *et al.* (2006) showed a 35% non compliance with the EU AFB_1 standard for fruits. There is need for AF surveys in commonly eaten fruits in Africa namely, oranges, banana, plantain, guava, dates, etc. in order to properly ascertain the extent to which fruits expose the African population to the toxins.

3.5 Roots and tubers

Roots and tubers are a major source of nutrition in the world after cereals. They are basic diets for about a billion people in the developing countries, providing 10% of world's food energy and protein intake (Shewry, 2007). They account for 40% of food eaten by half the population of SSA, contributing 20% of total energy intake in the continent (FAOSTAT, 2010). Cassava, potatoes, yam and taro form the bulk of roots and tubers consumed worldwide. Cassava and yam are not vulnerable to AF contamination (Bankole *et al.* 2006) and even the processed products such as cassava and yam chips and their flour have low contamination rates. Analysis of cassava and yam chips from Benin showed no contamination by AF (Gnonlonfin *et al.* 2008) neither was the toxins found in cassava products from Tanzania (Muzanila *et al.* (2000), Nigeria (Jimoh and Kolapo, 2008) and Ivory Coast (Kastner *et al.* 2010). However, data obtained from Cameroon (Table 1) show low levels of the toxins in stored cassava chips at levels greatly dependent on processing practices, storage facilities and duration of storage (Essono *et al.* 2009). Higher prevalence and contents of AF are observed in yam based products than in cassava products. Apart from the 100% frequency (Table 1) shown in Yam chips from Benin (Bassa *et al.* 2009), 23% of the 107 samples analyzed by Mestres *et al.* (2004) from the same country had AF levels over the 15 μg/kg CODEX standard value for total AF. AFB_1 and G_1 were detected in yam chips from Nigeria at levels ranging from 5-27μg/kg (Jimoh and Kolapo, 2008). AFB_1 was also found in 22% of samples from Nigeria (Bankole and Mabekoje, 2003) and a larger survey conducted later in same country found the toxin at a prevalence rate of 54.2% in the same products at toxicologically significant levels (range: 4–186 μg/kg; mean: 23 μg/kg), 32.3% with AFB_2 (range: 2-55 μg/kg), 5.2% were positive for AFG_1 (range: 4-18 μg/kg), while 2 samples contained AFG_2) (Bankole and Adebanjo, 2003). The present unwholesome practice of storing and marketing high moisture cassava and yam products is responsible for the contamination in Africa (Essono *et al.* 2009) and should be discouraged.

3.6 Animal products

3.6.1 Milk and milk products

Animals fed AFB_1 and B_2 contaminated feeds excrete into their milk the less toxic AFM_1 and M_2, respectively. AFM_1 is of particular interest being the hydroxylated metabolite of AFB_1 and is known to have 2-10% of the carcinogenic potency of the parent compound (Zinedine *et al.* 2007). The carryover of this carcinogen in cow at a transfer ratio (consumed AFB_1 to excreted AFM_1) of 200:1 (Smith and Moss, 1985) which could be as high as 40:0.05 (JECFA, 2001) into human and animal milk that are the main sources of nutrition for infants whose vulnerability due to undeveloped immune system is obvious, poses serious health concern. Its stability to heat, cold storage, freezing and drying (Yousef and Marth, 1985) during processing makes dairy products another important source of AFM_1 exposure. Milk and milk products are traditionally staple food commodities for the nomadic population of Africa. They are recognized by the elites as natural balanced diet and so are increasingly consumed by the urban populace in the continent. Therefore, they can no longer be ignored as they are among the main entry routes of AFM_1 into the human dietary system in Africa.

The natural occurrence of AFM_1 in raw cow milk has been reported in quite a number of African countries as reflected in Table 1. A regional variation has been demonstrated in the data with lower concentrations occurring in the drier North Africa (Egypt and Morocco) than in the more humid SSA (Nigeria, Sudan, Kenya, Cameroon and South Africa). Many of the milk samples from the region (7.4, 18.7, 40, 52.6, 61.9 and 83.3 % of samples from Morocco, Kenya, Cameroon, Egypt, South Africa and Sudan, respectively, had AFM_1

contents above the legislated levels (0.05µg/L) of several countries including those in the EU. Similarly, their recorded mean levels according to JECFA (2001) are higher than those reported for European (0.023 µg/L), Latin America (0.022 µg/L), Far Eastern (0.36 µg/L), Middle Eastern (0.005 µg/L) and African diets (0.0018 µg/L). Since AFM_1 content in milk is a good indicator of AFB_1 contamination in feeds, hence at a transfer rate of 200:1 for cows it can be estimated that diary animals in Africa are exposed to between 3.6 - 414 µg/kg of AFB_1 in their rations since the observed range of AFM_1 averages (Table 1) for Africa is 0.018 to 2.07 µg/L. Such high levels of AF in feeds in the continent have been shown in sections 3.1 and 3.2. The presence of AFM_1 in milk from breastfeeding mothers (Atanda *et al.* 2007) and other body fluids of diseased patients (Tchana *et al.* 2010) sometimes at above regulated levels, is confirmatory that humans are exposed to high AF levels in Africa and because milk is primary to infant nutrition it gives cause for considerable concern.

Raw milk is usually processed into dehydrated dairy products such as cream, butter, cheese and milk powder in order to extend its shelf-life. Yoghurt, another product of milk which has become part of human dietary system has a similar processing method as cheese making, only that the process is arrested before the curd forms. The separation of the components of milk during processing leads to distribution of AFM_1 into dairy products with consequent lower levels of the toxins in the individual products than in milk. In spite of this anticipated loss, AFM_1 has been found in African dairy products at unacceptable levels. High prevalence (75%) of AFM_1 with all positive samples having levels exceeding the legislated limit of 0.05 µg/L was seen in Libyan cheese (Elgerbi *et al.* (2004). Amer and Ibrahim (2010) found AFM_1 in 50/150 Egyptian cheese samples at levels between 0.051 to 0.182 µg/L and all the contaminated samples had levels beyond the EU regulated limit. Similarly, the toxin contaminates cheese, ice cream and yoghurt at intolerable amounts in Nigeria (Atanda *et al.* 2007). In order to determine how safe imported powdered milk samples are for human consumption, Makun *et al.* (2010) analyzed AFM_1 in 100 samples sold in the Lagos metropolis. As seen, 6 of the 19 positive samples had levels above the EU limit, however, all the tested samples were considered safe in Nigeria as levels of the toxin were below the country's maximum tolerable limit of 1 µg/kg. Clearly, the main strategy to reduce incidence of milk toxins is by feeding animals with AF free feeds.

3.6.2 Animal tissues

Relative to such mycotoxins as OTA, a much smaller proportion of AF is absorbed into animal tissues. The transfer ratios into tissues for AFB_1 can range between 1,000 and 14,000 (Smith and Moss, 1985). Any animal exposed to such high amount of mycotoxins would have shown some toxicity signs or even death and would likely not enter the human food chain in the developed world. Conversely, diseased animals are still eaten in Africa and as such, human exposure to AF via consumption of animal tissues can be a reality in the continent. There is a disparity between the AF content of beef and those of other edible organs with lower values obtained for the muscle tissue. This was shown by Oyero and Oyefolu (2010) when analysing fresh and sun-dried beef and edible organs (liver, kidney and heart). While, the dried animal tissues had lower levels of AFB (beef: 2.9, liver: 3.1, heart: 27.9, kidney: 75.8 ng/kg) and AFG (beef: 4.4, liver: 3.1, heart: 55, kidney: 141.3 ng/kg) than for fresh samples whereby AFB levels were for beef liver, heart, kidney were 21.7, 33.9, 55.9 and 85.2 ng/kg, respectively, and AFG levels in these organs recorded were respectively, 27, 41.4, 74.1 and 70.7 ng/kg. Accordingly, the levels of the toxins in the edible organs were consistently higher than those in beef with kidney being the most vulnerable. It is therefore, suggested that withdrawal of animals from contaminated feeds onto mycotoxin free diets for 3-4 weeks could have allowed for sufficient withholding period to clear the muscles and organs from the toxins. Despite the low carryover rates of the toxins into animal tissues, AF contaminate fresh and processed meat (especially liver and kidney) at toxicologically significant levels of up to 325µg/kg in Egypt (Aziz and Youssef, 1991; Abdelhamid, 2008). It is only proper to indicate further herein that consumption of animal visceral organs (*kayan chiki*) may constitute a major source of AF exposure than muscle.

The garnishing of beef with peanut paste to produce dried beef product (*kilishi*) and roasted beef (*Suya*) is a common practice in West Africa that elevates the AF content of these processed meat products to over 194µg/kg (Jones *et al.* 2001; Chukwu and Imodiboh, 2009) which is beyond any known accepted maximum. Smoke-drying is the commonest method of preservation of fish, another major source of protein in the African continent. The inadequacy of the method with regards to preservation from contamination of smoke-dried fish by aflatoxigenic fungi and AF (Table 1) has been demonstrated (Adebayo-Tayo *et al.* 2009).

The carry-over of AF from feed to poultry by-products including meat and eggs has been investigated and found to be quite low varying with the product. In a feeding trial, Hussain *et al.* (2010) demonstrated AF transfer ratios of 1: 914 and 1: 1939 for the liver and muscle of broilers respectively. In hen, the transfer ratios are 1: 1103 for edible organs (gizzard, kidneys and liver) and 1: 33,100 for breast muscle (Wolzak *et al.* 1986). The carryover into eggs occurs at obviously high ratio of 1:6633 (Wolzak *et al.* 1985). The corresponding ratio of AFB_1 in feeds to residual levels in egg yolk and albumen were shown to be 1:4615 and 1:3846, respectively, in chicken hen (Bintvihok *et al.* 2002). In light of these high ratios, the presence of AF in eggs from Egypt (Abdallah *et al.*, undated) and Cameroon (Tchana *et al.* 2010) at levels of up to 7.68 µg/kg (Table 1) indicates that poultry animals consume AF in feeds at alarming concentrations. Although the reported AF levels in both meat and meat products, and eggs from Africa seem insignificant, with chronic intake of such amount simultaneously occurring with other food borne toxicants can have deleterious health impact (Speijers and Speijers, 2004).

3.7 Fermented products

Although fermentation reduces mycotoxins in contaminated food products (Hell and Mutegi, 2011), there is ample evidence to suggest that fermented products in Africa contain significant levels of AF. Kpodo *et al.* (1996) detected AF at levels as high as 289µg/kg in fermented maize dough in Ghana. The presence of AF was observed in all samples of fermented yams and plantains analyzed from the Southern region of Nigeria (Jonathan *et al.* 2011) at levels ranging between 37.67 - 96.34 µg/kg. Detectable (5.2 - 14.5 µg/kg) amounts of the toxins were also found in fermented cassava products from Cameroon (Essono *et al.* 2009). Sorghum based traditional opaque beer from Malawi contained AF at levels above the CODEX permissible limit of 10 µg/kg (Table 1) (Matumba *et al.* 2011). In their review on mycotoxin problem in Africa, Wagacha and Muthomi (2008) reported incriminating levels of AF (200, 000 - 400, 000 µg/l) in 33% of traditionally brewed beer in South Africa. Levels of up to 50µg/kg were found in sorghum based local beer from Lesotho (Sibanda *et al.* 1997).

With these unsafe levels in our fermented products, it will only be proper to adhere to the advice of Pietri *et al.* (2010) that if raw materials comply with the legislated limits, contribution of a moderate daily consumption of beer to AFB_1 intake will not contribute significantly to exposure of the consumer.

3.8 Other foods

Plants and plant products used as medicinal herbs, tea and spices may be commonly contaminated by AF at significant levels of up 2,230µg/kg especially in the case of liver curative herbal medicine sold in India (Trucksess and Scott, 2007). According to these authors, contamination of the toxins has been observed in ginger, garlic and capsicum. A survey conducted for aflatoxigenic fungi and AF in spices from Egypt (Aziz and Youssef, 1991) found black and white pepper contaminated with AF at unhealthy levels (range: 22-35 µg/kg). Zinedine *et al.* (2006) found natural presence of AFB_1 in black pepper, ginger, red paprika and cumin from Morocco at average levels of 0.09, 0.63, 2.88 and 0.03µg/kg, respectively, with the highest level of contamination found in red paprika (9.68µg/kg). Although moulds are frequently isolated from African herbal plants (Bankole and Adebanjo, 2003), it seems the herbs are not prone to AF contamination (Katerere *et al.* 2008). Equally, a search on the literature failed to provide data on the incidence of AF in African tea.

An attempt is made in this chapter to provide an extensive review on AF contamination of commodities in Africa and further information on the subject is referred to the review of Sibanda *et al.* (1997), Shephard (2003), Bankole and Adebanjo (2003) Bankole *et al.* (2006) and Wagacha and Muthomi (2008).

4. Human AF exposure

An important part in elucidating the health effects of AF in humans is the estimation of exposure to the toxin that they receive. A tedious and rather inaccurate estimate of exposure level can be established by determining mean toxin level in the food people eat/amount eaten, either in the basic commodity, e.g. maize kernels and rice as in the case for Nigeria, or in its ready-to-eat form, e.g. porridge for those leaving in the rural area of Limpopo in South Africa. In order to approximate the level of exposure of Africans to AF, let's assume a Nigerian that eats 138kg of cereals annually (Bandyopadhyay *et al.* 2007) as a representative of Africa and whose main staples are maize and rice consumed in a ratio of about 3:1. From Table 1, the AFB_1 mean values for indigenous maize and rice destined for human consumption in the continent are approximately 258 and 83µg/kg, respectively. It can therefore be estimated that 29,500 µg/kg of AF is consumed yearly or let's say 81µg/kg daily by the subject from the two cereals. Even though food processing such as sorting, cooking and others factors will reduce the content (Hell and Mutegi, 2011), what might be left will still be incriminating causing certain chronic intoxications which are discussed subsequently in this chapter. Again, if 350 µg/kg is the mean level of AF in foods that can elicit acute symptoms (Azziz-Baumgartner *et al.*, 2005), it can rationally be deduced that Africans are exposed to sub acute doses of AF in maize and rice amongst other food crops. While, the inferences made herein are generalized for Africa with regards to AF intake, it is not without flaws and exceptions, as it can also be reasonably inferred that Africans are exposed to the toxins from virtually all the foods they consume as seen in Table 1. This corroborates with high incidence of the toxin and its biomarkers in their body tissues and fluids (Gong *et al.* 2003, Tchana *et al.* 2010). This is also in agreement with the fact that in many regions of the continent, the estimate on the frequency of human exposure to AF is about 98% (Wild, 1996).

More precise estimates of AF exposure among humans are best achieved by estimating the toxins and associated metabolites in bile, urine, faeces and hair as well as their distribution in blood, milk, liver, kidney and semen (Njobeh *et al.* 2010). Probably, the easiest one to use is the urine, which can conveniently be sampled and if a 24 hour sample is taken, it can give a fairly accurate content of AFB_1 ingested over a period from the measurement of levels of AFM_1 (Nyathi *et al.* 1987) or adducts (Groopman *et al.* 1992). Another useful method is to assay the amount in the blood and that bound to blood albumin. The latter can be regarded as a good biomarker for AFB_1 exposure (Wild *et al.* 1990) and in effect, removes AFB_1 from circulation. If one assumes that the half-life of albumin is about 20 days (Wild *et al.* 1990), then correlating the amount of conjugate with intake, an estimate of exposure that took place the previous week can be established. A similar case may be made for liver AFB_1-DNA adducts and this has been demonstrated in animals and by extrapolation, a relationship between the level of AFB_1 albumin and AFB_1-DNA conjugates is established (Wild *et al.* 1996).

5. Human health implications to AF exposure with reference to Africa

It is seen how there is extremely high degree of human exposure to AF in many parts of the African continent as previously stated in Section 4.0 of this chapter. Such level of exposure has enormous health and socioeconomic implications. Aspects of the effects of AF contamination of food and feed commodities in terms of the economy will also be highlighted in this review. In general, a disease caused by a mycotoxin is termed "mycotoxicosis" (aflatoxicosis in the case of AF). Aflatoxicosis seems a commoner in developing countries (Williams *et al.* 2004). Furthermore, for mycotoxins to exhibit disease symptoms of a chronic nature, they occur in feed or food at one part per billion (ppb)(μg/kg), the exception being FB_1 which is present at three parts per million (ppm) (mg/kg) mark before symptoms in farm animals begin to be manifest (Bucci and Howard 1996). The difficulty when considering levels of mycotoxins for intoxication in human is that, except for "natural" cases of mycotoxicoses, disease causing levels of mycotoxins cannot be derived and hence, extrapolations from animal experiments have to be made. From known cases of human aflatoxicosis (Azziz-Baumgartner *et al.* 2005), it would seem that acute symptoms are found when levels in food ingested were at a mean value of 350ppb. Again the difficulty with such estimates, however, is gauging the exact amount of the toxin ingested per kg body weight, because this will depend not only on how much is ingested at a specific time, but for how long. Acute doses of mycotoxins may be ingested at one time or over a short period, whilst chronic levels may be consumed over long periods. An example in point of the latter is AFB_1 in groundnut, a staple where rural people ingest this mycotoxin most of their life time and as a consequence, this may increase the incidence of liver cancer (hepatocarcinoma) proportionately (Peers and Linsell, 1973) in the region. A similar situation is found where FB_1 from maize as a staple is ingested but the consequence of a lifetime of exposure to this toxin is not as clearly defined as that of AFB_1, because FB_1 has a much lower toxicity but there are indications that it has a role in various diseases conditions (Dutton 2009). This is discussed subsequently in detail.

5.1 Toxicology of AFB_1

AFB_1 created great interest amongst the medical profession once it was established as a powerful carcinogen (Wogan, 1973). In fact it is claimed to be the most powerful naturally occurring carcinogen known. The toxin is also mutagenic and teratogenic (Raisuddin *et al.* 1993). Because of these toxic properties, several investigations have been carried out over the years since its discovery both *in vitro* and *in vivo* to elucidate its mode of action as a carcinogen, perhaps to the detriment of studying other toxic properties it exhibits as well as those of its congeners AFB_2, AFG_1 and AFG_2. To justify this obsession with AFB_1, it is fair to point out that it is the most commonly occurring and at the highest levels as well as being the most potent when compared with the other AF. However, it might be worthwhile investigating the toxic properties of AFB_2 more closely, as this has all the molecular attributes of AFB_1 apart from the bishydro-furano double bond that confers the carcinogenic property. This double bond makes all the difference, as it allows detoxifying systems (cytochrome P_{450}) in the body to convert it to an epoxide by cytochrome P_{450}s, specifically CYP3A4, CYP3A5, and/or CYP1A2 (Gallagher *et al.*, 1994; Wang *et al.*, 1999) to the exo-8, 9-epoxide, which ironically is the "activated" form of the molecule that can form adducts with deoxyribonucleic acid (DNA) leading to guanine nucleotide substitutions (Lilleberg *et al.*, 1992). In passing it is also important to note that cytochrome P450 types also convert AFB_1 to other derivatives e.g., AFM_1, AFP_1 and Q_1 (Campbell and Hayes 1976). The epoxide is a good alkylating agent and can react with bases such as those in DNA and RNA to form the AF- alkylated form. Obviously other factors are involved, as not all alkylating agents are carcinogens and further, AFB_1 has a tendency to specifically attack guanine, one of the four DNA bases (Taylor 1992). Because AFB_1 is somewhat non-polar, it passes though membranes and other lipid barriers easily and also has a slight water solubility so it passes from the aqueous phase at low concentrations and accumulates in fat soluble phases such as adipose tissues. Furthermore, the molecule itself, because it is primarily aromatic in nature, the main core is a coumarin structure, which is rather flat and therefore, can intercalate into DNA (Jones *et al.* 1998). The epoxide of course can react with other nucleophiles, including those acting as part of the detoxification system, e.g., glutathione (Gopalan *et al.* 1992) and proteins such as blood albumin.

Because of the interaction of AFB_1 with DNA, it is reasonable to ascribe it to be its main toxicological action, certainly at low chronic level of exposure. AFB_1 can form AFB_1-DNA adducts, DNA strand breaks, DNA base damage and oxidative damage that can lead to cancer (Wang and Groopman, 1999). This damage can be repaired by various mechanisms, e.g. base excision repair (Wood, 1999). However, certain mutations that occur due to AFB_1's action may interfere in these repair mechanisms, in particular, the xeroderma pigmentosum complementation gene group D (XPD) which is one of the groups encoding for groups in the nucleotide excision repair pathways. Recent findings suggest that two loci in this group are of particular importance in modulating the AFB_1 related development of liver cancer (Long *et al.* 2009). Probably of more direct importance is the action of AFB_1 on the p53 gene where it causes an AGG to AGT transverse mutation at codon 249 (Bressac *et al.* 1991). Gene p53 is responsible for producing the p53 protein which has an important role in the regulation of cell cycle and in suppressing genome mutation (May E. and May P., 1999).

5.2 Disease conditions in Africa linked to AF exposure

When considering mycotoxins and their effects in Africa there is a tendency to concentrate on those countries in SSA. The assumption being that those countries constituting the Sahara desert and border areas of the Mediterranean Sea are either more like European countries in terms of commercial food production and consumption; or the countries are so dry and unsuited to mass agrarian exploitation that they do not have the same problems of human mycotoxicoses as the rest within the continent. This is an erroneous supposition, as all foodstuffs at some stage may be contaminated with fungi and mycotoxins in these countries (Mokhles *et al,.* 2007, Ghali *et al.,* 2010) but it is fair to say that in SSA, apart from the major cities, the African population is rural, mainly relying on subsistence agriculture. Because this type of activity is unregulated and often is insufficient to maintain a proper nutritional supply to these populations, it is not unreasonable to suppose that they are exposed routinely to mycotoxins and often have little natural resilience to their effects. Consequently, several diseases, which can be correlated to exposure to mycotoxins or can be in part attributed to them, are known. As AF is ubiquitous in African commodities (Nyathi *et al.* 1989), it is not surprising that several of such conditions are caused or exacerbated by these toxins. However, as indicated above, the main problem is AFB_1 and these diseases will be discussed in terms of this toxin.

5.2.1 Acute toxicity

Although most naturally occurring cases of acute toxicity caused by AFB_1 are observed mainly in animals such as one that lead to the discovery of AF in turkey poults (Blount, 1961), there are several cases of human acute aflatoxicoses (most of which often go unnoticed). Since 1982, deaths caused by AF-contaminated maize have repeatedly occurred in the Eastern Province of Kenya (Probst *et al.* 2007). Similar cases have also previously been recorded at various times in India (Krishnamachari *et al.* 1975) and Malaysia (Lye *et al.* 1995). A recent case reported in Kenya is as a result of consuming contaminated maize (Nyikal *et al.* 2004) with 125 deaths out of 317 cases (Azziz-Baumgartner *et al.,* 2005) being recorded. The symptoms are anorexia, malaise and low grade fever leading to acute jaundice and lethal hepatitis. This outbreak was associated with high levels of AF in maize for human consumption (mean concentration of 355ppb) that led to increased levels of AF B_1 albumin adduct and higher hepatitis B titres in the patients than controls (Azziz-Baumgartner *et al.,* 2005). There seems to be little in the way of treatments for these cases, apart from straight forward strategies of antimicrobials and support for the damaged organs (Mwanza *et al.,* 2005) as the toxin acts rapidly and can be lethal.

5.2.2 Conditions of the liver

As the liver is the main organ of detoxification and the first major organ to be exposed to dietary intake of xenobiotics, it is not surprising to find that several liver conditions have been associated with AF, particularly in under developed countries, especially in Africa. The fact that many African crops, in particular, the staples such as ground nut and maize, can be routinely contaminated with AF, leads to an intuitive feeling that many diseases, including those of the liver, can be linked directly or indirectly, to AF, especially AFB_1. As already intimated, this is a dangerous supposition, as many chronic diseases are multi-factorial in nature, and as in any scientific hypothesis, needs strong evidence of support. This is not easily gathered in the rural areas of Africa where infrastructure is poor, health services varying from none to basic clinics and health centres with little time or inclination to gather usefully directed statistics. Even the analysis and quality control of staples is dependent upon external scientific studies, which often are one of investigations that merely provide a snapshot of a true situation. Consequently, the information available to investigators attempting to correlate disease conditions to mycotoxin exposure is sparse and patchy with respect to various African countries. In addition, the practise of comparing these conditions in Africa to those in developed countries may highlight interesting differences that can give clues to the aetiology of African disease but may also lead to western scientists dismissing explanations of such chronic diseases in Africa, because of well researched conditions in their own countries, which have explanations other than dietary ones. Clearly, people living in African rural areas have an environment that is completely different to those living in western cities and hence, elevated incidence of certain medical conditions may be explained by factors other than those appertaining let's say for example, in Europe or North America. In any case, there are readily available systems in developing countries particularly in Africa that could provide natural environments for human experimentations as it is in theory, possible to compare the chronic diseases in rural populations against those of urbanized populations of the same race group (Njobeh *et al.*, 2010) elsewhere.

5.2.2.1 Hepatitis

Hendrickse (1991) lists five possible roles for AFB_1 in human disease including fatal AFB_1 poisonings that can "masquerade as hepatitis". This condition is an inflammation of the liver cells caused by various agents including viruses e.g., hepatitis B virus, (HBV) and may be self healing or in extended chronic cases, lead to cirrhosis. Several cases of hepatitis have been reported in the literature including one in India that was attended by high mortality (Krishnamachari *et al.* 1975). The outbreak was associated with maize contaminated by AF and it was concluded that it was as a result of aflatoxicosis.

5.2.2.2 Cirrhosis

Cirrhosis of the liver is a well known disease condition usually related to alcohol consumption. However, it may be caused by exposure of the organ to toxic principles other than ethanol and can often be found in children because of their higher susceptibility to toxin exposure. In certain countries in Africa (e.g. Ethiopia) (Tsega, 1977) and India (Yadgiri *et al.*, 1970), the condition has been described and efforts made to correlate it with AF intake (Tandon *et al.*, 1978). The whole issue is somewhat clouded because of the commoner occurrence of hepato-cellular carcinoma (HCC) together with cirrhosis, in areas of high AF exposure and hence, a relationship has been suggested by the two conditions; simply this would be that the onset of childhood and other cirrhoses leading to liver cancer (Lata, 2010). There is no real evidence, however, that there is a link (Kew and Popper 1984) and further, that the roles of HBV, hepatitis C (HVC) or AFB_1 is by no means clear. In some cases of cirrhosis, there is an over expression of the p53 gene and this has been related to simple cellular stress or mutation of the gene, which occurs in many of the subjects exposed to HBV and/or AFB_1 (Livni *et al.*, 1995). More recent work on cirrhosis in The Gambia (Kuniholm *et al.* 2008) tentatively concludes that health effects due to exposure to AFB_1 could include this disease. It would seem that, because of various factors that African populations face, including alcohol abuse, it is difficult to tease out any particular one and ascribe a role to it, other than it does make a contribution to the overall condition.

5.2.2.3 Hepatocellular Carcinoma (HCC)

This chronic disease is a major global health problem, causing over 600,000 cases per annum (Ferenci *et al.* 2010) and accounts for over 70% of all liver carcinomas (Lata, 2010). In Africa, the most likely cause or promoter of such a cancer is AFB_1 (it has been estimated that AFB_1 may play a causative role in 4.6-28.2% of all global HCC cases) (Liu and Wu, 2010), which is a primary carcinoma in the case of areas where factors such as hepatitis and AF are found. In other situations, the cancer may be of secondary type, i.e., an infiltration of a metastasised type from other parts of the body. Thus the situation is, not as clear cut as might have been thought. Areas in Africa where groundnut is consumed regularly, which is one of the principal sources of the mycotoxin, such as Sudan (Omer *et al.* 1998); Mozambique and Kenya, have high levels of liver cancer which have been shown to be correlated to AFB_1 levels (van Rensburg *et al.*, 1985). These areas also have a high incidence of viral diseases that can also affect the liver, e.g., the highly contagious HBV plus HIV infection (Kew, 2010a); HCV (Ashfaq *et al.* 2011); and iron over load (Kew and Asare, 2007) which tends to cause the issue. Hence it has been argued that these viruses are the cause of the high liver cancer incidence with other factors such as alcoholism and mycotoxins being of secondary importance (Stoloff, 1969; Perz *et al.*, 2006). Given the multi-factorial nature of cancer, these claims are dangerous ones to make and also apply to the school of thought that says HCC is mainly due to AFB_1. Nevertheless, hepatitis viruses do seem to play a major role in the development of HCC in African countries (Kirk *et al.*, 2004; Ocama *et al.*, 2009). It is, however, known that the risk posed by AFB_1 is independent of that of HBV (Blondski *et al.*2010) although where both factors occur together, the risk is increased. It has been suggested, for example that cirrhosis of the liver, as previously discussed, may be a precursor to liver carcinoma (Kew, 2010a; Lata, 2010). It seems, therefore that HCC is multi-factorial, although some factors may contribute more than others, depending upon circumstances, e.g., in Africa where AFB_1 is commonly found in food, it may play a bigger role here than elsewhere, where it is not found to any extent. Further, in simple terms one thinks of a cancer initiator and a cancer promoter. It may well be that the highest liver cancer areas have a combination of both cancer forming factors plus others such as genetic ones. This certainly is the case for AFB_1 as it has been seen that AFB_1 needs to be activated by the cytochrome P_{450} system in particular the CYP 3A4 version plus CYP3A5 and 3A7 (Kamdem *et al.*, 2006). Hence those persons with genetics dictating less expression of these forms of cytochrome P_{450} would, in theory, have less chance of developing the cancer and vice versa. As earlier mentioned, AFB_1 may cause a missense mutation, where a guanine residue is converted to a thymine in the 249 codon of the p53 gene and this is considered to be an important marker in the promotion of HCC by AFB_1 (Bressac *et al.*, 1991). In a study that was conducted on Chinese subjects, this mutation was detected in the blood and liver of HCC patients (Jackson *et al.*, 2001). However, not all the patients had this mutation in their livers, indicating yet again a multi-factorial situation, although some of the liver negative subjects did have the mutation in their plasma samples. It was concluded that the detection of the codon 247 mutation may present a method of an early diagnosis of HCC.

5.2.3 Kwashiorkor

Kwashiorkor is a malnutrition condition which is in essence a protein deficiency of the young like marasmus but unlike it, the patient has sufficient calories. It is characterized by oedema, anorexia, dermatitis and an enlarged liver with fatty infiltrates (Bhattacharyya,

1986). Because of the liver's involvement, it was suggested that AFB_1 may be involved (Hendrickse *et al.*, 1983). This appeared to be supported by other observation, e.g., relating peak prevalence for kwashiorkor with climatic conditions such as high humidity (de Vries and Hendrickse, 1988); general occurrence of AFB_1 in children's excreta (de Vries *et al.* 1987, 1990); liver (Lamplugh and Hendrickes, 1982); serum (Coulter *et al.*, 1986; Hatem *et al.*, 2005); and its reduced product, aflatoxicol, in livers of Ghanaian children (Apeagyei *et al.*, 1986).

Other work, however, does not support the hypothesis that AF have a role in Kwashiorkor (Househam and Hundt, 1991). What AFB_1 exposure seems to influence in kwashiorkor patients is that their recovery time in hospital is lengthened and this is linked to a difference in the way AFB_1 is metabolised as compared to control non-kwashiorkor infants (Ramjee *et al.*, 1992). It could, therefore, be argued that kwashiorkor is protein deficiency in children who are receiving sufficient carbohydrate such as diet on maize gruel given to weaned children. Maize is not a balanced source of protein but has plenty of starch and hence its use as a staple would fit this hypothesis. Oedema would arise due to lack of sufficient blood proteins, i.e., hypoalbuminemia (Waterlow, 1984). This view has been questioned as it has been observed that the oedema symptoms can disappear before the blood albumin levels return to normal (Golden *et al.*, 1980). An alternative mechanism suggested was that the condition was caused by oxidative species not being regulated by anti-oxidants, e.g., vitamin E, which were depleted because of the poor diet (Golden and Ramdath, 1987). In order to test this, children in Malawi were treated with antioxidants such as riboflavin, vitamin E and selenium (Ciliberto *et al.*, 2005) but both the treated and control group given placebos showed similar levels as kwashiorkor as evidenced by oedema. More recent work seems to suggest that the incidence of kwashiorkor in Africa may be dropping, as evidenced by a Nigerian study (Oyelami and Ogunlesi, 2007). This was put down to better management of diarrhoeal diseases. An interesting point that was made in an earlier publication (Enwonwu, 1984) is that protein deficiency affects the cytochrome P_{450} system, which would allow for the accumulation of AFB_1 in kwashiorkor children due to lack of conversion products. Because of the lack of the AFB_1 epoxide production, its carcinogenic property would be lost, which, ironically, would be re-instituted if protein supplementation was given. It would seem that the underlying causes of kwashiorkor still remains a mystery, is multi-factorial (Oyelami *et al.*, 1995) and unrelated to the lack of any particular food or nutrient (Lin *et al.*, 2007) although potassium supplement appeared to help (Manary and Brewster, 1997). However, it is still tempting to evoke some toxic factor in the diet of children that interacts with other factors to produce the condition. Kwashiorkor has been referred to at a certain time as a "maize disease" (Adhikari *et al.*, 1994). This label has been used in other circumstances with reference to *Fusarium* mycotoxins, in particular the FB, i.e., FB_1 (Dutton, 2009). It is not impossible that these mycotoxins rather than AFB_1 have some role in the development of kwashiorkor.

5.2.4 Miscellaneous

5.2.4.1 Reye's syndrome

This can be a fatal disease that affects mainly children and is typified by fatty liver, swelling of the brain (encephalopathy) and hypoglycemia. The liver becomes enlarged and firm but with no jaundice. There is also evidence of kidney damage. It is not particularly associated with Africa and indeed many cases were reported in first world countries (Harwig *et al.*,

1975). When first described (Reye *et al.*, 1963), the cause was unknown but it often occurred after a mild respiratory tract infection and, because of involvement of the liver, as in kwashiorkor and similar cases in Thailand, it was suggested that AFB_1 may be involved (Becroft, 1966; Olson *et al.*, 1971). This was supported by further studies where materials resembling AF were isolated from the liver of a girl of 8 months of age (Becroft and Webster, 1972) and AFB_1 in the livers of 5 children (Stora *et al.* 1983) but other investigations showed that the presence of AF in Reye's syndrome livers was variable and probably, the syndrome was the result of multiple interrelated factors (Ryan *et al.*, 1979; Rogan *et al.*, 1985). Nevertheless, concern was raised with respect to the presence of AF in the serum and urine of children with Reye's syndrome and it was concluded that this was of general public health importance (Nelson *et al.*, 1980). Because there was no clear cut evidence that AF played a central role in Reye's syndrome, other factors were considered. One of these *inter alia*, was the use of salicylates (Aspirin) in treating children with various infections, e.g., respiratory viral infections (Trauner, 1984). It has been claimed that since aspirin was abolished as a medication for influenza and similar diseases for patients under 18 years old, Reye's syndrome has become very rare (Kimura, 2011) and is considered a secondary mitochondrial disease.

5.2.4.2 Lung cancer

Oyelami and co-workers (1997) found that children with various diseases, including pneumonia in Nigeria had high levels of AF in their lungs. The route of acquisition of this toxin was not clear but there is evidence to suggest that the lung can be exposed to AFB_1 by inhalation (Hayes *et al.*, 1984) and that this may give rise to lung cancer (Dvorackova *et al.* 1981). Human lung microsomes cytochrome P_{450} does not seem to be well expressed, the CYP1A2 form is not expressed at all (Wheeler *et al.* 1990) although some CYP1A2 activity is (Kelly *et al.* 1997). The exposure of lung cells, however, to polycyclic aromatic hydrocarbons induces cytochrome P_{450}s that may activate AFB_1 (van Vleet *et al.*, 2001). A study (Donnelly *et al.*, 1996) was conducted on human lung tissue fractions to investigate the possible role of other oxidative systems in the activation of AFB_1 and its deactivation by glutathione S-transferase. The study concluded that AFB_1 activation in human lung was primarily due to lipoxygenase and prostaglandin H synthase activity and that low conjugation activity contributed to human pulmonary susceptibility to AFB_1. Similar studies (van Vleet *et al.*, 2002a) showed that cytochrome P_{450} in lung was capable of activating AFB_1 (van Vleet *et al.*, 2002b) and CYP1A2 has greater importance in lung tissue in activating AFB_1. Later work has indicated that cytochrome P_{450} CYP2A13 is highly expressed in human bronchial cells, indicating that this cytochrome may be responsible for AFB_1 activation in the lung (Zhu *et al.*, 2006). Irrespective of what activates AFB_1, it seems that exposure of humans to AFB_1 inhalation may result in an increased risk of lung cancer (van Vleet *et al.*, 2002b). In the case of rural Africans, this is then a possibility as they process their maize manually (partly by winnowing), often in enclosed conditions where they may be exposed to dust from contaminated cobs.

5.2.4.3 Immuno-suppression and reproductive problems

An often neglected aspect of chronic mycotoxin exposure in humans is the effect of these on the immune system. With the arrival of infections such as HIV, which also attacks this system and is a major problem in SSA, the extra burden of immuno-suppressors may affect the course of the immune deficiency syndrome (Bondy and Pestka, 2000; Dutton, 2004; Jiang *et al.*, 2008). There is a strong evidence to show that mycotoxins can promote "secondary infections" of organisms that are generally commensal in an animal due to immuno suppression, e.g. OTA in pigs (Stoev *et al.*, 2000). Several mycotoxins have been shown to be immuno-suppressive or have a potential to be so (Sharma, 1993) and these include AFB_1 (Cusumano *et al.*, 1996). These effects may be compounded by the interaction of more than one mycotoxin, which may occur in substrates (Hussein and Brasel, 2001) that may cause synergism, although doubts have been expressed on assessing these effects (Speijers and Speijers, 2004). There is, however, evidence from *in vitro* experiments that this may occur (Creppy *et al.*, 2004, Luongo *et al.*, 2006; Del Rio Garcia *et al.*, 2007; Orsi *et al.*, 2007; Smith *et al.*, 1997) and it is reasonable to suppose that these effects may be extrapolated to humans.

The action of AFB_1 on the human reproductive system and gestation and birth defects is experienced by animals (Ibeh and Saxena, 1997) so it not unreasonable to suppose that these may occur in humans and, from a recent review on the subject, it would seem that AFB_1 exposure in humans has several effects on reproduction (Shuaib *et al.*, 2010a). Several studies have measured the blood albumin-AFB_1 conjugate levels in pregnant African women and have found substantial levels as might be expected (Turner *et al.*, 2007). In one study, there was a correlation between these levels and anaemia in pregnant women in Ghana, if iron deficiency anaemia was excluded (Shuaib *et al.*, 2010b). In the case of men, sperm abnormalities have been associated with AFB_1 in their semen (Ibeh *et al.*, 1994). Other effects are low birth weight (de Vries *et al.*, 1989) and jaundice in neonates (Abulu *et al.*, 1998) which was correlated with AFB_1 in cord blood and effects on their immune system (Turner *et al.*, 2003). AFM_1 is a cytochrome P_{450} mediated hydroxylated product of AFB_1 found in most mammal secreted milks, previously exposed to AFB_1 (Motawee *et al.*, 2009). It does have toxic and carcinogenic properties, although not as marked as AFB_1 (Hsieh *et al.*, 1984). It has been found in breast milk of women from several African countries (Coulter *et al.*, 1984; Zarba *et al.*, 1992). Its presence in cow's milk, particularly in that produced in African rural areas (Mwanza, 2007; Tchana *et al.*, 2010) is of great concern, considering its use in the nutrition of children.

To summarize on the health implications associated with human exposure to AF from the African viewpoint, it is but normal to state that of the AF, AFB_1 is considered to be the most important and much likely to be involved in human diseases. Although most work has concentrated on its role, sight must not be lost of the fact that its congeners, AFG_1 and M_1, which are also present in the environment, are capable of being converted to the active epoxide derivatives by the cytochrome P_{450} system. AFM_1 is of particular importance, because of its occurrence in milk from dairy cattle fed feeds contaminated with AFB_1. With AFB_1 itself, as discussed herein, it is very difficult to tease out its exact role in various human disease conditions. In animal trials, it has been shown to be a powerful carcinogen and there is no reason to suspect that it does not have a carcinogenic effect in humans. A strong claim that the role of mycotoxins in human disease has been ignored was recently made (Wild and Gong, 2010) and interestingly, AFB_1 and FB_1 were cited as major culprits, because of their common occurrence in staple foods, often in combination. However, because of the complicating issues in the human environment and lack of direct experimentation on humans, it is difficult to assign a definitive role in any disease condition. In the case of HCC for example, at best we can claim that AFB_1 is responsible for 5-28% of all cases in Africa, although the detection of specific point base mutations in liver cancer cells may allow for an estimation of the contribution of AFB_1 to human liver cancer. Whatever the precise medical role of mycotoxins in human diseases is , one cannot help but have a strong empathy with the pleas of Wild and Gong (2010) that they should be taken very seriously in Africa and other developing parts of the world.

6. Possible intervention control strategies for AF in Africa

Considering the action and toxic nature of AF, it may have been thought that attempting to completely eliminate their actions in humans or perhaps, animals would be rather futile. However, several approaches have been or can be taken, either during pre or post-exposure that can assist in alleviating or moderating the actions of AF, particularly those of AFB_1. Whatever actions are taken or have been adopted, do require financial considerations and because the current global financial situation is not in a good stead, make this even much more difficult, but of great importance. In order to assess "overall disease burden" (ODB), the concept of "disability-adjusted life year" (DALY) is used (Murray, 1994). It is defined as the number of years lost due to early death, ill health or disability and this is deduced by the addition of two variables, i.e. "years of life lost" (YLL) to "years lived with disability" (YLD). It follows that one DALY is equal to one year of healthy life lost. In order, therefore, to calculate the cost effectiveness of a programme to prevent HCC for example, the DALYs for this disease must be determined followed by computing the likely lowering of this figure. Against this, is the cost of reducing mycotoxins like AFB_1 and FB_1 in a food staple such as maize. This in turn depends upon regulations governing the health risk factors, which in order to reduce risk to health, must become more stringent resulting in economic losses in the crop due to it not meeting the recommended standard. For example, it is claimed that export losses from AF in groundnut may exceed US$ 450 million, if the level were 4ppb AF (European Union regulation) as opposed to the level imposed in USA of 20ppb which would result to about $100 million loss (Wu, 2004). An important aspect of this approach is the fact that WHO has in the past not recognised AF as a high priority problem within the top 6 health problems but it has been argued that AF not only has an impact on HCC, but probably modulates several of the other top 6 problems (Williams *et al.*, 2004). In order to determine the cost effectiveness of various interventions, it is necessary to establish the cost effectiveness ratio (CER), which is the gross domestic product (GDP) multiplied by DALY saved per unit cost. In a study involving AF in maize in Nigeria and groundnut in Guinea, two strategies were compared viz: pre-harvest control; and post harvest interventions (Wu and Khlangwiset, 2010a). Accordingly, it was shown that the cost of both interventions exceeded the monetised values of lives saved and quality of life gained by reducing HCC, if applied nationwide. Furthermore on this study of Wu and Khlangwiset (2010a), CER for biocontrol in Nigeria ranged from 5.1 to 24.8 and for post harvest intervention for groundnut in Guinea from 0.21- to 2.08. Any intervention with a CER <1.0 is considered very cost effective and more that >0.33 as just simply cost effective (Wu and Khlangwiset, 2010a). The implications of such calculations cannot be under estimated, as they do not only indicate to governments and world bodies their value, but also have an ethical dimension in terms of human and animal sufferings.

When considering interventions, several routes may be taken (Wu and Khlangwiset, 2010b). The best approach is that of prevention which is always better than cure. One such intervention is that of releasing non-aflatoxigenic strains of *Aspergillus flavus* into the agricultural environment and such a commercial product called Afla-Guard® is available commercially. This results in suppression of naturally occurring aflatoxigenic strains (Abbas *et al.*, 2011). Another is the introduction of genetically modified variety of crops, e.g., genetically modified (GM) Bt maize which inhibits insect damage and hence fungal infection (Wu, 2006). Another preventive measure is feeding of animals with amino acids and vitamins particularly lysine and vitamin C that have protective actions against mycotoxins (Obidoa and Gugnani, 1992; Smith *et al.*, 2000). A more traditional way is the use of fungicides and pesticides, although current preference is not in favour of this. The use of natural predators (cats and dogs) at fields and storage sites to deter rodents, birds and monkeys is a very practicable preventive control strategy for Africa.

Post harvest treatments are a little more difficult due to the persistence of AF in commodities even after processing. Early harvesting, effective drying (to moisture level of less than 14%), cleaning, removal of damaged produce (sorting), e.g., small and discoloured groundnuts (Chiou *et al.*, 1994); good storage facilities with controlled humidity (Kew, 2010b) and packaging can all contribute to lowering the level of the mycotoxin in the final product. While these remain the most effective post harvest control measures for Africa, other alternative but less effective measures include reduction of storage time, use of chemical and botanical preservatives, and detoxification of contaminated produce. The most commonly used chemical preservatives are the organic acids; formic, acetic, propionic, sorbic and benzoic acids. Nonetheless, they are ineffective in foods that contain basic components that neutralize these acids (Smith and Moss, 1985). Alkalis, strong acids and oxidizing agents are quite effective in detoxifying AF but because they could drastically change the properties of the products, ammoniation is still the most preferred and developed detoxification procedure. But the changes in chemical compositions and organoleptic properties of ammoniated meals makes them unfit for human consumption nevertheless good enough for animals. Commercialization of the ammoniation procedure in Africa by governments and private companies as has been successfully done in the USA, could help provide livestock farmers in developing countries with relatively safer feeds in the face of highly contaminated feedstuffs and shortage of feeds.

The toxic and 'off flavour' products of chemical preservation and detoxification processes has led scientists to search for natural, safer and environment-friendly fungicidal products. Among such African based studies, *Lippia multiflora* leaf extract has been shown to have fungistatic effect on *A. flavus* (Anjorin *et al.*, 2008). More intense field trials of such promising plant products and their subsequent formulation into botanical fungicide would be impressive for the continent. Gamma irradiation of AF contaminated foods lowers both the toxicity (Ogbadu and Bassir, 1979) and production (Ogbadu, 1979; Ogbadu, 1980a; Ogbadu, 1981; Ogbadu, 1988) of the toxin in irradiated foods and so this could be a good post harvest, processing and packaging treatment option for African countries if suitable infrastructures are put in place. Hazard Analysis Critical Control Point (HACCP), a proactive management system in which food safety is maintained through the analysis and control of biological, chemical, and physical hazards from raw material production, procurement and handling, to manufacturing, distribution and consumption of the finished product has become a priceless tool for mycotoxin control (FAO, 2003).

Clinical applications in the control of conditions such as HCC have been applied with varying degree of success. This range from preventative measures such as the use of Novasil clay being added to the diet to bind AF (Afriyie-Gyawu *et al.*, 2008); and HVB immunisation (Kew 2005). The use of drugs can be considered in two parts, one that blocks cytochrome P_{450}s responsible for the activation of AFB_1 to an active (its epoxide) form, e.g. oltipraz (Langouet *et al.*, 1995; Wang *et al.*, 1999) and natural foods, e.g. Brassicas (Manson *et al.*, 1997); and those which may have some other non-clear cut effect such as the use of plant extracts as protective agents (Kotan *et al.*, 2011); boric acid (Turkez and Geyikoglu, 2010); sorafenib, a blocker for signalling pathways involved in HCC (Dank, 2010).

7. Legislation

In order to protect consumers against the hazards of mycotoxins, many countries including 15 of those from Africa (Sibanda *et al.*, 1997; Fellinger, 2006; Njobeh *et al.*, 2010) have instituted legislation against some mycotoxins notably AF. According to these authors, the maximum tolerable limits for AF in human foods in Africa is between 5-20 ppb, while for animal feeds is from 5 to 300 ppb with infant foods having the least regulated levels (0.05-10 ppb) (0.05 ppb for AFM_1 in the case of South Africa). While these maximum allowable limits would protect citizens from the dangers of AF, the biggest challenge in regulating mycotoxins in the continent is the lack of enforcement of legislation partly attributed to the presence of informal food market systems operating in most countries. Under this market structure, raw agricultural produce from farms and storage barns are sold directly to consumers without being screened for mycotoxins neither are they subjected to inspection for spoilage. Furthermore, government agencies charged with the responsibility to regulate mycotoxins are non-existent in many of these countries and even when available, they are dysfunctional as they are composed of deplorable infrastructures and logistics. An effective mycotoxin surveillance and food quality control unit which ensures that all foods and feeds destined for human and animal consumption are devoid of mycotoxins at harmful levels must be in place to implement mycotoxin legislation in the continent.

8. Prevention by surveillance and awareness campaign

Assessing the levels of mycotoxins and indeed other food toxicants is paramount to evaluating food safety. In line with the recommendation for effective mycotoxin survey and food and feed inspection for implementation of legislation for food safety, African governments must build or strengthen already existing regional laboratories to monitor mycotoxins in foods and feeds on regular basis. And to ensure that they are in compliance with set standards. Invariably, Africa must reinforce its food quality control agencies and this can only be achievable if professionals working in such establishments possess the academic as well as technical capacity for mycotoxin management which calls for the inclusion of courses on mycotoxins in the curricula for training of agriculturists, medical personnel and laboratory based scientists. Awareness on the adverse impact of mycotoxins should not be limited to professionals in the food and feed related industries, but to the entire consumers. Public awareness campaign on impact and prevention of mycotoxins especially the notorious AF via electronic and print media and other information dissemination modes is therefore an imperative. Such scientific and public enlightenment interventions require concerted national and international multidisciplinary strategies (WHO, 2006). It is but imperative to engage both national and international bodies to partner with one another to effectively manage mycotoxins. For example in Nigeria, experts in devised but related fields from academia and research institutions formed the Nigerian Mycotoxin Awareness and Study Network in 2005 (NMASN-www.ngmycotoxin.org) with a common goal of offering scientific and technical support towards managing mycotoxins in the country. In doing so, the network organizes annual workshops for stakeholders. Similarly this year (2011), the International Society of Mycotoxicology organised a world conference in Cape Town, South Africa on mycotoxin reduction (www.mycoredafrica2011.co.za) which brought scientists from all over the globe to not only share knowledge and expertise but to establish research collaborations towards strengthening the capacity of the African mycotoxicologists and laboratories. European Union (Leslie *et al.*, 2008) and World Health Organisation (WHO, 2006) had earlier organized such international conferences in 2005 and 2006 in Ghana and Congo Brazzaville, respectively. It is only pertinent now to encourage scientists and institutions involved in mycotoxin research in Africa to collaboratively seek accessible research grants from the AU, EU and other foreign funding agencies for more effective investigations and control of mycotoxins.

9. Summary

Aflatoxins are toxic secondary metabolites produced notably by *Aspergillus flavus* and *A. parasiticus* that frequently invade foods and feedstuffs before and after harvest. The four major aflatoxins include aflatoxin B_1, B_2, G_1 and G_2 with aflatoxin B_1 recognized as the most prevalent and toxic of all aflatoxins. Their presence in foods and feeds is inevitable and as such, humans and animals are exposed to them on a continuous basis leading to a wide array of health complications. Particularly aflatoxins B_1, they have been directly linked to hepatocarcinoma and deaths among humans and animals. Although this may be the case worldwide, the situation in sub-Saharan Africa is very severe as increased levels of exposure to this group of mycotoxins is a common phenomenon since it presents suitable environmental conditions for aflatoxin concentration in various food and feed materials. Again, the problem is further exacerbated by increased prevalence in the continent, of such endemic diseases as malaria, hepatitis and HIV/AIDS. In Africa recently, we have experienced the most fatal aflatoxin-poisoning outbreaks including two episodes in Kenya and one in Nigeria. In view of the significance therefore, of aflatoxin exposure, this chapter reviews the disparity in aflatoxin contamination of food and feeds worldwide with particular emphasis on Africa. It has also expounded briefly on those factors that influence the distribution of aflatoxins in various food and feeds. Additionally, an in-depth review is provided on the negative public health problems and the impact in the economy associated with this notorious group of secondary metabolites with particular reference made from the African context, while also discussing those control strategies possible within the continent's technological capacity.

10. Keywords

Aflatoxins, Occurrence, Agricultural commodities, Health, Control, Africa.

11. References

Abalaka, J.A (1984). Aflatoxin distribution in edible oil-extracting plants and in poultry feed mills. *Food and Chemical Toxicology* 22 (6): 461-463

Abbas, H. K., Zablotowicz, R. M., Horn, B. W., Phillips, N. A., Johnson, B. J., Jin, X. and Abel, C. A. (2011) Comparison of majo control strains of non-toxigenic *Aspergillus flavus* for the reduction of aflatoxins and cyclopiazonic acid in maize, *Food Additives and Contaminants Part A 28*, 198-208.

Abdallah, M.I.M.; Manal, M.A.*; Dawoud, A.S. And Marouf, H.A. Occurrence of aflatoxins in table eggs sold in Damietta City regarding its health significance. Retrieved on 15th July, 2011 from www.nu.edu.sa/uploads/sss/2/7/1/3/7.pdf

Abdelhamid, A.M (2008). Thirty Years (1978 - 2008) of Mycotoxins Research at Faculty of Agriculture, Almansoura University, Egypt. Retrieved on 1st June, 2011 from www. engormix.com

Abulu, E.O., Uriah, N., Aigbefo, H.S., Oboh, P.A. and Agbonlahor, D.E. (1998) Preliminary investigation on aflatoxin in cord blood of jaundiced neonates *West African Journal of Medicine 17*, 184-187.

Adebayo-Tayo, B. C., Adegoke, A. A. and Akinjogunla, O. J. (2009). Microbial and physico-chemical quality of powdered soymilk samples in Akwa Ibom, South Southern Nigeria. *African Journal of Biotechnology* 8 (13): 3066-3071

Adebayo-Tayo, B. C. and Ettah, A E. (2010). Microbiological quality and aflatoxin B1 level in poultry and livestock feeds. *Nigerian Journal of Microbiology,* 24(1): 2145 - 2152

Adebayo-Tayo, B.C., Onilude, A.A., Ogunjobi, A.A., Gbolagade, J.S. and M.O. Oladapo (2006). Detection of fungi and aflatoxin in shelved bush mango seeds (*Irvingia* spp.) stored for sale in Uyo, Nigeria *African Journal of Biotechnology* 5 (19):1729-1732

Adebayo-Tayo, B.C., Onilude, A.A. and Ukpe, G.P. (2008) Mycofloral of Smoke-Dried Fishes Sold in Uyo, Eastern Nigeria *World Journal of Agricultural Sciences* 4 (3): 346-350

Adhikari, M., Ramjee, G. and Berjak, P. (1994) Aflatoxin, kwashiorkor and morbidity *Natural Toxins 2*, 1-3.

Afla-guard, (2005). Aflatoxin in Africa. Retrieved on 11th May, 2011 from www.circleoneglobal.com/aflatoxin_africa.htm.

Afriyie-Gyawu, E. Ankrah, N.A., Huebner, H.J., Ofosuhene, M., Kumi, J., Johnson, N.M., Tang, L., Xu, L., Jolly, P.E., Ellis, W.O., Ofori-Adjei, D., Williams, J.H., Wang, J.S., Phillips, T.D. (2008) Novasil clay intervention in Ghanaians at high risk for aflatoxicosis. I. Study design and clinical outcomes *Food Additives and Contaminants Part A 25,* 76-87.

Agboola S.D (1992). Post harvest technologies to reduce mycotoxin contamination of food crops. In Z.S.C Okoye (ed) Book of proceedings of the first National Workshop on Mycotoxins held at University Jos, on the 29th November, 1990, 73-88.

Akinmusire, O.O. (2011). Fungal species associated with the spoilage of some edible fruits in Maiduguri, Northern Eastern Nigeria *Advances in Environmental Biology* 5(1): 157-161

Akrobortu, D.E. (2008). Aflatoxin contamination of maize from different storage locations in Ghana. An M.Sc. Thesis submitted to the Department of Agricultural Engineering, Kwame Nkrumah University of Science and Technology, Ghana. 27-32

Amer, A.A. and Ibrahim, M.A.E. (2010) Determination of aflatoxin M1 in raw milk and traditional cheeses retailed in Egyptian markets *Journal of Toxicology and Environmental Health Sciences* 2 (4): 50-53

Anjorin, S.T, Makun, H.A, and Iheneacho, H.E (2008). Effect of *Lippia Multiflora* leaf extract and *Aspergillus flavus* on germination and vigour indices of *Sorghum Bicolor* [L] (Moench). *International Journal of Tropical Agriculture and Food System,* 2 (1): 130 - 134.

Apeagyei, F., Lamplugh, S. M., Hendrickse, R. G., Affram, K. and Lucas, S. (1986) Aflatoxin in the livers of children with kwashiorkor in Ghana *Tropical and Geographical Medicine 38*, 273-276.

Aroyeun, S.O and Jayeola, C.O (2005). Mycotoxins in cocoa. A paper presented at the Regional Workshop on Mycotoxins organized by National Agency for Food and Drug Administration and Control (NAFDAC) in collaboration with International Atomic Energy Agency (IAEA), Held at Meidan Hotels, Victoria Garden City, Lagos, Nigeria between 7th and 11th February, 2005.

Ashfaq, U. A., Javed, T., Rehman, S., Nawaz, Z. and Riazudding, S. (2011) An overview of HCV molecular biology, replication and immune response *Virology Journal 8*, In Press.

Atehnkeng, J., Ojiambo, P.S., Donner, M., Ikotun, K.,Sikora, R.A.,Cotty, P.J. and Bandypadhyay, R. (2008). Distribution and toxicity of *Aspergillus* species isolated from maize kernels from three agro-ecological zones of Nigeria. *International Journal Food Microbiology. 122*, 74–84.

Atanda, O., Oguntubo, A., Adejumo, O., Ikeorah, J. and Akpan, I. (2007). Aflatoxin M1 contamination of milk and ice cream in Abeokuta and Odeda local governments of Ogun State, Nigeria *Chemosphere* 68, 1455–1458

Aziz, N.H. and Youssef, Y.A (1991). Occurrence of aflatoxins and aflatoxin-producing moulds in fresh and processed meat in Egypt *Food Additives and Contaminants.* 8 (3):321-31

Azziz-Baumgartner, E., Lindblade, K., Gieseker, K., Rogers, H. S., Kieszak, S., Njapau, H., Schleicher, R., McCoy, L. F., Misore, A., DeCock, K., Rubin, C. and Slutsker, L. (2006) Case control study of an acute aflatoxicosis outbreak, Kenya 2004 *Environmental Health Perspectives 113*, 1779-1783.

Baiyewu, R.A., Amusa, N. A., Ayoola, O.A. and Babalola, O.O. (2007). Survey of the post harvest diseases and aflatoxin contamination of marketed pawpaw fruit (*Carica papaya* L) in South Western Nigeria *African Journal of Agricultural Research* 2 (4):178-181.

Bandyopadhyay, R., Kumar, M. and Leslie, J.F. (2007). Relative severity of aflatoxin contamination of cereal crops in West Africa Food *Additives and Contaminants* 24 (10):1109-14.

Bankole, S.A and Adebanjo, A. (2003). Mycotoxins in food in West Africa: current situation and possibilities of controlling it. *African Journal of Biotechnology* 2 (9): 254-263.

Bankole, S.A. and Mabekoje, O.O (2004). Mycoflora and occurrence of aflatoxin B1 in dried yam chips from markets in Ogun and Oyo States, Nigeria *Mycopathologia.* 157(1):111-5

Bankole, S.A., Ogunsanwo, B.M. and Eseigbe, D. A (2005). Aflatoxins in Nigerian dry-roasted groundnuts. *Food Chemistry* 89: 503–506.

Bankole, S.A., Ogunsanwo, B.M., Osho, A. and Adewuyi, G.A (2006). Fungal contamination and aflatoxin B1 of _egusi_melon seeds in Nigeria. *Food Control* 17: 814–818.

Bankole,S Schollenberger, M. and Drochner, W. (2006). Mycotoxins in food systems in Sub Saharan Africa: A review *Mycotoxin Research* 22 (3): 163-169

Barkai-Golan, R. and Paster, N. (2008). Mycotoxins in Fruits and Vegetables. Academic Press, San Diego, USA. 3-11

Bassa, S., Mestres, C., Champiat, D., Hell, K., Vernier, P. and Cardwell, K. (2009) First report of aflatoxin in dried yam chips in Benin. *Plant Disease* 85 (9): 1032

Becroft, D.M.O. (1966) Syndrome of encephalopathy and fatty degeneration of viscera in New Zealand children *British Medical Journal* 2, 1351.

Becroft, D. M. O. and Webster, D. R. (1972) Aflatoxins and Reye's disease *British Medical Journal 14th October*, p117.

Beucker, S., Raters, M. and Matissek, R. (2005). Mycotoxin Study III: MycoDONA Deoxynivalenol, Ochratoxin A and Aflatoxins in Cocoa and Cocoa-containing Products Analysis, Situation Assessment, and Monitoring. Retrieved on 15th July, 2011 from http://www.kakao-stiftung.de/pdf/Projekt_38e.pdf

Bhattacharyya, A. K. (1986) Protein-energy malnutrition (Kwashiorkor-Marasmus syndrome): terminology, classification and evolution *World Review of Nutrition and Dietetics 47*, 80-133.

Bilgrami, K.S., Prasad, T., Misra, R.S. and Sinha, K.K.1981. Aflatoxin contamination in maize under field conditions.*Indian Phytopatho*logy. *34*, 67–68.

Bintvihok, A., Thiengnin, S., Doi, K. and Kumagai, S. (2002). Residues of aflatoxins in the liver, muscles and eggs of domestic fowls *Journal of Veterinary Medical Science* 64 (11): 1035-1037

Blondski, W., Wojciech, Kotlyar, D. S. and Forde, K. A. (2010) Non-viral causes of hepatocellular carcinoma *World Journal of Gastroenterology 16*, 3603-3615.

Blount W.P. (1961) Turkey X disease *Turkeys (Journal of the British Turkey Association) 9*, 55-58.

Bondy, G. S. and Peska, J. J. (2000) Immunomodulation by fungal toxins *Journal of Toxicology and Environmental Health B 3*, 109-143.

Bressac, B., Kew, M., Wanda, J. and Ozturk, M. (1991) Selective G to G mutations of p53 gene in hepatocellular carcinoma from southern Africa *Nature 350*, 429-431.

Bucci, T. J. and Howard, P. C. (1996) Effect of fumonisin mycotoxins in animals *Journal of Toxicology Toxin Reviews 15*, 293-302.

Brimer, L (2011). Chemical food safety of traditional grains. Retrieved on 1st June, 2011 from http://www.sik.se/traditionalgrains/workshop/proceedings/Leon_Brimer_proc.pdf

Campbell, T.C. and Hayes, J.R. (1976) The role of aflatoxin metabolism in its toxic lesions *Toxicology and Allied Pharmacology 35*, 199-222.

Center for Disease Control and Prevention (CDC), 2004. Outbreak of aflatoxin poisoning—Eastern and Central provinces, Kenya, January-July, 2004. Retrieved on 20th June, 2011 from http://www.cdc.gov/mmwr/preview/mmwrhtml/mm5334a4.htm.

Chandrashekar, A., Bandyopadhyay, R., and Hall, A.J. (eds.). 2000. Technical and institutional options for sorghum grain mold management: proceedings of an international consultation, 18-19 May 2000, ICRISAT, Patancheru, India. (In En. Summaries in En, Fr.) Patancheru, 502324, Andhra Pradesh, India: International Crops Research Institute for the Semi- Arid Tropics. 299 pp. ISBN 92-9066-428-2. Order code CPE 129.

Chiou, R. Y. Y., Wu, P. Y. and Yen, Y. H. (1994) Color sorting of lightly roasted and de-skinned peanut kernels to diminish aflatoxin contamination in commercial lots *Journal of Agricultural Food Chemistry 42*, 2156-2160.

Chukwu, O. and Imodiboh, L.I. (2009) Influence of Storage Conditions on Shelf-Life of Dried Beef Product (*Kilishi*) *World Journal of Agricultural Sciences* 5 (1): 34-39

Ciliberto, H., Ciliberto, M., Briend, A., Ashorn, P., Bier, D. and Manary, M. (2005) Antioxidant supplementation for the prevention of kwashiorkor in Malawian children: randomised, double blind, placebo controlled trial *British Medical Journal 330*, 1109-1114.

Codex Alimentarius Commission. 2004. Code of Practice for the Prevention and Reduction of Aflatoxin Contamination in Peanuts. Retrieved 13th May, 2011 from http://www.codexalimentarius.net/download/standards/10084/CXC_055_2004e.pdf.

Codex Alimentarius Commission. (2011). Discussion paper on mycotoxins in Sorghum. Joint FAO/WHO Food Standards Programme CODEX Committee on contaminants in foods' 5th Session held in The Hague, The Netherlands on 21 - 25 March 2011.

Coulter JB, Lamplugh SM, Suliman GI, Omer MI and Hendrickse RG. (1984) Aflatoxins in human breast milk *Annals of Tropical Paediatrics 4,* 61-66.

Coulter, J. B., Hendrickse, R. G., Lamplugh, S. M., MacFarlane, S. B., Moody, J. B., Omer, M. I., Suliman, G. I., and Williams, T. E. (1986) Aflatoxins and kwashiorkor: clinical studies in Sudanese children, *Transactions of the Royal Society of Tropical Medicine and Hygiene 80,* 945-951.

Creppy, E. E., Chiarappa, P., Baudrimont, I., Borracci, P., Moukha, S. and Carratu, M. R. (2004) Synergistic effects of fumonisin B_1 and ochratoxin A: are *in vitro* cytotoxicity data predicitive of *in vivo* acute toxicity? *Toxicology 201,* 115-123.

Cusumano, V., Rossano, F., Merendino, R. A., Arena, A., Costa, G. B., Mancuso, G., Baroni, A. and Losi, E. (1996) Immunobiological activities of mould products: functional impairment of human monocytes exposed to aflatoxin B_1 *Research in Microbiology 147,* 385-391.

Dank, M. (2010) Treatment of primary hepatocellular carcinoma *Orv Hetil 151,* 1445-1449. (In Hungarian; abstract PubMed 20739261).

Del Rio Garcia, J. C., Moreno, R., C., Pinton, P., Mendoza, E. S. and Oswald, I. P. (2007) Evaluation of the cytotoxicity of aflatoxin and fumonisins on swine intestinal cells *Rev Iberoam Micol 24,* 136-141. (In Spanish; abstract PubMed 17604433).

de Vries, H. R. and Hendrickse, R. G. (1988) Climatic conditions and kwashiorkor in Mumias: a retrospective analysis over a 5-year period *Annals of Tropical Paediatrics 8,* 268-270.

de Vries, H. R., Lamplugh, S. M. and Hendrickse, R. G. (1987) Aflatoxins and kwashiorkor in Kenya: a hospital based study *Annals of Tropical Paediatrics 7,* 249-251.

de Vries, H.R., Maxwell S.M. and Hendrickse, R.G. (1989) Foetal and neonatal exposure to aflatoxins *Acta Paediatric Scandanavia 78,* 373-378.

de Vries, H. R., Maxwell, S. M. and Hendrickse, R. G. (1990) Aflatoxin excretion in children with kwashiorkor or marasmic kwashiorkor - a clinical investigation *Mycopathologia 110,* 1-9.

Donnelly, P.J., Stewart, R.K., Ali, S.L., Conlan, A.A., Reid, K.R., Petsikas, D. and Massey, T.E. (1996) Biotransformation of aflatoxin B_1 in human lung *Carcinogenesis 17,* 2487-2494.

Dvorakova, I., Stora, C. and Ayraud, N. (1981) Evidence for aflatoxin B_1 in two cases of lung cancer in man *Journal of Cancer Research and Clinical Oncology, 100* 221-224.

Dutton, M. F (2004) Chapter 13 Fumonisin B_1 in animal and human health. In Recent Researches on Fungi Ed. R.K.S. Kushwaha, Scientific Publishers (India) Jodhpur.

Dutton, M. F. (2009) The African *Fusarium*/maize disease, *Mycotoxin Research 25,* 29-39.

Dutton, M., Mwanza, M., de Kock, S., Khilosia, L. (2012). Mycotoxins in South African foods: a case study on aflatoxin in milk. *Mycotoxin Research* 28: 19-25

Elgerbi, A. M., Aidoo, K. E. Candlish, A. A. G. and Tester, R. F. (2004). Occurrence of aflatoxin M_1 in randomly selected North African milk and cheese samples *Food Additives & Contaminants: Part A,* 21 (6) : 592 - 597

Elzupir AO, Younis M.H, Himmat Fadul M, Elhussein AM (2009) Determination of Aflatoxins in Animal Feed in Khartoum State, Sudan. *Journal of Animal and Veterinary Advances* 8 (5): 1000-1003

Enwonwu, C. O. (1984) The role of dietary aflatoxin in the genesis of hepatocellular carcinoma in developing countries, *Lancet 2 (8409)*, 956-958.

Essono, G., Ayodele, M., Akoa, A., Foko, J., Filtenborg, O. and Olembo, S. (2009). Aflatoxin-producing Aspergillus spp. and aflatoxin levels in stored cassava chips as affected by processing practice *Food Control* 20 648–654

Fapohunda, S.O (2011).Impact of Mycotoxins on Sub-Saharan Africa : Nigeria as a Case Study. European Mycotoxin Awareness Network Retrieved on 7th July, 2011 from http://www. services.leatherheadfood.com/mycotoxins/index.asp

F.A.O. (2003) Manual on the application of HACCP system to mycotoxin control. Retrieved on 13th May, 2011 from http://www.fao.org/docrep/005/y1390e/y1390e00.htm).

F.A.O (1983). Post harvest losses in quality of food grains. Food and Agriculture Organisation (Food and Nutrition Paper No 29, pg. 103

FAOSTAT (2006). Production statistics. Retrieved on July, 2010 from www.faostat.fao.org

FAOSTAT, (2010) Crops primary equivalent. Retrieved on 16th May, 2011 from www.faostat.fao.org

FDA (2011), Hazard Analysis Critical Control Point. Retrieved on 13th May, 2011 from http://www.fda.gov/food/foodsafety/hazardanalysiscriticalcontrolpointshaccp/default.htm)

Fellinger, A. (2006). Worldwide mycotoxin regulations and analytical challenges. World Grain Summit: Foods and Beverages, September 17–20, 2006, San Francisco, California, USA.

Ferenci, P., Fried ,M., Labrecque, D., Bruix, J., Sherman, M., Omata, M., Heathcote, J., Piratsivuth, T., Kew, M., Otegbayo, J.A., Zheng, S.S., Sarin, S., Hamid, S., Modawi, S.B., Fleig, W., Fedail, S., Thomson, A., Khan, A., Malfertheiner, P., Lau, G., Carillo, F.J., Krabshuis, J., Le Mair, A., World Gastroenterology Organisation Guidelines and Publications Committee (2010) World gasteroenterology organisation guideline. Hepatocellular carcinoma (HCC): a global perspective, *Journal of Gastrointestinal Liver Disease 19*, 311-317.

Fernández-Cruz, M.L., Mansilla, M.L. and Tadeo, J.L. (2010). Mycotoxins in fruits and their processed products: Analysis, occurrence and health implications *Journal of Advanced Research* 1, 113–122

Frisvad, J.C. and Samson, R.A (2004). *Emericella venezuelensis*, a New Species with Stellate Ascospores Producing Sterigmatocystin and Aflatoxin B1 *Systematic and Applied Microbiology*. 27, 672-680.

Frivad, J.C., Samson, R.A. and Smedsgaard, J. (2004) *Emericella astellata*, a new producer of aflatoxin B1, B2 and sterigmatocystin. *Letters in Applied Microbiology* 38, 440–445

Frisvad, J. C., Skouboe, P., and Samson, R. A. (2005). Taxonomic comparison of three different groups of aflatoxin producers and a new efficient producer of aflatoxin B1, sterigmatocystin and 3-*O*-methylsterigmatocystin, *Aspergillus rambellii* sp. nov. *Systematic AppliedMicrobiology, 28*, 442–453.

Gallagher, E.P., Wiekers, L.C., Stapleton, P.L., Kunze, K.L and Eaton, D.L. (1994) Role of human microsomal and human complementary DNA expressed cytochromes P450 1A2 and 3A4 in the bioactivation of aflatoxin B_1 *Cancer Research 54*, 101-108.

Gbodi, T.A. (1986). Studies of mycoflora and mycotoxins in Acha, maize and cotton seed in plateau state, Nigeria. A Ph. D thesis, submitted to Department of Physiology and Pharmacology, Faculty of Veterinary Medicine, A.B.U, Zaria, pg. 1-213.

Ghali, R., Belouaer, I., Hdiri, S., Ghorbel, H., Maaroufi, K. and Hedilli, A. (2009) Simultaneous HPLC determination of aflatoxins B1, B2, G1 and G2 in Tunisian sorghum and pistachios. *Journal of Food Composition and Analysis* 22, 751–755

Ghali, R., Khlifa, K. H., Ghorbel, H., Maaroufi, K. and Hedilli, A. (2008). Incidence of aflatoxins, ochratoxin A and zearalenone in Tunisian foods. *Food Control* 19, 921–924

Ghali, R., Khlifa, K. H., Ghorbel, H., Maaroufi, K. and Hedilli, A. (2010) Aflatoxin determination in commonly consumed foods in Tunisia, *Journal of the Science of Food and Agriculture 90*, 2347-2351.

Golden, M.H.N., Golden, B.E. and Jackson, A.A., (1980) Kwashiorkor and nutritional oedema *Lancet 1 (8180)*, 114-116.

Golden, M.H.N. and Ramdath, D. (1987) Free radicals in the pathogenesis of kwashiorkor *Proceedings of Nutrition Society 46*, 53-68.

Gong, Y.Y., Egal, S., Hounsa, S., Hall, A.J., Cardwell, K.F., Wild, C.P., (2003). Determinants of aflatoxin exposure in young children from Benin and Togo, West Africa: the critical role of weaning. *International Journal of Epidemiology* 32, 556–662.

Gong, Y., Hounsa, A., Egal, S., Sutcliffe, A.E., Hall, A.J., Cardwell, K.F., Wild, C.P., (2004). Postweaning exposure to aflatoxin results in impaired child growth: a longitudinal study in Benin, West Africa. *Environmental Health Perspectives* 112, 1334–1338.

Gopalan, P., Jensen, D. E. and Lotikar, P. D. (1992) Glutathione conjugation of microsome mediated and synthetic aflatoxin B1-8,9-oxide by purified glutathione S-transferase from rats *Cancer Letters 64* 225-233.

Gnonlonfin, G.J.B., Hell, K., Fandohan, P. and Siame, A.B. (2008). Mycoflora and natural occurrence of aflatoxins and fumonisin B1 in cassava and yam chips from Benin, West Africa *International Journal of Food Microbiology*

Groopman, J. D., Hall, A. J., Whittle, H., Hudson, G. J., Wogan, G. N., Montesano, R. and Wild, C. P. (1992) Molecular dosimetry of aflatoxin-N7-guanine in human urine obtained in the Gambia, West Africa *Cancer Epidemiological Biomarkers and Prevention 1*, 221-227.

Harwig, J., Przybylski, W. and Moodie, C. A. (1975) A link between Reye's syndrome and aflatoxins? *Canadian Medical Association Journal 113*, 281.

Hatem, N. L., Hassab, H. M., Abd Al-Rahman, E. M., El-Deeb, S. A. and El-Sayed Ahmed, R. L. (2005) Prevalence of aflatoxins in blood and urine of Egyptian infants *Food and Nutrition Bulletin 26*, 49-56.

Hayes, R.B., Van Nieuwenhuize, J.P., Raatgever, J.W. and Ten Kate, F.J.W. (1984) Aflatoxin exposures in industrial setting: an epidemiological study of mortality *Food Chemistry and Toxicology 22*, 39-43.

Hell, k. and Mutegi, C. (2011). Aflatoxin control and prevention strategies in key crops of Sub Saharan Africa. *African Journal of Microbiology Research* 5 (5):459-466

Hell, K., Cardwell, K.F., Setamou, M., Poehling, H.M., (2000). The influence of storage practices on aflatoxin contamination in maize in four agroecological zones of Benin, West Africa. *Journal of Stored Products Research* 36,365–382.

Hell, K., Gnonlonfin, .G.J., Kodjogbe, G., Lamboni, Y. and Abdourhamane, I.K. (2009). Mycoflora and occurrence of aflatoxin in dried vegetables in Benin, Mali and Togo, West Africa *International Journal of Food Microbiology* 135, 99–104

Hendrickse, R. G. (1991) Clinical implications of food contaminated by aflatoxins *Annals of Academic Medicine Singapore 20*, 84-90.

Hendrickse, R. G., Coulter, J. B., Lamplugh, S. M., MacFarlane, S. B., Williams, T. E., Omer, M. I., Suliman, M. I., and El-Zorganui, G. A. (1983) Aflatoxins and kwashiorkor. Epidemiology and clinical studies in Sudanese children and findings in the autopsy liver samples from Nigeria and South Africa *Bulletin of the Exotic Pathology Society 76*, 559-566

Holbrook, C.C., Guo, B.Z., Wilson, D.M., Kvien, C., (2004). Effect of drought tolerance on preharvest aflatoxin contamination in peanut. Proceedings of the 4th International Crop Science Congress Brisbane, Australia, 26 Sep–1 Oct 2004

Househam, K. C. and Hundt, H. K. (1991) Aflatoxin exposure and its relationship to kwashiorkor in African children *Journal of Tropical Pediatrics 37*, 300-302.

Hsieh, D. P. H., Cullen, J. M. and Ruebner, B. H. (1984) Comparative hepatocarcinogenicity of aflatoxins B_1 and M_1 in the rat *Food and Chemical Toxicology 22*, 1027-1028.

Hussain, Z., Khan, M.Z., Khan, A., Javed, I., Saleemi, M.K., Mahmood, S. and Asi, M.R. (2010). Residues of aflatoxin B1 in broiler meat: Effect of age and dietary aflatoxin B1 levels *Food and Chemical Toxicology* 48 (12): 3304-3307

Hussein, H. S., and Brasel, J. M. (2001) Toxicity, metabolism and impact of mycotoxins on humans and animals, *Toxicology 167*, 101-134.

International Crops Research Institute for the Semi-Arid Tropics (ICRISAT) (2010). Assessing occurrence and distribution of aflatoxins in Malawi. Project Final Report (Grant No. 08-598). Retrieved on 1st June, 2011 from http://mcknight.ccrp.cornell.edu/program_docs/project_documents/SAF_0642_groundnut_breeding/Assessing%20Occurrence%20and%20Distribution%20of%20Aflatoxin%20%20final2.pdf.

Ibeh, I. N. and Saxena, B. N. (1997) Aflatoxin B_1 and reproduction. I. Reproductive performance in female rats, *African Journal of Reproductive Health 1*, 79-84.

Ibeh IN, Uraih N, Ogonar JI. (1994) Dietary exposure to aflatoxin in human male infertility in Benin City, Nigeria. *International Journal of Fertility and Menopausal Studies 39*, 208-214

IPCC, (2007) International Panel on Climate Change. Fourth Assessment Report. Retrieved on 16th May, 2011 from www.ipcc.ch/publications_and_data/publications_and_data_reports.shtml

Ito, Y., Peterson, S. W., Wicklow, D. T., & Goto, T. (2001). *Aspergillus pseudotamarii*, a new Aflatoxin producing species in *Aspergillus* section *Flavi*. *Mycological Research, 105*, 233–239.

Jackson, P. E., Qian, G.-S., Friesen, M. D., Zhu, Y.-R., Lu, P., Wang, J.-B., Wu, Y., Kensler, T. W., Vogelstein, B. and Groopman, J. D. (2001) Specific p53 mutations detected in plasma and tumors of hepatocellular carcinoma patients by electrospray ionisation mass spectrometry, *Cancer Research 61*, 33-35.

Jarvis, B (1971). Factors affecting the production of mycotoxins *Journal of Applied Bacteriolog.* 34(1):199-213.

Jarvis, B. (1976). Mycotoxins in food. In: Skinner, F.A and Carr, J.G (Eds), *Microbiology in Agriculture, Fisheries and Food*, Academic Press, London, pp. 251-267.

JECFA, (2001). WHO Food additives series:47 safety evaluation of certain food additives and contaminants: Aflatoxin M1. Joint FAO/WHO Expert Committee on Food Additives

Jiang, Y., Jolly, P. E., Preko, P., Wang, J. S., Ellis, W. O., Phillips, T. D. and Williams, J. H. (2008) Aflatoxin related immune dysfunction in health and in human immunodeficiency virus disease, *Clinical and Developmental and Immunology 2008*, 12 pages. (On Line Free access).

Jimoh, K.O. and Kolapo, A.L. (2008). Mycoflora and aflatoxin production in market samples of some selected Nigerian foodstuffs. *Research Journal of Microbiology* 3 (3): 169-174

Jonathan, G., Ajayi, I. and Omitade, Y.(2011). Nutritional compositions, fungi and aflatoxins detection in stored 'gbodo' (fermented *Dioscorea rotundata*) and 'elubo ogede' (fermented *Musa parasidiaca*) from South western Nigeria. *African Journal of Food Science* 5(2): 105 - 110.

Jonathan, S. G. and Esho E.O. (2010) Fungi and aflatoxin detection in two stored oyster mushroom (*Pleurotus ostreatus and Pleurotus pulmonarius*) from Nigeria *Electronic Journal of Environmental, Agricultural and Food Chemistry* 9 (11): 1722-1736

Jones, W. R., Johnston, D. S. and Stone, M. P. (1998) Refined structure of the doubly intercalated d(TAT^{afb}GCATA)$_2$ aflatoxin B_1 adduct *Chemical Research in Toxicology 11*, 873-881.

Jones, M. J., Tanya, V. N., Mbofiing, C. M.F., Fonkem, D. N. and Silverside, D. E. (2001) A Microbiological and nutritional evaluation of the West African dried meat product, Kilishi *The Journal of Food Technology in Africa*, 6 (4), 126-129

Juan, C., Zinedine, A., Molto, J.C., Idrissi, L. and Man˜es, J. (2008). Aflatoxins levels in dried fruits and nuts from Rabat-Sale´ area, Morocco *Food Control* 19, 849–853

Kaaya, N.A and Kyamuhangire, W. (2006). The effect of storage time and agroecological zone on mould incidence and aflatoxin contamination of maize from traders in Uganda. *International Journal of Food Microbiology* 110, 217–223

Kaaya, N.A. and Warren, H.L. (2005). A review of past and present research on aflatoxin in Uganda *African Journal of Food Agriculture Nutrition and Development* 2005, 18 pages (On Line Free Access)

Kamdem, L. K., Meineke, I., Godtel-Armbrust, U., Brockmoller, J. and Wojnowski, L. (2006) Dominant contribution of P450 3A4 to the hepatic activation of aflatoxin B_1 *Chemical Research in Toxicology 19*, 577-586.

Kamika, I. and Takoy, L.L (2011). Natural occurrence of Aflatoxin B1 in peanut collected from Kinshasa, Democratic Republic of Congo. *Food Control.* In press.

Kang.ethe, E.K. and Lang.a, K.A. (2009). Aflatoxin B1 and M1 contamination of animal feeds and milk from urban centers in Kenya. *African Health Sciences* 9 (4): 218-226

Kastner, S., Kandler, H., Hotzb, K., Bleisch, M., Lacroix, C, Meile, L.(2010). Screening for mycotoxins in the inoculum used for production of attie´ke´ a traditional Ivorian cassava product. *Food Science and Technology* 43: 1160–1163

Katerere , D.R., Stockenstrom, S., Thembo, K.M., Rheeder, J.P., Shephard, G.S. and Vismer, H.S. (2008). A preliminary survey of mycological and fumonisin and aflatoxin contamination of African traditional herbal medicines sold in South Africa *Human and Experimental Toxicology* 27 (11): 793-798

Kelly, J.D., Eaton, D.L., Guengerich, F.P. and Colombe (1997) Aflatoxin B_1 activation in human lung *Toxicology and Applied Pharmacology 144*, 88-95.

Kew, M. C. (2005) Prevention of hepatocellular carcinoma, *HPB 7*, 16-25.

Kew, M. C. (2010a) Hepatocellular carcinoma in African blacks: recent progress in etiology and pathogenesis *World Journal of Hepatology 2*, 65-73.

Kew, M. C. (2010b) Prevention of hepatocellular carcinoma *Annals of Hepatology 9*, 120 - 132.

Kew, M. C. and Asare, G.A. (2007) Dietary iron overload in the African and hepatocellular carcinoma *Liver International 27*, 735-741.

Kew, M. C. and Popper, H. (1984) Relationship between hepatocellular carcinoma and cirrhosis *Seminars in Liver Disease 4*, 136-146.

Kimura, A. (2011) Reye's syndrome and Reye-like syndrome, *Nippon Rinsho 69*, 455-459. (In Japanese, PubMed 21400838).

Kirk, G. D., Lesi, O. A., Mendy, M., Akano, A. O., Sam, O., Goedert, J. J., Hainaut, P., Hall, A. J., Whittle, H. and Montesano, R. (2004) The Gambia liver cancer study: infection with hepatitis B and the risk of hepatocellular carcinoma in west Africa *Hepatology 39*, 211-219.

Kotan, E., Alpsoy, L., Anar, M., Aslan, A. and Agar, G. (2011) Protective role of methanol extracts of *Cetrartia islandica* against oxidative stress and genotoxic effects of aflatoxin B_1 in human lymphocytes *in vitro Toxicology and Industrial Health 2011*, In press.

Kpodo, K., Sorensen, A.K. and Jakobsen, M.(1996). The occurrence of mycotoxins in fermented maize products *Food Chemistry* 56 (2):147-153.

Krishnamachari, K. A., Bhat, R. V., Nagarajan, V. and Tilak, T. B. (1975) Hepatitis due to aflatoxicosis. An outbreak in Western India *Lancet 1(7915)*, 1061-1063.

Kuniholm, M. H., Lesi, O. A., Mendy, M., Akano, A. O., Sam, O., Hall, A. J., Whittle, H., Bah, E., Goedert, J. J., Hainaut, P. and Kirk, G. D. (2008) Aflatoxin exposure and viral hepatitis in the etiology of liver cirrhosis in the Gambia West Africa *Environmental Health Perspectives 116*, 1553-1557.

Lamplugh, S. M. and Hendrickse, R. G. (1982) Aflatoxins in the livers of children with kwashiorkor *Annals of Tropical Paediatrics 2*, 101-104.

Lane, K.S, (2005). New support for FDA regulation of tobacco. Retrieved on July, 2006 from www.Tobacco.org.

Langouet, S., Coles, B., Morel, F., Becquemonet, L., Beaune, P., Guengrich, P., F., Ketterer, B. and Guillouzo, A. (1995) Inhibition of CYP1A2 and CYP3A4 by olitpraz results in reduction of aflatoxin B_1 metabolism in human hepatocytes in primary culture *Cancer Research 55*, 5574-5579.

Lata, J. (2010) Chronic liver disease as tumor precursors, *Digestive Diseases 28*, 596-599.

Leslie, J.F., Bandyopadhyay, R. and Visconti, A. (2008), Mycotoxins: Detection methods, management, public health and agricultural trade. Cromwell Press, Trowbridge, UK. Pp 476

Lilleberg, S. L., Cabonce, M. A., Raju, N. R., Wagner, L. M. and Kier, L. D. (1992) Alterations in the structural gene and expression of p53 in rat liver tumors induced by aflatoxin B_1 *Molecular Carcinogenesis 6*, 159-172.

Lin, C. A., Boslaugh, S., Ciliberto, H. M., Maleta, K., Ashorn, P., Briend, A. and Manary, M. J. (2007) A prospective assessment of food and nutrient intake in a population of Malawian children at risk for kwashiorkor *Journal of Paediatric and Gastroenterological Nutrition 44*, 487-493.

Liu, Y. and Wu, F. (2010) Global burden of aflatoxin-induced hepatocellular carcinoma: a risk assessment *Environmental Health Perspectives 118*, 818-824.

Livni, N., Eid, A., Ilan, Y., Rivkind, A., Rosenmann, E., Blendis, L. M., Shouval, D. and Galun, E. (1995) p53 expression in patients with cirrhosis with and without hepatocellular carcinoma *Cancer 75*, 2420-2426.

Long, X. D., Ma, Y., Zhou, Y. F., Yao, J. G., Ban, F. Z., Huang, Y. Z. and Huang, B. G. (2009) XPD codon 312 and 751 polymorphisms and aflatoxin B_1 exposure and hepatocellular carcinoma risk *BMC Cancer 9*, 400-409.

Lubick, N (2010) Examining DDT's urogenital effects. *Environmental Health Perspectives 118*, A18

Luongo, D., Severino, L., Bergamo, P., De Luna, R., Lucisano, A. and Rossi, M. (2006) Interactive effects of fumonisin B_1 and alpha-zearalenol on proliferation and cytokine expression in Jurkat T cells *Toxicology in Vitro 20*, 1403-1410.

Lye, M. S., Ghazali, A. A., Mohan, J., Alwin, N. and Nair, R. C. (1995) An outbreak of acute hepatic encephalopathy due to severe aflatoxicosis in Malaysia *American Journal of Tropical Medical Hygiene 53*, 68-72.

Maenetje, P.W. and Dutton, M.F (2007). The incidence of fungi and mycotoxins in South African barley and barley products *Journal of Environmental Science and Health*, Part B: 42, (2) 229 - 236

Makun, H.A, Anjorin, S.T., Moronfoye, B., Adejo, F.O., Afolabi, O.A., Fagbayibo, G., Balogun, B.O. and Surajudeen, A.A. (2010). Fungal and aflatoxin contaminations of some human food commodities in Nigeria. *African Journal of Food Sciences*. 4 (4): 127 - 135

Makun, H. A., Dutton, M.F., Njobeh, P.B., Mwanza, M. and Kabiru A.Y. (2011) Natural multi- mycotoxin occurrence in rice from Niger State, Nigeria *Mycotoxin Research*. 27 (2): 97-104.

Makun HA, Gbodi TA, Akanya HO, Sakalo AE, Ogbadu HG (2007) Fungi and some mycotoxins contaminating rice (Oryza sativa) in Niger state, Nigeria. *African Journal of Biotechnology* 6(2):99–108

Makun HA, Gbodi TA, Akanya HO, Salako EA, Ogbadu GH (2009) Fungi and some mycotoxins found in mouldy Sorghum in Niger State, Nigeria. *World Journal of Agricultural Sciences* 5(1):5–17

Manary, M. J. and Brewster, D. R. (1997) Potassium supplement in kwashiorkor *Journal of Paediatric and Gasteroenterological Nutrition 24*, 194-201.

Manson, M. M., Ball, H. W. L., Barrett, M. C., Clark, H. L., Jujah, D. J., Williamson, G. and Neal, G. E. (1997) Mechanism of action of dietary chemoprotective agents in rat liver: induction of phase I and II drug metabolizing enzymes and aflatoxin B_1 metabolism *Carcinogenesis 18*, 1729-1738.

Matumba, L., Monjerezi, M., Khonga, E. B., Lakudzala, D.D. (2011). Aflatoxins in sorghum, sorghum malt and traditional opaque beer in southern Malawi. *Food Control* 22, 266-268

Mashinini, K. and Dutton, M.F. (2006) The incidence of fungi and mycotoxins in South Africa wheat and wheat-based products *Journal of Environmental Science and Health Part B*, 41:285-296

Maxwell, D., Levin, C., Armar-Klemesu, M., Ruel, M., Morris, S., Ahiadeke, C. (2000). Urban Livelihoods and Food and Nutrition Security in Greater Accra, Ghana. IFPRI, Washington, p. 172.

May, P. and May, E. (1999) Twenty years of p53 research: structural and functional aspects of the p53 protein *Oncogene, 18* 7621–36.

Martin, P.M.D, and Gilman, G.A. (1976). A consideration of the mycotoxin hypothesis with special reference to the mycoflora of maize, sorghum, wheat and groundnut. *Rep. Trop. Prod. Inst.* G105, Vil=112pg. 1-63.

Mazen, M.B, El-Kady, I.A and Saber, S.M (1990). Survey of the mycoflora and mycotoxins of cotton seeds and cotton seed products in Egypt. *Mycopathologia* 110 (3):133-138.

MRC (Medical Research Council) (2006). Report on aflatoxins in groundnuts and peanuts products retrieved on 14th July, 2011 from www.doh.gov.za/department/foodcontrol/docs/nmp.html

Mestres, C., Bassa, S., Fagbohoun, E., Nago, M., Hell, K., Vernier, P., Champiat, D., Hounhouigan, J. and K. F. Cardwell. (2004). Yam chip food sub-sector: hazardous practices and presence of aflatoxins in Benin. *Journal of Stored Products Research* 40 (5): 575-585

Miller, J.D. (1995). Fungi and mycotoxins in grains: Implications for stored product research. *J. Stored. Prod. Res.* 31 (1): 1-16

Miller, J.D. (1996). Mycotoxins. In: Cardwell, K.F. (Ed.), Proceedings of the Workshop on Mycotoxins in Food in Africa. November 6–10, 1995, Cotonou, Benin. International Institute of Tropical Agriculture, Cotonou,Benin, pp. 18–22.

Mircea, C., Poiata, A., Tuchilus, C., Agoroae, L., Butnaru, E. and Stanescu, U. (2008) Aflatoxigenic fungi isolated from medicinal herbs *Toxicology Letters*. 180: 32-246

Motawee, M.M., Bauer, J. and McMahon, D.J. (2009) Survey of aflatoxin M_1 in cow, goat, buffalo and camel milks in Ismailia-Egypt. *Bulletin of the Environmental and Contamination Toxicology, 83,* 766-769.

Mokhles, M., Abd El Wahhab, M. A., Tawfik, M., Ezzat, W., Gamil, K. and Ibrahim, M. (2007) Detection of aflatoxin among hepatocellular carcinoma in patients in Egypt *Pakistan Journal of Biological Sciences 10,* 1422-1429.

Mngadi, P.T., Govinden, R. and Odhav, B. (2008). Co-occurring mycotoxins in animal feeds. *African Journal of Biotechnology* 7 (13): 2239-2243.

Muhammad, S., Shehu, K., Amusa, N.A., (2004). Survey of the market diseases and aflatoxin contamination of tomato (Lycopersicon esculentum Mill) fruits in Sokoto, northwestern Nigeria. *Nutrition and Food Science* 34, 72–76.

Murray, C. J. L. (1994) Quantifying the burden of disease: the technical basis for disability-adjusted life years *Bulletin of the World Health Organisation 72*, 429-445.

Mutegi, C.K., Ngugi, H.K., Hendriks, S.L. and Jones, R.B.(2009). Prevalence and factors associated with aflatoxin contamination of peanuts from Western Kenya *International Journal of Food Microbiology* 130, 27–34.

Muthomi, J.W., Ndung'u, J.K., Gathumbi, J.K., Mutitu, E.W. and Wagacha, J.M. (2008). The occurrence of Fusarium species and mycotoxins in Kenyan wheat. *Crop Protection* 27, 1215– 1219.

Muzanila, Y.C., Brennan, J.G. and King, R.D. (2000) Residual cyanogens, chemical composition and aflatoxins in cassava ¯our from Tanzanian villages *Food Chemistry* 70, 45-49

Mwanza, M. (2007) A survey of fungi and mycotoxins with respect to South African domestic animals in the Limpopo Province. Master of Technology Dissertation, University of Johannesburg. http://152.106.6.200:8080/dspace/bitstream/10210/884/1/Mwanza%20M%20tech%20dissertation.pdf

Mwanza, O. W., Otieno, C. F. and Omonge, E. (2005) Acute aflatoxicosis: case report *East African Medical Journal 82*, 320-324.

Nelson, D. B., Kimbrough, R., Landrigan, P. S., Hayes, A. W., Yang, G. C. and Benanides, J. (1980) Aflatoxin and Reye's syndrome: a case study *Pediatrics 66*, 865-869.

Njobeh, B.P., Dutton, M.F., Koch, S.H. and Chuturgoon, A. (2009) Contamination with storage fungi of human foods from Cameroon *International Journal of Food Microbiology* 135, 193-198.

Njobeh B. P., Dutton F. M., Makun, H.A (2010). Mycotoxins and human health: Significance, prevention and control In: Ajay K. Mishra, Ashutosh Tiwari, and Shivani B. Mishra (Eds) 'Smart Biomolecules in Medicine' VBRI Press, India 132-177

Nyathi, C. B., Mutiro, C. P., Hasler, J. A. and Chetsanga, C. J. (1987) A survey of urinary aflatoxin in Zimbabwe *International Journal of Epidemiology 16*, 516-519.

Nyathi, C. B., Mutiro, C. F., Hasler, J. A. and Chetsang, C. J. (1989) Human exposure to aflatoxins in Zimbabwa *Central African Journal of Medicine 35*, 542-545.

Nyikal, J., *et al.* (2004) Outbreak of aflatoxin poisoning - Eastern and Central Provinces, Kenya January-July 2004, *Morbidity and Mortality Weekly Control 53*, 790-793.

Obidoa, O and Gugnani, H.C. (1992). Mycotoxins in Nigerian foods: causes, Consequences and remedial measures. In Z.S.C Okoye (ed) Book of proceedings of the first National Workshop on Mycotoxins held at University Jos, on the 29th November, 1990, 95-114.

Obilana AB. 2002. 'Overview: importance of millets in Africa', Retrieved on 1st June, 2011 from Website: www.afripro.org.uk/papers/Paper02Obilana.pdf

Ocama, P., Nambooze, S., Opio, C. K., Shields, M. S., Wabinga, H. R. and Kirk, G. D. (2009) Trends in the incidence of primary liver cancer in central Uganda. 1960-1980 and 1991-2005 *British Journal of Cancer 100*, 799-802.

Odoemelam, S. A and Osu, C.I (2009). Aflatoxin B1 contamination of some edible grains marketed in Nigeria. *E-Journal of Chemistry* 6 (2):308-314.

Ogbadu, G. (1979) Effect of low gamma irradiation on the production of aflatoxin B1 by *Aspergillus flavus* growing on *Capsicum annuum Microbios letters* 10, 139-142.

Ogbadu, G. and Bassir, O. (1979) Toxicological study of γ-irradiated aflatoxins using the chicken embryo *Toxicology and Applied Pharmacology* 51, 379-382.

Ogbadu G (1980a) Influence of gamma irradiation of aflatoxin B1 production by *Aspergillus flavus* growing on some Nigerian foodstuffs. *Microbios.*; 27(107):19-26.

Ogbadu, G. (1981) Ultra structural changes in gamma-irradiated *Aspergillus flavus* spores *Cytobios* 30, 167-171.

Ogbadu, G. (1988) Use of gamma irradiation to prevent aflatoxin B1 production in smoked dried fish *International Journal of Applied Instrumentation* 31, 207-207.

Ogunsanwo, B.M., Faboya, O.O., Idowu, O.R., Lawal, O.S. and Bankole, S.A (2004). Effect of roasting on the aflatoxin contents of Nigerian peanut seeds. *African Journal of Biotechnology* 3 (9): 451-455

Okoye, Z. S. C. (1987). Carryover of aflatoxin B1 in contaminated substrate corn into Nigerian native beer. *Bulletin of Environmental Contamination and Toxicology* 37 (4) 482-489

Olson, L.C., Bourgeois, C.H., Cotton, R.B., Harikul, S., Grossman, R.A. and Smith, T.J. (1971) Encephalopathy and fatty degeneration of the viscera in northeastern Thailand. Clinical syndrome and epidemiology *Paediatrics 47*, 707- 716.

Omer, R. E., Bakker, M. I., van't Veer, P., Hoogenboom, R. L., Polman, T. H., Alink, G. M., Idris, M. O., Kadaru, A. M. and Kok, F. J. (1998) Aflatoxin and liver cancer in Sudan *Nutrition and Cancer 32*, 174-180.

Ominski, K.H., Marquardi, R.R., Sinha, R.N and Abramson, D (1994). Ecological aspects of growth and mycotoxin production by storage fungi. In: Miller, J.D and Trenholm, H.L (1994). Mycotoxins in grains: Compounds other than aflatoxins. Eagan Press, St. Paul Minnesota, USA. 287-314.

Opadokun, J.S (1992). Occurrence of aflatoxin in Nigerian food crops In Z.S.C Okoye (ed) Book of proceedings of the first National Workshop on Mycotoxins held at University Jos, on the 29th November, 1990, 95-114.

Orsi, R. B., Oliveira, A. A., Dilkin, P., Xavier, J. G., Direito, G. M. and Correa, B. (2007) Effects of oral administration of aflatoxin B_1 and fumonisin B_1 in rabbits (*Oryctiolagus cuniculus*) *Chemical and Biological Interactions 170*, 201-208.

Otsuki, T., Wilson, J.S. and Sewadeh, M. (2001) What price precaution? European harmonization of aflatoxin regulations and African groundnut exports. *European Review of Agricultural Economics* 28: 263-283.

Oyelamin, O. A., and Ogunlesi, T. A. (2007) Kwashiorkor - is it a dying disease?, *South African Medical Journal 97*, 65-68.

Oyelami, O. A., Maxwell, S. M., Aladekomo, T. A. and Adelusola, K. A. (1995) Two unusual cases of kwashiorkor: can protein deficiency explain the mystery? *Annals of Tropical Paediatrics 15*, 217-219.

Oyelami, O. A., Maxwell, S. M., Adelusola, K. A. and Oyelese, A. O. (1997) Aflatoxins in the lungs of children with kwashiorkor and children with miscellaneous diseases in Nigeria *Journal of Toxicology and Environmental Health 51*, 623-628.

Oyero, G.O and Oyefolu, A.B (2010) Natural occurrence of aflatoxin residues in fresh and sun-dried meat in Nigeria *Pan African Medical Journal.* 7:14

Paterson, R.R.M and Lima, N (2010) How will climate change affect mycotoxins in foods? *Food Research International* 43, 1902 - 1914

Peers, F. G. and Linsell, C. A. (1973) Dietary aflatoxins and human liver cancer. A population based study in Kenya *British Journal of Cancer 27*, 473-484.

Perz, J. F., Armstrong, G. L., Farrington, L. A., Hutin, Y. J. and Bell, B. P. (2006) The contribution of hepatitis B virus and hepatitis C virus infection to cirrhosis and primary liver cancer world wide *Journal of Hepatology 45*, 529-538.

Peterson, S.W., Ito, Y., Horn, B.W., Goto, T. (2001) Aspergillus bombycis, a new aflatoxigenic species and genetic variation in its sibling species, A. nomius. *Mycologia* 93, 689–703

Pietri, A., Bertuzzi, T., Agosti, B. and Donadini, G. (2010). Transfer of aflatoxin B_1 and fumonisin B_1 from naturally contaminated raw materials to beer during an industrial brewing process. *Food Additives & Contaminants:* Part A: Chemistry, Analysis, Control, Exposure & Risk Assessment. 27 (10): 1431 - 1439

Pildain, M.B., Frisvad, J.C., Vaamonde, G., Cabral, D., Varga, J. and Samson, R.A. (2008). Two novel aflatoxin-producing *Aspergillus* species from Argentinean peanuts *International Journal of Systematic and Evolutionary Microbiology* 58, 725-735.

Probst, C., Njapau, H., and Cotty, P. J. (2007) Outbreak of an acute aflatoxicosis in Kenya in 2004: identification of the causal agent, *Applied and Environmental Microbiology 73*, 2762-2764.

Rehana, F. and Basappa, S.C. (1990). Detoxification of aflatoxin B1 in maize by different cooking methods. *Journal Food Science Technology*. 27: 379-399.

Raisuddin, S., Singh, K. P., Zaidi, S. I. A., Paul, B. N. and Ray, P. K. (1993) Immunosuppressive effects of aflatoxin in growing rats *Mycopathologia 124*, 189-194

Ramjee, G., Berjak, P., Adhikari, M. and Dutton, M. F. (1992) Aflatoxins and kwashiorkor in Durban, South Africa *Annals of Tropical Paediatrics 12*, 241-247.

Reiter EV, Dutton MF, Mwanza M, Agus A, Prawano D, Häggblom P, Razzazi-Fazeli E, Zentek J, Andersson G, Njobeh PB (2011) Quality control of sampling for aflatoxins in animal feedingstuffs: Application of the Eurachem/CITAC guidelines. *Analyst.* 136 (19): 4059-4069

Reye, R.D.K., Morgan, G. and Baral, J. (1963) Encephalopathy and fatty degeneration of the viscera: a disease entity in childhood Lancet 2: 749-752.

Riba, A., Bouras, N., Mokrane, S., Mathieu, F., Lebrihi, A. and Sabaou, N (2010). Aspergillus section Flavi and aflatoxins in Algerian wheat and derived products. *Food and Chemical Toxicology* 48, 2772–2777

Robertson, A. (2005). Risk of aflatoxin contamination increases with hot and dry growing conditions. Integrated Crop Management 185–186. Retrieved on 11th May, 2011 from http://www.ipm.iastate.edu/ipm/icm/2005/9-19/aflatoxin.html.

Rogan, W. J., Yang, G. C. and Kimborough, R. D. (1985) Aflatoxin and Reye's syndrome: a study of livers from deceased cases *Archives of Environmental Health 40*, 91-95.

Ryan, N. J., Hogan, G. R., Hayes, A. W., Unger, P. D. and Sirai, M. Y. (1979) Aflatoxin B_1; its role in the etiology of Reye's syndrome *Paediatrics 64*, 71-75.

Sahar, N., Ahmed, M., Parveen, Z., Ilyas, A. and Bhutto, A. (2009). Screening of mycotoxins in wheat, fruits and vegetables grown in Sindh, Pakistan *Pakistan Journal Botany.*, 41(1): 337-341

Sánchez-Hervás, M., Gil, J.V., Bisbal, F., Ramón, D and Martínez-Culebras, P.V (2008). Mycobiota and mycotoxin producing fungi from cocoa beans. *International Journal of Food Microbiology* 125:336–340.

Schindler, A.F. (1977) Temperature limits for production of aflatoxin by twenty-five isolates of *Aspergillus flavus* and *Aspergillus parasiticus*. *Journal of Food Protection*. 40:39–40.

Sharma, R. P. (1993) Immunotoxicity of mycotoxins *Journal of Dairy Science 76*, 892-897.

Shuaib, F. M., Ehiri, J., Abdullahi, A., Williams, J. H., and Jolly, P. E. (2010a) Reproductive health effects of aflatoxins: a review of the literature, *Reproductive Toxicology 29*, 262-270.

Shuaib, F. M., Jolly, P. E., Ehiri, J., Jiang, Y., Ellis, W. O., Stiles, J. K., Yatich, N. J., Funkhouser, E., Person, S. D., Wilson, C. and Williams, J. H. (2010b) Association between anemia and aflatoxin B_1 biomarker levels among pregnant women, *American Journal of Tropical Medical Hygiene 83*, 1077-1083.

Sibanda L, Marovatsanga LT, Pestka JJ (1997) Review of mycotoxin work in sub-Saharan Africa. *Food Control* 8:21–29 647.

Sinha, R.N (1984). *Journal of Economic Entomology* 77: 1463-1488.

Smith, T.K., Mehrdad, M. and Ewen, J.M. (2000) Biotechnology in the Feed Industry *.Proceedings of Alltech's 16th Annual Symposium*, Pp 383–390.

Smith, T. K., McMillan, E. G. and Castillo, J. B. (1997) Effect of feeding blends of *Fusarium* mycotoxin-contaminated grains containing deoxynivalenol and fusaric acid on growth and feed consumption of immature swine *Journal of Animal Science 75*, 2184-2191.

Smith, J.E and Moss M.O. (1985). Mycotoxins: formation, analysis and significance. John Wiley & sons. Chichester, Britain, 83-103.

South African Department of Health (2004a) Government Gazette 6th March 2009. Regulations governing tolerance for fungus-produced toxins in foodstuffs. Foodstuff, Cosmetics and Disinfectants Act 1972 (Act 54 of 1972).

South African Department of Health (2004b). Government Gazette 6th March 2009. Regulations governing tolerance for fungus-produced toxins in foodstuffs. Foodstuff, Cosmetics and Disinfectants Act 1972 (Act 54 of 1972) 1.

Speijers, G. J. and Speijers, M. H. (2004) Combined toxic effects of mycotoxins *Toxicology Letters 153*, 91-98.

Stoev, S. D., Goundasheva, D., Mirtcheva, T. and Mantle, P. G. (2000) Susceptibility to secondary bacterial infections in growing pigs as an early response in ochratoxicosis *Experimental Toxicology and Pathology 52*, 287-296.

Stoloff, L. (1989) Aflatoxin is not a probable human carcinogen: the published evidence is sufficient *Regulatory Toxicology and Pharmacology 10*, 272-283.

Stora, C., Dvorackova, I. and Ayraud, N. (1983) Aflatoxin and Rey's syndrome *Journal of Medicine 14*, 47-54.

Stössel, P. (1986). Aflatoxin contamination in soybeans: role of proteinase inhibitors, zinc availability, and seed coat integrity. *Applied and Environmental Microbiology* 52, 68–72.

Sylla, A., Diallo, M.S., Castegnaro, J., Wild, C.P (1999). Interaction between hepatitis V virus and exposure to aflatoxin in the development of hepatocellular carcinoma: a molecular epidemiological approach. *Mutation Research* 428:187-196

Tandon, H. D., Tandon, B. N. and Ramalingaswami, V. (1978) Epidemic toxic hepatitis in India of possible mycotoxin origin *Archives of Pathology and Laboratory Medicine 102*, 372-376.

Taylor, E. R. (1992) Aflatoxin B_1 and DNA adducts. Proposed model for surface noncovalent and covalent complexes with N7 of guanine *Journal of Biomolecular Structure and Dynamics 10*, 533-550.

Tchana, A. N., Moundipa, P. F. and Tchouanguep, F. M. (2010) Aflatoxin contamination in food and body fluids in relation to malnutrition and cancer status in Cameroon *International Journal of Environmental Research and Public Health 7*, 178-188.

Trauner, D. A. (1984) Reye's syndrome *The Western Journal of Medicine 141*, 206-209.

Trenk, H.L., and Hartman, P.A. (1970) Effects of moisture content and temperature on aflatoxin production in corn. *Applied Microbiology*. 19:781–784.

Trucksess, M. W. and Scott, P. M.(2008) 'Mycotoxins in botanicals and dried fruits: A review', *Food Additives & Contaminants*: Part A, 25: 2, 181 — 192

Tsega, E. (1977) Current views on liver diseases in Ethiopia *Ethiopian Medical Journal 15*, 75-82.

Turkez, H., and Geyikoglu, F. (2010) Boric acid: a potential chemoprotective agent against aflatoxin B_1 toxicity in human blood, *Cytotechnology 62*, 157-165.

Turner, P.C., Moore S.E., Hall, A.J., Prentice A.M. and Wild C.P. (2003) Modification of immune function through exposure to dietary aflatoxin in Gambian children *Environmental Health Perspectives 111*, 217-220.

Turner, P.C., Collinson, A.C., Cheung, Y.B., Gong, Y., Hall, A.J., Prentice, A.M. and Wild, C.P. (2007) Aflatoxin exposure *in utero* causes growth faltering in Gambian infants *International Journal of Epidemiology 36*, 1119-1125.

Udoh, J.M., Cardwel, K.F., Ikotun, T. (2000) Storage structures and aflatoxin content of maize in five agro-ecological zones of Nigeria. *Journal of Stored Products Research* 36, 187–201.

US Grain Council (2000) Manual pp 1-42

Van Rensburg, S. J., Cook-Mozaffari, P., Van Schalkwyk, J. J., Van der Watt, J. J., Vincent, T. J. and Purchase, I. F. (1985) Hepatocellular carcinoma and dietary aflatoxins in Mozambique and Transkei, *British Journal of Cancer 51*, 713-726.

Van Vleet, T. R., Klein, P. J. and Coulombe, R. A. (2001) Metabolism of aflatoxin B_1 by normal bronchial epithelial cells *Journal of Toxicology and Environmental Health 63*, 525-540.

Van Vleet, T. R., Klein, P. J. and Coulombe, R. A. (2002a) Metabolism and cytotoxicity of aflatoxin B_1 in cytochrome P_{450} expressing human lung cells *Journal of Toxicology and Environmental Health A 65*, 853-867.

Van Vleet, T. R., Mace, K. and Coulombe, R. A. (2002b) Comparative aflatoxin B_1 activation and cytotoxicity in human bronchial cells expressing cytochromes P_{450} 1A2 and 3A4 *Cancer Research 62*, 105-112.

Wagacha, J.M., Muthomi, J.W. (2008) Mycotoxin problem in Africa: current status, implications to food safety and health and possible management strategies. *International Journal Food Microbiology*. 124, 1–12.

Wang, J. S. and Groopman, J. D. (1999) DNA damage by mycotoxin *Mutation Research 424*, 167-181.

Wang, J.-S., Shen, X., He, X. H., Zhu, Y.-R., Zhang, B.-C., Wang, J.-B., Qian, G.-S., Kuang, S.-Y., Zarba, A., Egner, P. A., Jacobson, L. P., Munoz, A., Helzlsouer, Groopman, J. D. and Kensler, T. W. (1999) Protective alterations in Phase 1 and 2 metabolism of aflatoxin B_1 by oltipraz in residents of Qidong, People's republic of China *Journal of the National Cancer Institute 91*, 347-354.

Waterlow, J.C. (1984) Kwashiorkor revisited: the pathogenesis of oedema in kwashiorkor and its significance. Transactions of the Royal Society of Tropical Medical Hygiene *78*, 436-441.

Wheeler, C.W., Park, S.S. and Guenthner, T.M. (1990) Immunological analysis of cytochrome P450 1A1 homologue in human lung microsomes *Molecular Pharmacology 38*, 634-643.

World Health Organization (WHO) (2006) Mycotoxins in African foods: implications to food safety and health. AFRO Food Safety Newsletter. World Health Organization Food safety (FOS), Issue No. July 2006. www.afro.who.int/des.

Wild, C. P. (1996). Summary of data on aflatoxin exposure in West Africa. In K. F. Cardwell (Ed.), Proceedings of the workshop on mycotoxins in food in Africa, November 6–10, 1995 (p. 26). Cotonou, Benin: International Institute of Tropical Agriculture

Wild, C. P. and Gong, H. Z. (2010) Mycotoxins and human disease: a largely ignored global health, *Carcinogenesis 31*, 71-82.

Wild, C. P., Jiang, Y. Z., Sabbioni, G., Chapot, B. and Montesano, R. (1990) Evaluation of methods for quantitation of aflatoxin-albumin adducts and their application to human exposure assessment *Cancer Research 50*, 245-251.

Wild, C. P., Hasegawa, R., Barraud, L., Chutimataewin, S., Chapot, B., Ito, N. and Montesano, R. (1996) Aflatoxin-albumin adducts: a basis for comparative carcinogenesis between animals and humans *Cancer Epidemiology, Biomarkers and Prevention 5*, 179-189.

Williams, J. H., Phillips, T. D., Jolly, P. E., Stiles, J. K., Jolly, C. M. and Aggarwal, D. (2004) Human aflatoxicosis in developing countries a review of toxicology, exposure potential health consequences and intervention *American Journal of Clinical Nutrition 80*, 1106-1122.

Wogan, G. (1973) Aflatoxin carcinogenesis *Methods in Cancer Research 7*, 309-344.

Wolzak,A., Pearson, . A. M., Coleman, T. H., Pestka, J. J. and Gray, J. I. (1985). Aflatoxin deposition and clearance in the eggs of laying hens *Food and Chemical Toxicology*, 23 (12): 1057-1061

Wolzak, A., Pearson, A. M., Coleman, T. H., Pestka, J. J., Gray, J. I. and Chen, C. (1986) Aflatoxin carryover and clearance from tissues of laying hens *Food and Chemical Toxicology* 24, 37-41

Wood, R.D. (1999) DNA damage recognition during nucleotide excision repair in mammalian cells, *Biochemie 81*, 39-44.

Wu, F. (2004) Mycotoxin risk assessment for the purpose of setting international regulatory standards *Environmental Science and Technology 38*, 4049-4055.

Wu, F. (2006) Mycotoxin reduction in Bt corn: potential economic, health and regulatory impacts *Transgenic Research 15*, 277-289.

Wu, F. and Khlangwiset, P. (2010a) Health economic impacts and cost-effectiveness of aflatoxin reduction strategies in Africa: case studies in biocontrol and post harvest interventions *Food Additives and Contaminants 27*, 496-509.

Wu, F. and Khlangwiset, P. (2010b) Evaluating the technical feasibility of aflatoxin risk reduction strategies in Africa *Food Additives and Contaminants 27*, 658-676.

Yadgiri, B., Reddy, V., Tulpule, P. G., Srikantia, S. G. and Goplan, C. (1970) Aflatoxin and Indian childhood cirrhosis *The American Journal of Clinical Nutrition 23*, 94-98.

Yousef, A.E. and Marth, E.H. (1985) Degradation of aflatoxin M_1 in milk by ultraviolet energy. *Journal of Food Protection*, 48, 697–698.

Zain, M.E (2010). Impact of mycotoxins on humans and animals. Journal of Saudi Chemical Society. Retrived on 16th

Zarba A., Wild C.P., Hall A.J., Montesano R, Hudson G.J. and Groopman JD. (1992) Aflatoxin M1 in human breast milk from The Gambia, west Africa, quantified by combined monoclonal antibody immunoaffinity chromatography and HPLC *Carcinogenesis 13*, 891-894.

Zhu, L. R., Thomas, P. E., Lu, G., Reuhl, K. R., Yang, G. Y., Wang, L. D., Wang, S. L., Yang, C. S., He, X. Y. and Yan, H. J. (2006) CYP2A13 in human respiratory tissues and lung cancers: an immunohistochemical study with a new peptide specific antibody *Drug Metabolism and Deposition 34*, 1672-1676.

Zinedine, A., Juan, C., Soriano, J.M., Moltó, J.C., Idrissi, L. and Mañes, J. (2007a). Limited survey for the occurrence of aflatoxins in cereals and poultry feeds from Rabat, Morocco *International Journal of Food Microbiology* 115, 124–127

Zinedine, A., González-Osnaya, L., Soriano, J.M., Moltó, J.C., Idrissi, L., Mañes, J. (2007b). Presence of aflatoxin M1 in pasteurized milk from Morocco *International Journal of Food Microbiology* 114, 25–29

Zinedine, A., Brera, C., Elakhdari, S., Catano, C., Debegnach, F., Angelini, S., De Santis, B.,Faid, M., Benlemlih, M., Minardi, V., Miraglia, M. (2006) Natural occurrence of mycotoxins in cereals and spices commercialized in Morocco. *Food Control* 17, 868–874.

11

Study of Evaluation Machinability of a Stainless Steels and Accompanying Phenomena in the Cutting Zone During Machining

Jozef Jurko[1], Anton Panda[1] and Tadeusz Zaborowski[2]
[1]Technical University of Košice,
[2]IBEN Wlkp
[1]Slovak Republic
[2]Poland

1. Introduction

The production process - cutting process (working materials) is currently still one of the fundamental technologies of production parts. It is therefore necessary that the cutting process was effective, the products are competitive mainly in terms of production costs. If we consider the technological system M-machine, T-tool, W-workpiece and F-fixture, each element of this system has a different percentage of production costs. But the most important reason for the need of productive machining production time is that it is necessary to always use the best. This is a guide to success for a large part of production processes, ie maximum number of products per unit of time. Especially during operation, for variations in technological progress in production, there are significant changes in the cost per piece produced parts. Currently we can apply several alternatives for increasing productivity cutting process:

a. Alternative: Investment in new CNC machine to ensure the production program is especially important that this machine can significantly increase performance and speed aside, to improve the conditions of production and thereby ensuring returns in the coming years, and profit.
b. Alternatives: Improving the cutting process is a way of improving the cutting process continuously, provided the correct application of cutting tools - for defined machining conditions and for a defined set of M-F-W technology. In comparison with a) this solution is inexpensive (compared to the price of a new machine).

Developments in the machining method is dynamic and applied a decade or five years ago are largely already non-progressive. Seeing similar developments in the production of advanced cutting tools (for a defined structure for the defined basic materials for defined surface layer). On this basis, it is possible to achieve a productive day as production. one operating the machine, especially the change of applied cutting tools. Non-use benefits of development and inability to correct application of advanced cutting tools may be one outcome, namely: untapped potential for increasing its profits and competitiveness. Nature of the cutting process is that there is a complete plastic deformation separated workpiece (chip-form) by interaction T (cutting tool) and W (workpiece), defined in terms of a defined technological system M-F. Application of advanced cutting tools to be mainly in terms of economy: faster return on investment (new machinery), improve performance (old machine) and labor productivity service. Currently one of the cost items, which affects the production costs for 1 piece parts is heading: the cost of cutting tool. Every company in the total cost to produce one piece of different components defines the percentage of the cost of cutting tool. Although the costs of cutting tool, the proportion of the total cost to produce one piece of a few parts percentage, have a significant impact mainly on:

1. Cutting process - interaction (cutting tool-workpiece) - impact on the overall result (damage, surface roughness, defects, errors, change in mechanical properties), machined surfaces of the parts.
2. Hours of downtime of the machine - damaged cutting tool (its durability, respectively life, to ensure logistics system - storage, maintenance, service.
3. The number of necessary operations - change in technological progress, quality progressive tool can provide cost savings (production of precise holes - drilling, reaming).
4. Time, which parts must remain in the company.
5. Number of cutting tools, which must be available (logistics, availability).

At the cost mainly affects: labor, machinery, tooling, material, overhead and savings. Automated production of, in the sense of, machine production has characteristic features: a reduction of production costs, stimulation of the development of cutting tools, and changes in the construction of machine tools, all of which work against the creation of optimal technological methods, which thrusts the technological process of cutting into a more important position. These trends confirm that the cutting process remains one of the basic manufacturing technologies. A condition of the economic usage of modern, automated programmed machines is the optimal course of the cutting process, i.e. the use of optimal work conditions. A summary of optimal work conditions requires knowledge of the laws of cutting theory and knowledge of the practical conditions of their application.

2. Stainless steels

Stainless steels are distinguished by many features, including:

- inherent resistance to corrosion and heat under given conditions,
- aesthetic factors, hygienic characteristics, ease of cleaning and sterilization,
- a high ratio of strength to weight,
- low magnetic permeability.

Stainless steels are fundamentally subdivided by their chemical composition and metallographic structure. Austenitic steels are the most extensive and thus the most important category of stainless steels. Several kinds of these steels are known, which differ among themselves in their carbon, nickel, and sometimes in their titanium content. Titanium is an important element, which increases the steel's resistance to intercrystalline corrosion.

Current developments in stainless steels is divided into three directions:

1. Development of several species containing up to 0.02% carbon
2. Development of new types of nitrogen added as an effective alloying elements
3. Development of steel with a further increase in the level of corrosion resistance.

The basic chemical elements of austenitic steels are resistant to high temperatures and loads chromium-nickel-chromium and nickel. To increase the added resistant to high temperatures and loads tungsten-W (about 2%), molybdenum, Mo (1-3%), ever-Nb niobium, titanium Ti, vanadium V, N-nitrogen. Chromium content varies from 12 to 20%. Chromium belongs to the group the ferrite formed elements to completely conclude the gamma region. In a purely binary system of iron (Fe) - chromium (Cr) achieves homogeneous area of gamma at 1075 ° C up to 10.6% chromium content. In technical alloys may be limits to move to higher chromium content, because it had a small amount of carbon and nitrogen gamma significantly expand the area. At high content of Cr in the binary system Fe - Cr appears fragile intermediate sigma phase with variable composition and hardness about 100 HV, which is stable up to temperatures of 815 ° C. When cooling is falling apart at 460 ° C eutectoid response to a phase α, α' the solid solution of Fe - Cr with higher iron content and higher content of chromium. Today, most assumed that the fragility of the high-chromium steels causes the coexistence of these phases, and not the sigma phase. The nickel content is higher than in austenitic steels 18/9, it must σbalance the impact the ferrite formed elements and avoid elimination phase which reduces impact strength. For alloys Fe-Ni could be envisaged with a large thermal hysteresis. For a content of about 7% nickel austenite transformation takes place even under normal conditions of cooling (especially the cooling rate) free diffusion, ie sliding mechanism similar to alloys of iron and manganese. According by Lavesa Fe_2Mo phase.σMolybdenum increases resistant to high temperatures and loads, however, supports the formation of undesirable phases and Vanadium increases the stress-rupture, especially at lower temperatures. At 700 ° C has little influence vanadium. The additive is especially effective in the presence of nitrogen, since the dislocation creep precipitated vanadium nitrides, whose gross even after long soaking at a temperature only slightly. Carbon inhibits secretion of delta ferrite and increases the mechanical properties. If it is excluded in the basic matrix precipitates as M_{23} (C, N)$_6$, or M_{23} (C, B)$_6$, slows the flow. In an unstable steels inhibits formation of sigma phase, which is produced only when the greater part is excreted as carbon carbide. Carbon suppresses the formation phase $Fe_{18}Cr_6Mo$ or chi (Fe, Ni)$_{18}Cr_9Mo_4$. Some Heat resistant austenitic steel containing a small amount of niobium or titanium with carbon are stable carbide. The highest stress-rupture of steel to the ratio Ti/C from 0.8 to 1.2. Similarly affects resistant to high temperatures and loads niobium, the optimal ratio nNb:NC is lower and more dependent on test temperature and time. The disadvantage of steel with niobium is their tendency to crack in the weld area and a decrease in toughness after long lasting working temperature. To increase the yield strength of hot, ultimate strength, respectively. stress-rupture is added to austenitic steels unstabilised small amounts of nitrogen, which slows down the diffusion rate of carbon and thus the elimination of coagulation and carbides $M_{23}C_6$ and other intermetallic phases. Because at the same time reduces the rate of substitutional alloying elements in grain boundaries, extended incubation period of precipitation. Solubility of nitrogen in the steel matrix with a maximum of 0.003% C at 1150 °C is about 0.30% at 600 to 650 ° C decreases to 0.005%. Admixture of nitrogen to 0.14% increases resistant to high temperatures and loadsaustenitic steels containing up to 0.03% as well, that their properties aligned carbon steel containing 0.06 to 0.09%.

2.1 Categories of stainless steels

Austenitic-A family of alloys containing chromium and nickel (and manganese and nitrogen when nickel levels are reduced), generally built around the type 302 chemistry of 18% Cr, 8% Ni, and balance mostly Fe. These alloys are not hardenable by heat treatment.

Ferritic - This group of alloys generally containing only chromium, with the balance mostly Fe, are based upon the type 430 composition of 17% Cr. These alloys are somewhat less ductile than the austenitic types and again are not hardenable by heat treatment.

Martensitic - The members of this family of stainless steels may be hardened and tempered just like alloy steels. Their basic building block is type 410 which consists of 12% Cr, 0.12% C, and balance mostly Fe.

Precipitation-Hardening - These alloys generally contain Cr and less than 8% Ni, with other elements in small amounts. As the name implies, they are hardenable by heat treatment.

Duplex - This is a stainless steel alloy group, or family, with two distinct microstructure phases, ferrite and austenite. The Duplex alloys have greater resistance to chloride stress corrosion cracking and higher strength than the other austenitic or ferritic grades. Classification of stainless steels defines also diagram on base equivalent Ni and Cr by Fig.1. Equations 1 and 2 described of equivalent Ni and Cr.

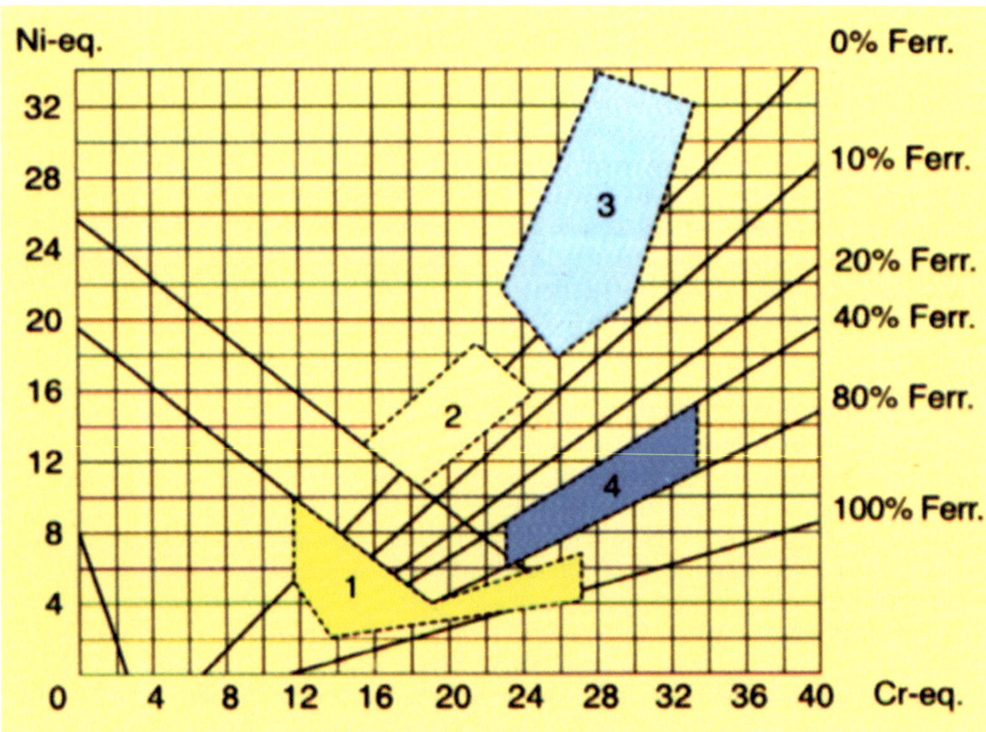

Fig. 1. Schäeffler diagram for Stainless steels, zone 1-structure ferrite, martenzite, zone 2-structure duplex (austenite, ferrite), zone 3-structure austenite, zone 4-structure ferrite

$$E_{Ni} = \%\,Ni + 30x\,\%\,C + 0{,}5x\,\%\,Mn \quad (1)$$

$$E_{Cr} = \%\,Cr + \%\,Mo + 1{,}5x\,\%\,Si + 0{,}5x\%\,Nb \quad (2)$$

In Table 1 are mentioned mostly applied stainless steels.

Alloy	C	Mn	P	S	Si	Cr	Ni	Mo	Others
301	0.15	2.00	0.045	0.030	1.00	17.00	7.00	-	-
302	0.15	2.00	0.045	0.030	1.00	18.00	9.00	-	-
303	0.15	2.00	0.20	0.15	1.00	18.00	9.00	0.60	-
303Se	0.15	2.00	0.20	0.06	1.00	18.00	9.00	0.60	0.15Se
304	0.08	2.00	0.045	0.030	1.00	19.00	9.25	-	-
304L	0.03	2.00	0.045	0.030	1.00	19.00	10.0	-	-
309S	0.08	2.00	0.045	0.030	0.75	23.00	13.5	-	-
310S	0.08	2.00	0.045	0.030	1.50	25.00	20.5	-	-
316	0.08	2.00	0.045	0.030	1.00	17.00	12.0	2.5	-
316L	0.03	2.00	0.045	0.030	1.00	17.00	12.0	2.5	-
317	0.08	2.00	0.045	0.030	1.00	19.00	13.0	3.5	-
317L	0.03	2.00	0.045	0.030	1.00	19.00	13.0	3.5	-
321	0.08	2.00	0.045	0.030	1.00	18.00	10.5	-	Ti 5 X C
329	0.10	2.00	0.045	0.030	1.00	27.50	4.5	1.50	-
347	0.08	2.00	0.045	0.030	1.00	18.00	11.0	-	Cb+Ta 10 X C
409	0.08	1.00	0.045	0.045	1.00	11.50	-	-	Ti 6 x C
410	0.15	1.00	0.040	0.030	1.00	12.50	-	-	-
416	0.15	1.25	0.040	-	1.00	13.00	-	0.60	S =0.15 min.
416Se	0.15	1.25	0.060	0.060	1.00	13.00	-	-	0.15 Se
420	0.15 min.	1.00	0.040	0.030	1.00	13.00	-	-	-
430	0.12	1.00	0.040	0.030	1.00	17.00	-	-	-
440C	1.00	1.00	0.040	0.030	1.00	17.00	-	-	-
904L	0.02	2.00	0.045	0.035	1.00	21.00	25.5	4.5	Cu 1.5
17-4 PH	0.07	1.00	0.045	0.035	1.00	16.5	5.5	-	Cu 3-5, 0.4 Al
17-7 PH	0.09	1.00	0.045	0.035	1.00	17.0	7.0	-	0.75-1.5 Al

Table 1. Chemical composition of the Stainless steels

3. Machining of stainless steels

Austenitic stainless steels are produced with graded carbon content. The content of chromium steels in this group of about 18% nickel content is tailored to the requirement that the steel structure was largely austenite. Minor phases present in the structure are made of ferrite and carbides of chromium δ mainly $M_{23}C_6$ type. Austenite is in this group of steels stable even at temperatures well below freezing. Mechanical properties in the solvating annealing, where part of the solution passes into the carbide, depending on the content carbon. Machining austenitic steels is more difficult compared with the low and medium alloy steels.

High strength, low thermal conductivity, high ductility and a tendency to high firming austenitic stainless steels are the main factors that make their machinability difficult. Most difficult place to drill deep holes with small diameter, shown Fig.2. Any method of cutting process leaves the surface of certain characteristics and form of a specific surface condition. On the surface of the cut and formed macroprofil and microprofil of surfaces. Force effects a working tool for cutting a thin surface layer beneath the surface of the deformed section. As a result of deformation heating and heat, which always accompanies the process of cutting, are the tensions in this layer and change its physical and mechanical properties.

Fig. 2. Drilling process

Problem of machining austenitic steels-austenitic stainless steels are characterized by high toughness, low thermal conductivity and a high degree of hardening of machined surface after machining. In terms the machinability of stainless steel is of great importance especially bonding surface finish. Low thermal conductivity causes adverse creation and shaping of particles in the editing plane, and therefore the material of cutting tools should be applied to sintered carbides. Austenitic steels are more prone to plastic deformation due to the large number of slip systems. Hardening of the machined surface (evaluated by measuring hardness) is the result of transformation of metastable austenite. For working metal, should be seen as a process of plastic deformation. Plastic deformation in relation to the cut surface of an object can take different forms and plastic (hardening), elastic-plastic or elastic. For a group very hard materials, this form of plastic deformation only. On the surface of the section, is of key importance, the thickness of plastically deformed material surface. The various levels of plastically deformed material thickness is a form of distortion varies. Aim of this paper is to present research findings determine the size of the zone of plastic deformation near the surface of the holes drilled by methods of light and electron microscopy and characterized by measuring the intensity of hardening the microhardness values of plastically deformed zone. The paper presents results of deformation of material under the layer surface finish when drilling austenitic stainless steel 1.4303.

According to different authors, very hard materials have ductility that is fifty percent greater than convetional materials, but handle heating more poorly. The specific electrical resistance depends on the chemical composition, not on the structure of the material. For very hard materials it is also characteristic that they have a low thermal conductivity, which is mostly a function of the chemical composition. It is important to stress that, as to their mechanical, physical and chemical characteristics, each very hard materials should be studied independently. Working is the result of a mutual interaction of the cutting tool and the workpiece that is accompanied by a number of phenomena that produce the synergic effect. Chip production is described through the plasticity theory. The field of slip lines intervenes with the area of the plastic deformation, the machined surface and the chip. Slip lines represent a durable, highly intense deformation. During the interaction of the cutting elements and the workpiece and due to heat, wearing takes place in the cutting area, because friction depends on the interaction of pure metal surfaces between the frontal surface of the cutting part of the tool and the chip, as it is stated. Wearing is a synergic effect of factors that cause the change of weight, volume or the change of dimensions on the cutting part of the tool. The area of the slip contact has an important role during cutting. The creation of flow zone on the frontal and back surface of the tool called the scab, is one of phenomena resulting from the interaction of the workpiece and the cutting edge of the tool. The order of phases of the scab as a periodic phenomenon that mostly negatively influences the process of cutting. Figure 3 and Figure 4 each photograph includes the information of the thermo-plastic deformation. The paper described verification of CAD systems applied by analysis of drilling tools. Analysis of tool life is very important for proces effectivity.

There is a tendency in the field of machined materials towards stainless steels with higher strength and higher corrosion resistance. Duplex stainless steels are often applied (Schintlmeister & Wallgram). High-efficiency machining and the machinability of high-strength stainless steels are considered to be the most important future trends affecting machining operations. Duplex stainless steels have a lower nickel content than austenitic stainless steels, and an austenite plus ferrite structure. Increased strength with enhanced properties for service in a corrosive environment provides difficulties from a machinability point of view. The characteristics of stainless steels raised from the austenitic structure are high toughness, low thermal conductivity and high workhardening co-efficient (Peckner & Bernstein). From a machinability point of view the most important characteristic is the workhardening. Because of the low thermal conductivity, the chips are formed on the basis of catastrophic failure in narrow shear surfaces (Dolinšek). When carbide tools are used these characteristics cause the formation of BUE and low values of tool life. Cutting forces are also increased and the unfavourable formation of tough chips appears (Dolinšek). Tool materials, such as CVD-coated (Chemical Vapour Deposition) with hard Al_2O_3 coatings, are often preferred (Belejchak). The need for a hard tool surface coating is especially required when HIPed stainless steels containing hard inclusions are to be machined. Near-equiatomic nickel-titanium alloys (NiTi) have many attractive properties for engineering applications, such as pseudo-elasticity and good cavitation resistivity, in addition to their more well-known shape memory properties (Li & Sun, Starosvetsky & Gotman). In drilling stainless steel with a pseudo-elastic coating material, machinability difficulties are involved with the pseudo-elastic properties of the coating material. Nevertheless, both technical and commercial limitations arise when NiTi is considered as a material for large engineering components. Consequently, interest in NiTi-coating technologies, for example for stainless steels, is on the rise. The cutting process of NiTi-based shape memory alloys is influenced by their high ductility and high degree of workhardening, and the unconventional strain-stress behaviour (Weinert & Petzoldt). The investigation of machining austenitic stainless steels in different cutting processes has been initiated by industry, where the need for effective tools and demands for reliable data on cutting parameters extends far beyond the experiences or recommendations given by tool producers (Dolinšek). The machinability studies are often carried out by v_cT-tests in turning, milling and drilling operations. Tool wear is studied by using optical microscopy to define the amount of flank and crater wear. The interaction between tool and chip can be effectively studied using SEM. There are several tendencies affecting the technology and methods used in the metalworking industry. Highly efficient machining strategies are used, and HEM (High Efficiency Machining) is used as a machining method. In HEM machine tools, modern tools are used with sufficient cutting parameters. HEM focuses on optimising cutting efficiency to maximise material removal rate. Compared to HSM (High Speed Milling), lower spindle speeds and increased chip thicknesses are used. The modern tools and tool materials available for this research were specifically designed cutting tools for HEM machine tools. The machine tool reliability and productivity is controlled by optimising machining parameter selection and acceptable and adequate sufficient parameters are used. Also, nowadays modern machine tools are very complex mechatronical systems and their capability and efficiency are mainly determined by their kinematics, structural dynamics, computer numerical control system and the machining process (Altintas et al., and Weck et al.).

For materials of the cutting tools are allocated to three variables, which can determine the choice of material, cutting tool in a machining operation:

- wear resistance,
- toughness, respectively. resistance to fracture and deformation,
- wear resistance at elevated temperature.

4. Examples results from experiments

4.1 Example 1

The material were selected for the purposes of the experiment X4Cr19Ni9 stainless steel, which were then compared to X03Cr16Ni8 steel. The chemical composition of machined materials is reported in Table 2.

Element	C	Cr	Ni	Mn	Ti	Mo	Si	N	P	S
X03Cr16Ni8	0.3	16.0	7.5	1.8	-	-	0.36	-	0.04	0.03
X4Cr19Ni9	0.04	19.0	9.0	1.2	0.4	0.09	0.30	0.04	0.03	0.03

Table 2. Chemical composition of stainless steels in [wt.%]

X03Cr16Ni8 steel microstructure contains larger quantities of complex carbides.; these are generated along grain edges when compared to X4Cr19Ni9 steel. For the purposes of the experiments, the applied technical system was: Machine: Chiron FZ12 CNC. Tool: Screw drill with diameter d=7.0 mm - new cutting area structure in sintered carbide. Tool fixture: high-precision hydraulic clamping head. Workpiece: the following materials were employed for the purposes of experimental measurements: X03Cr16Ni8 steel, X4Cr19Ni9 steel with low carbon content. Samples with the following dimensions were used for the purposes of experimental measurements: $bxhx\ell$ (40x40x200) mm. Cutting conditions: cutting speed in interval v_c= 40-60 m per min, feed in interval f=0.01-0.08 mm per rev. Machining method:

Dry Machining. With regard to cutting tool life, the following criterion was applied: VB_k=0.2 mm. Measurement was carried out by means of an optical microscope without extracting the cutting tool from the clamping fixture. The cutting tools and fragments were analyzed using an electron microscope (SEM).

The result of cutting tool life assessment is T-v_c dependence. Analysis of T-v_c dependence used a tool with a new cutting area; when drilling X03Cr16Ni8 steel at 40 m per min cutting speed, the following drilling length was achieved: 4.5 m (at feed f=0.065 mm per rev.). Due to the increased cutting speed of 60 m per min, drilling length was reduced to 0.9 m. Tool life was reduced from 16 minutes to 4.5 minutes. When drilling X4Cr19Ni9 steel at 40 m per min cutting speed, the following drilling length was achieved: 4.0 m (at feed f=0.065 mm per rev.). Due to the increased cutting speed of 60 m per min, drilling length was reduced to 1.0 m. Tool life was reduced from 9 minutes to 3.0 minutes. When comparing the execution of holes for cutting tools into X03Cr16Ni8 steel, we observed that tool life was increased by 36 % compared to X4Cr19Ni9 steel. The higher percentage of carbon does influence the increase of carbide formation, especially along grain edges; this fact has an impact on cutting tool lifes. By decreasing carbon content we eliminate carbide formation and therefore we increase cutting tool life.

The tendency to generate built-up edges was more significant in the case of X4Cr19Ni9 steel than in the case of X03Cr16Ni8 steel. One example of BUE formation for X4Cr19Ni9 steel during drilling followed these cutting conditions: cutting speed v_c=45 m per min, feed f=0.065 mm per rev. A cutting speed increase has an impact on BUE formation over the front area. By comparing the steels it was ascertained that in the case of X03Cr16Ni8 steel the corner was not worn (unlike X4Cr19Ni9 steel). BUE formation over the face area and cutting part shape (geometry) influence the chipping on the cutting edge. An electron microscope was used for wear analysis for sintered carbide cutting tools (SEM - Scanning Electron Microscopy). Qualitative and quantitative workpiece modifications do occur in the cutting zone. With regard to circumstances in the cutting zone, it is important to be aware of the input features of interacting objects (tool and workpiece) as well as of conditions influencing such interaction from the point of view of machining equipment. Output elements generated from the cutting: a workpiece with a machined surface, and tool with possible cutting area damage. For the purposes of analysis of the creation and shaping of chips, the samples were examined using an SEM according to the following cutting conditions: cutting speed v_c=40 and 60 m per min. X03Cr16Ni8 steel chips; X4Cr19Ni9 steel chips are illustrated; These figures illustrate the chip surfaces from their concave and convex sides. Cutting tool wear increased when cutting speed rose from 40 to 60 m per min. At this cutting speed interval it was possible to observe a tendency to make larger BUE in the case of X03Cr16Ni8 steel compared to X4Cr19Ni9 steel. Here we can observe traces/grooves as a result of the interaction between the chip and the front area of the tool. The concave side of the chip (dished chip) reported in Fig.3 shows a modification of the thickness of the elements/flakes (lamellae) when cutting speed increased. The higher tendency of BUE formation with X03Cr16Ni8 steel can be compared against the fragment surface (convex shape). Worse (reduced) machinability (resulting from the comparison of X03Cr16Ni8 and X4Cr19Ni9) could be caused by the irregular tooth-like endings on the convex (protuberant) sides of the chip. When comparing the chips with X03Cr16Ni8 and X4Cr19Ni9, it is possible to observe grooves (depressions) and protruding material. Chip shaping is influenced by structurally unstable effects. Tool wear occurred continuously for X03Cr16Ni8 steel with the use of a cutting tool in dry machining. In this respect, at increasing cutting speeds tool plastic deformation takes place with gradual laminar flaking of the surface on the cutting tool (Fig. 3a) and with destruction of the coat (frittering) over the front area (Fig.3b); tool wear is influenced by the formation of built-up edges and by coat flaking.

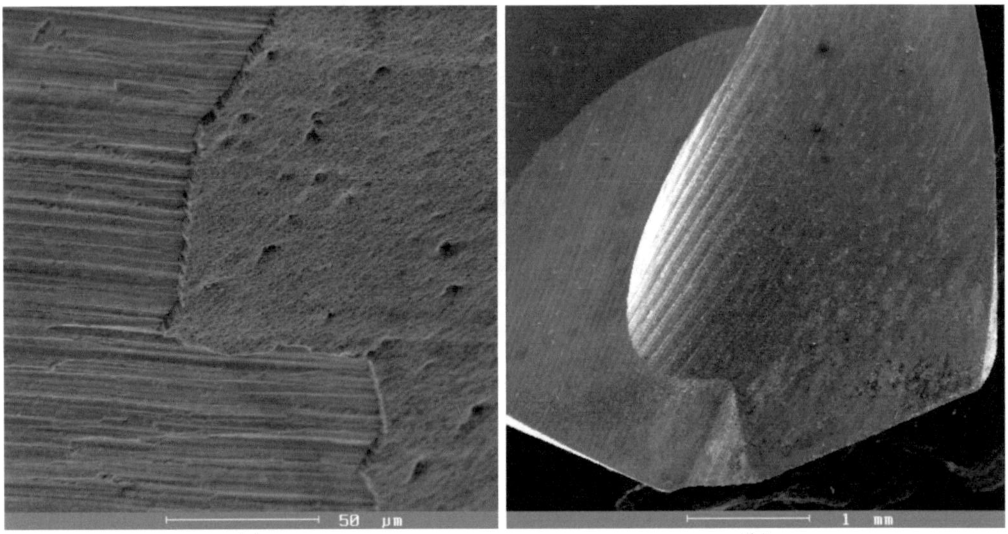

Fig. 3. Helical drill - characteristical wear in cutting part. (a) Coat damage on dorsal area - gradual coat laminar flaking on cutting tool. (b) Coat damage - destruction (frittering) on front area

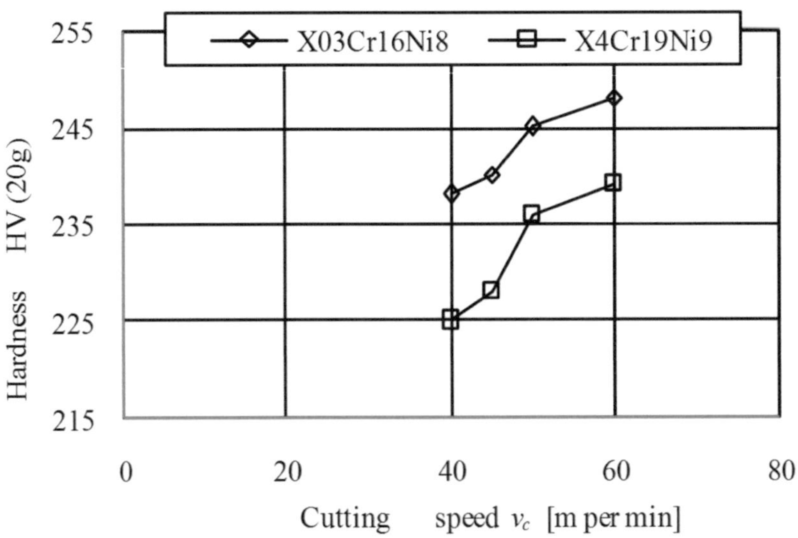

Fig. 4. Relation between machined workpiece micro-hardness and cutting speed when drilling holes into X03Cr16Ni8 and X4Cr19Ni9 steels, with feed 0.065 mm per rev.

Stainless steels are influenced by charging due to intensive mechanical reinforcement during machining. The examination of reinforced surfaces can be carried out by measuring the micro-hardness of the bottom part of the fragment; indeed, the bottom part of the fragment can be considered as the most deformed fragment zone. The results of micro-hardness examination are reported in Fig. 4, these results are as follows: X03Cr16Ni8 steel: fragments are strongly deformed compared to X4Cr19Ni9 steel. Austenite fragment, bottom part, v_c=60 m per min, 310 HV (20 g), if the bottom part of the fragment is measured: austenite 240 HV (20 g). 2. X4Cr19Ni9 steel: v_c=45 m per min, austenite microhardness 230 HV (20 g), if the bottom part of the fragment is measured: austenite 234 HV (20 g). The results presented in this article can be summarized as stated in the following main conclusions for the sake of comparison:

1. Machinability of X03Cr16Ni8 steel and X4Cr19Ni9 steel is influenced by the formation of BUE. X03Cr16Ni8 steel tends to form more BUE than X4Cr19Ni9 steel
2. Tool life (for the applied cutting tool) is between 4.5 min. and 12 min., when drilling X4Cr19Ni9 steel.
3. BUE formation is caused by adhesive wear; in terms of cohesion, this fact indicates that the above mentioned mechanism is likely to be the predominant mechanism in the damaging process of sintered carbide tools when drilling into X4Cr19Ni9 steel.

4.2 Example 2

During the drilling simulation, the cutting edges of the drill bit are shearing the workpiece material at high speeds which separate the material from the workpiece by chip formation. The material separation criterion for machining has been a topic of interest in the development of the theory of finite element modeling of machining. Initially, a parting line model was assumed to simplify the simulation process. This model assumed a small crack existed in the material and the chip was separated from the workpiece in a predetermined "unzipping" fashion. Eventually, the maximum plastic strain model was proposed and this criterion has been adopted by most FEM models. This maximum plastic strain model assumes that material separation occurs when an element reaches a critical plastic strain for the material model of the workpiece. The element is then split into two elements and a chip is formed. One can argue whether drilling actually produces smooth separation. Regardless, the maximum plastic strain criterion has been implemented and this has been the most accepted method of failure criteria to model burr formation in drilling. Historically, the two standard FEM meshes are Eulerian and Lagrangian. There are also combinations such as the Arbitrary Lagrangian Eulerian (ALE) and the Coupled Eulerian Lagrangian (CEL) meshes. Although the Lagrangian mesh is not as comprehensive as the Eulerian mesh, it has much better simulation cycle times as a result. Lagrangian mesh in simulating drilling processes is the ability to know the entire time history of the key variables at every point during the simulation. That means, if a simulation crashes for any reason, a new simulation can start where the crashed simulation stopped. This is particularly useful because nearly every simulation has some sort of problem during the run. This is possible because the Lagrangian mesh is reformulated at nearly every time step, in order to manage the deformation of the material. Several different types of machining operations can be accomplished with

Proengineer including drilling, turning, and milling. If a tool geometry can be modeled, the machining operation can be simulated. One of the most difficult problems faced with modeling drilling operations is obtaining an accurate model of a drill bit. More authors both present how this can be done and the latter developed a program to do this quickly and easily. Fig. 5,Fig. 6, Fig.7., Fig. 8., Fig.9., Fig. 10. each photograph includes the information of the thermo-plastic deformation. The paper described verification of CAD systems applied by analysis of drilling tools. Analysis of tool life is very important for proces effectivity.

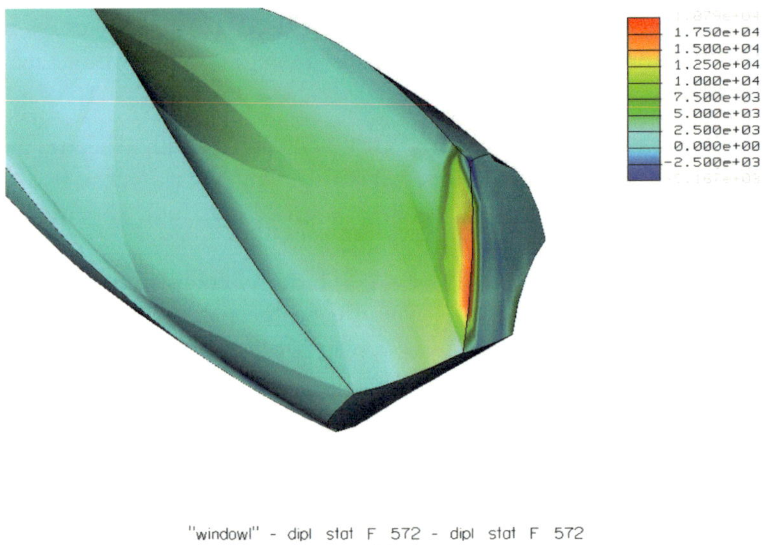

(a)

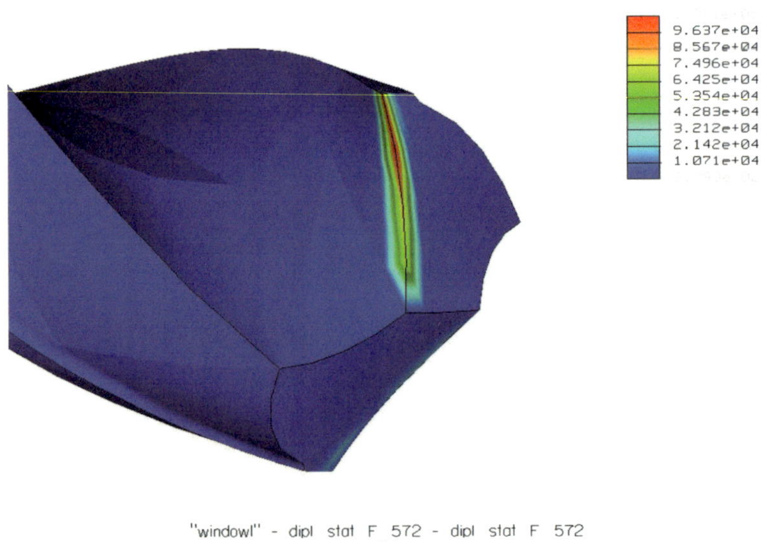

(b)

Fig. 5. Modeling of screw drill. (a) Tool wear- stress on the cutting edge. (b) Tool wear - plastic deformation on the cutting edge

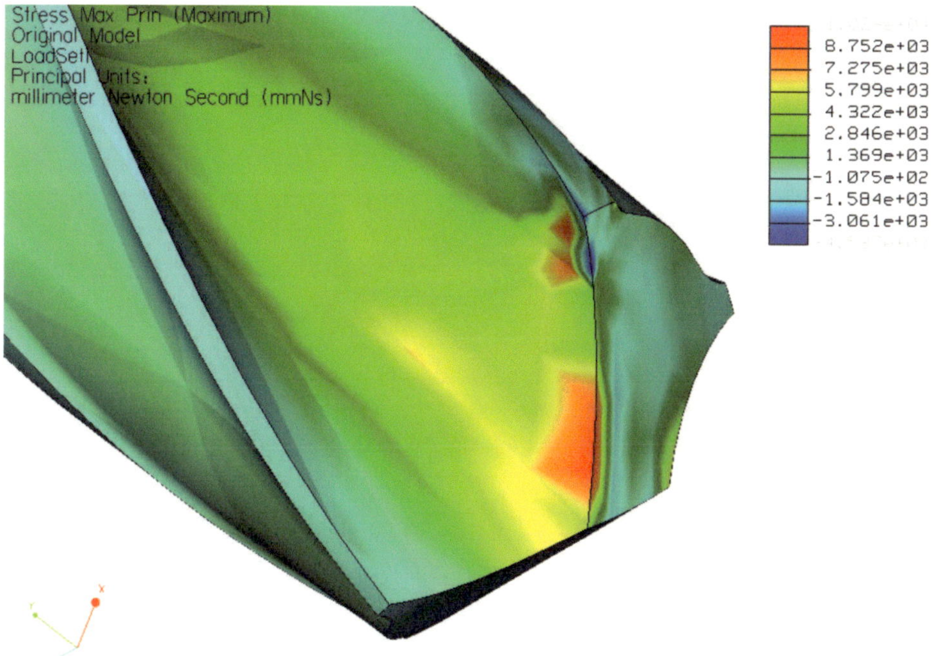

(a)

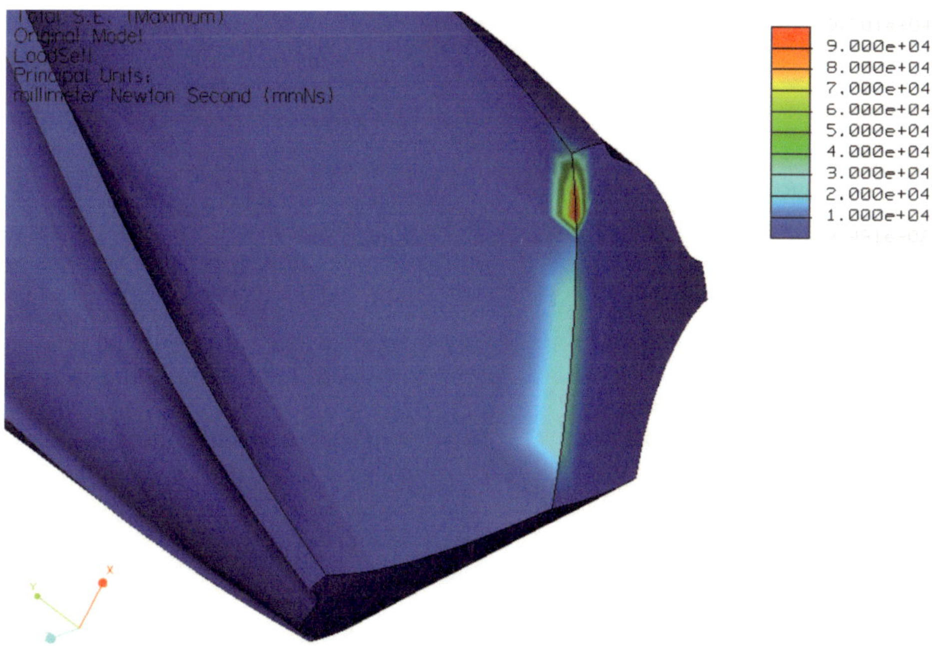

(b)

Fig. 6. Modeling of screw drill with two margins. (a) Tool wear- stress on the cutting edge. (b) Tool wear - plastic deformation on the cutting edge

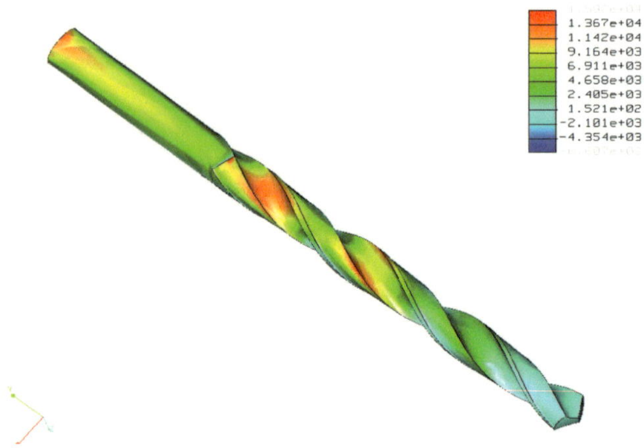

Fig. 7. Modeling of screw drill - Tool wear- stress on the cutting edge

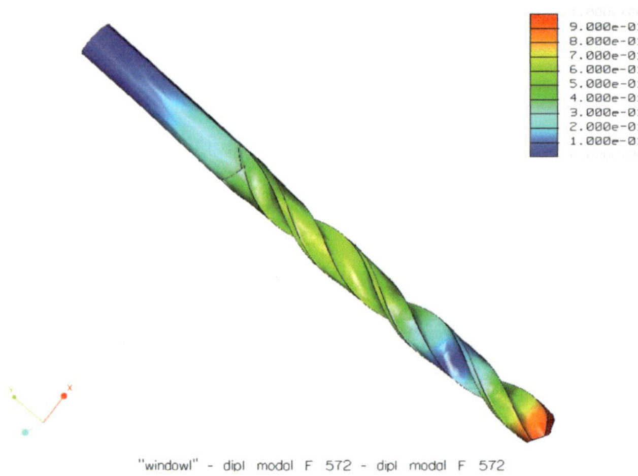

Fig. 8. Modeling of screw drill - Load on the cutting part

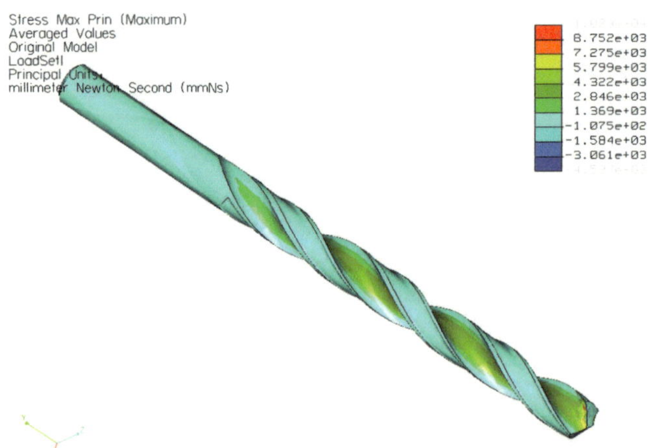

Fig. 9. Modeling of screw drill with two margins - stress on the cutting part

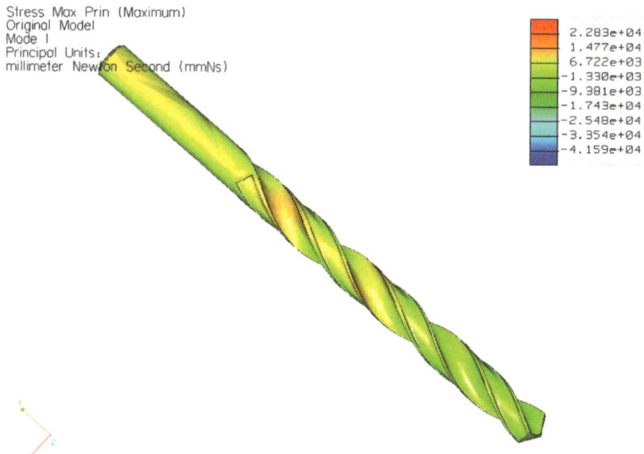

Fig. 10. Modeling of screw drill with two margins - Load on the cutting part

The cutting edge of an insert in a finishing operation is worn out when it can no longer generate a certain surface texture. Not a lot of wear is needed along a very small part of the insert nose for the edge of an insert to need changing. In a roughing operation wear develops along a lot longer part of the edge and considerably more wear can be tolerated as there are no surface texture limitations and accuracy is not close. The tool-life may be limited when the edge looses its chip control ability or when the wear pattern has developed to a stage when the risk for edge breakdown is imminent.

4.3 Example 3

For the experiments was applied technological system machine-tool-workpiece-fixture: CNC Chiron FZ12, helical drill with a diameter d=7.0 mm, the new design of the cutting edge with uncoated cemented carbide. Fixture for the cutting tool: high precision hydraulic clamping head, the workpiece fixture: mechanical vise. Workpiece: stainless steel-austenitic X4Cr17Ni8 steel. The materials to be machined were type of a new austenitic stainless steels with chemical composition listed in Tab.3. Experiments were used for sample size *hxbxl* (30x30x200) mm, for drilling holes 15 mm in depth and definition to the evaluation of internal machined surface. Cutting conditions: cutting speed of v_c=interval of 20-60 m/min, feed f=interval from 0.01 to 0.08 mm per revolution, depth of cut a_p=3.5 mm. Method of machining and dry machining. Workpieces of X4Cr17Ni8 steel, obtained by drilling, were analyzed for electron microscopy. To deformed the steel-hardening layer, was rated a hard layer width from the cut surface, a hard layer microhardness, and surface finish has been rated *Ra* surface roughness and surface morphology.

C	Cr	Ni	Mn	Si	P	Ti	S
0.04	17.2	8.2	1.6	0.6	0.04	0.06	0.03

Table 3. Chemical composition of stainless steels in wt [%]

The parameters of the process of drilling holes and dimensional characteristics of the drill are: T1(feed is 0.01 mm, cutting speed is 20 m/min., d=7 mm, angle $2\kappa_r$=135°), T2 (feed is 0.065 mm, cutting speed is 35 m/min., d=7 mm, angle $2\kappa_r$=135°), T3 (feed is 0.08 mm, cutting speed is 40 m/min., *d*=7 mm, angle $2\kappa_r$=135°). Workpieces for metallographic analysis were collected spark so that the plane metallographic sections axis hole. Subsequently, the workpieces were prepared and ready in dentacryle conventional metallographic procedures. For purposes of analysis has been used light microscopy experimental technique: an inverted metallographic microscope for observation in polarized light and differential interference contrast using (DIC). To verify the value of the depth of plastically deformed zones around the drilled holes (determined by light microscopy) technique was used for scanning electron microscopy - with a JEOL JSM 7000F autoemission nozzle. Microhardness plastically deformed zone was set at a distance of 60 μm for austenitic steel X4Cr17Ni8 steel. Methods of light and scanning electron microscopy was specified range plastically deformed zones in drill holes under the surface of the material X4Cr17Ni8 steel. Local hardening plastically defor-med zone around the drill holes was characterized by measuring the microhardness LM 700 AT micro-hardness with a load of 0.2452 N. For more information about local hardening deformed zone in the E position drill holes were obtained by measuring the indentation hardness apparatus Nanoindentation Hardness Tester - TTX NHT2 CSM method, where the grow monotonous load is superimposed small sinusoidal load. Measurement parameters Berkovich indenter types were: feed rate indenter 3000 nm/min, a maximum load of 100 mN, loading and unloading speed 200 mN/min, sinus frequency 20 Hz, amplitude sinusoidal load of 5 mN, stay to the maximum load of 10 s. Indentation measurements were in the position E hole. Analyses of samples plastically deformed layers under the surface of machined steel X4Cr17Ni8 steel can confirm these allegations. Autors indicates that for X4Cr17Ni8 steel, resizing a layer of plastic deformation is related mainly to the material structure and properties of austenitic grain size.

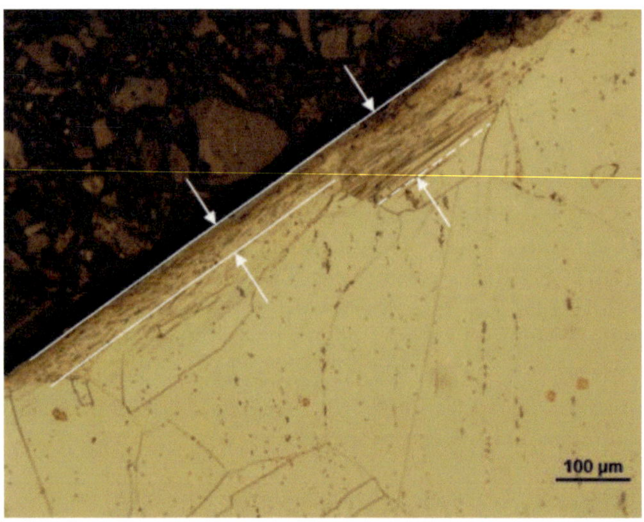

Fig. 11. Plastic deformation of the surface, local plastic deformation in austenitic grain-hardening surface, zone of hole F

Hardening uneven layer width from 30 μm to 550 μm, example as shown in Fig.11, generated by the cutting conditions of 50 m/min to 60 m/min and feeds on larger than 0.06 mm. Plastic deformation of the machined surface has the character in Fig.12.

Fig. 12. Plastic deformation of the surface to a depth of 220 µm, local plastic deformation in the austenitic grain, zone of hole H

To plastic deformation occurs under the surface finish and the local austenite grains. At a cutting speed of 30 m/min to 45 m/min and feeds from 0.04 mm to 0.08 mm width was measured to strengthen the layers from 20 µm to 68 µm, which is plastically deformed smallest layer, as shown Fig.13. Under these conditions, the cutting width was measured to be uniform solidified layer. At the same time claim autor[2] confirmed that this is true for smaller austenite grains (~to 80 µm). Hardness HV (20g) in plastically deformed layer after drilling X4Cr17Ni8 steel Fig.14 shows. For the evaluation of machined surface was measured roughness Ra=0.85 µm at a cutting speed of 50 m/min and feed is 0.04 mm.

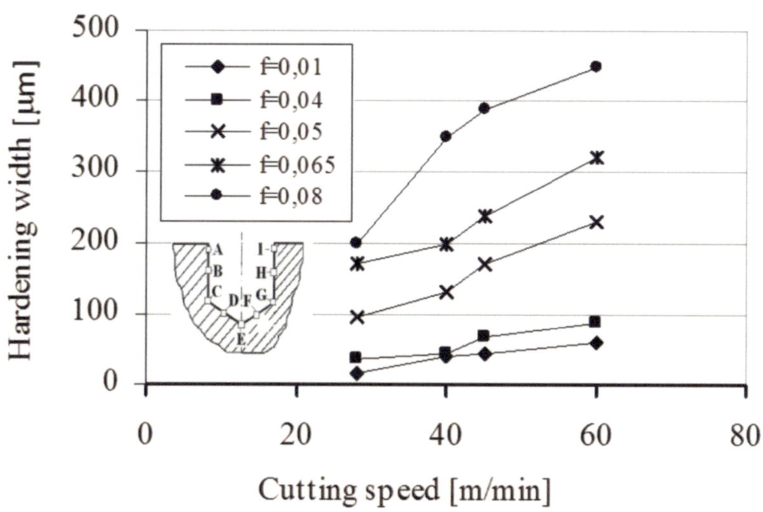

Fig. 13. Dependence of width of the deformed area of cutting speed and different feed, zone of hole F

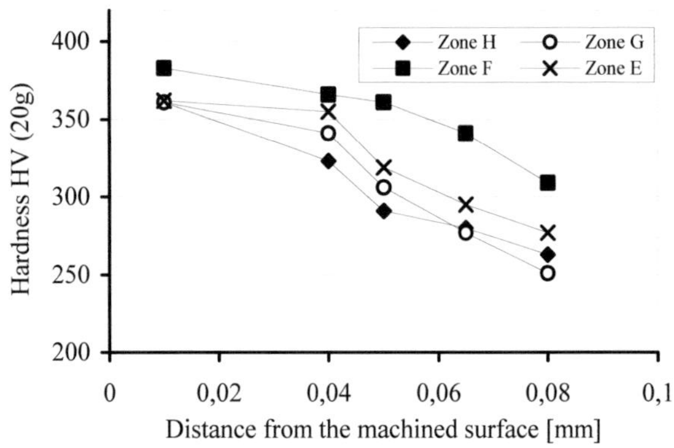

Fig. 14. Dependence of the hardness of the deformed distance from the machined surface

When the measured surface roughness Ra=0.85 μm (minimum of surface roughness) width was measured to strengthen the layers from 20 μm to 48 μm. This result was achieved, the plastically deformed layer can be taken further by removing (example: at increasing of holes diameter, or threadings) and by extending tool life by 20 to 30 %, and thus eliminate the catastrophic damage of tool. For evaluation of the cutting tools life were applied criterion VB_k=0.1 mm. Measurements were made using an optical microscope without removing the cutting tool from the fixture. Cutting tools and the chips were analyzed for electron microscopy (SEM). Plastic deformation initiates the so-called cutting edge wear. chipping, especially for the discontinuous method of machining such as drilling. This phenomenon is called plastic lowering the cutting edge. Light microscopy and SEM was measured amount of plastic strain in the top spot (position E) 18 to 152 μm. In these areas was measured indentation hardness HIT, depending on the depth of indenter penetration. The measurement matrix was 3x3 indentation impress whose mutual distance was 18 μm. The minimum distance from the indenter impresses edge of the sample was 18 μm. Maximum indentation hardness values were measured in the top spot at the sample X4Cr17Ni8 steel. In this position he was microscopically measured maximum value of local plastic deformation of ~ 152 μm.

4.4 Example 4

Study of cutting tool wear with regard to the elimination of occurrence of poor-quality holes when drilling into new austenitic ELC (Extra Low Carbon) stainless steel. The problem of drilling holes with diameter D=5 to 8 mm resides in the fact that 30 to 35% of these holes do not comply with prescribed requested requirements. The cutting tools (helical drills as monoliths) get damaged and wear out. The result of the damage is very often the unforeseen destruction of the cutting tools; therefore their operational tool life is reduced. On the basis of practical experience and experiments carried out in the past 15 years, we have observed that the operational tool life of helical drills is reduced by 30 to 40%. This article presents the results of experiments focusing on the study of the damage process in helical drills with diameter d=8.0 mm when drilling into new austenitic ELC stainless steel. This study also includes an analysis of accompanying phenomena in the cutting zone by measuring some selected parameters. The results of the experiments were compared with X2Cr18Ni8 steel and then verified when drilling holes into specific products. Production technology is the fundamental aspect in terms of requested values for the production of components. The relationship between the quality of a machined component surface and the applied technical machining method can be assessed using the following factors (Abrao, A.M. & Aspinwall, D.K.):

- Machined surface in the space (e.g. morphology, texture and surface roughness)
- Features of machined surface (e.g. dimension and hardness)
- Impact of the technical machining method over the features of the machined surface and their direct influence on the function of the component (e.g. surface defects, impurities and inclusions)

During the machining process, the following phenomena occur on the machined surface:

- Plastic deformation of the machined surface,
- Structural modification of the machined surface,
- Modification of mechanical features of the machined surface,
- Modification of the integrity of the machined surface,
- Modification of residual tension under the machined surface,
- Modification of chemical features of the machined surface.

Austenitic stainless steels are characterised by high strength, low heat conductivity, and a high degree of hardening of the machined surface after machining (Belluco, L. & De Chiffre, L.), (Peckner, D. & Bernstein, I.M.). With regard to the machinability of stainless steels, the hardening of the machined surface is a very significant aspect. Low heat conductivity causes negative formation and shaping of chips in the shear plane (Brinksmeier, E.W. & Reucher, A. & Griet-Solter, J.), (Dolinšek, S.); therefore, as far as cutting tool materials are concerned, it is necessary to apply sintered carbides. When machining stainless steels, we often note the occurrence of built-up edge; this phenomenon results in a reduction of tool life (Ceretti, E. & Giardini, C. & Filice, A. & Rizzuti, L. & Umbrello, S.D.), (Jurko, J. & Panda, A. & Gajdoš, M.) Cutting tool wear is the result of the combination of various wear mechanisms: abrasive, adhesive, diffuse, and chemical (oxidation), (Grzesik, W.). Generally speaking, with regard to sintered carbide tools the most dominant types of phenomena are in the following order: abrasive, adhesive, diffuse, and chemical wearing mechanism (Jurko, J.), (Jurko, J. & Panda, A.). As for tool cutting area materials, three parameters are selected, through which it is possible to define the choice of tool cutting area material with regard to a given machining operation:

- Resistance against wear
- Strength, i.e. resistance against breaking and deformations
- Resistance against wear at increased temperature.

According to (Liew, W. & Ngoi, B. & Lu, Y.): "Wear is an undesired modification of the surface or of the dimensions of solid bodies; such modification is caused either by the mutual action of functional surfaces or by mutual action of the functional surface and the environment - generating wear during their mutual relative movement".

Cutting tool wear is the result of a combination of encumbering factors influencing the cutting area of a tool. Wear is then the interaction between the cutting tool, workpiece material, and cutting and machining conditions (Nam, P. Suh). In terms of the technical method for drilling, mutual interaction is generated between the cutting area of the tool and the workpiece, according to the following steps (Jurko, J. & Panda, A. & Gajdoš, M.):

- First contact occurs when the elements - edges of the surface of the peak of the helical drill - are pushed into workpiece material (chisel cutting interaction edge/workpiece),
- Progressive incision of the chisel cutting edge into the workpiece over a length $\ell_{sw,2}$
- Progressive incision of the cutting edges into the workpiece over a length ℓ_s (length of the major cutting edge); this data is calculated according to equation 3

$$\ell_s \cong \frac{d}{2.\sin\kappa_r} \text{ [mm]} \tag{3}$$

where

ℓ_s– length of the major cutting edge in [mm],

d- drill diameter in [mm],

κ_r - tool cutting edge angle in [°],

$\ell_{s,1}$, $\ell_{s,2}$- lengths of cutting edges on a double-wedge helical drill in [mm]

and the following equation 4 is valid:

$$\ell_{s,1} = \ell_{s,2} \equiv \ell_s \text{ [mm]} \tag{4}$$

provided that the cutting edge of the tool is symmetrically reground.

- The progressive incision of the cutting edges into workpiece material ends at point H (corner of tool cutting part). The process of cutting edge incision is different in each single point of this cutting edge; this happens because of modifications concerning the kinetic rates for each single cutting edge point. Cutting speed v_c changes from point V (peak of tool cutting part) – where v_c=0 m per min – up to point H (corner in the tool cutting part), where maximum cutting speed is defined by this equation 5:

$$v_{c,\max} = \frac{\pi.d.n_c}{1000} \tag{5}$$

in m per min. As a result of such kinetic rates, during incision we observe various forms of damage in the elements of tool cutting area – specifically it happens in the following way for one cutting wedge: at peak V, or at the transversal cutting edge, with continuous progressive incision of the main cutting edge, we observe damage in the major flank area and in the face area; the damage process continues on at point H, proceeding towards the adjacent cutting edge and towards the minor flank areas. In Fig.15. we reported the main elements and sectors on the helical drill in terms of damage (successively we can exactly measure and assess these elements and sectors). The type and course of wear originated on the cutting area of the tool provide important information about the course of machining.

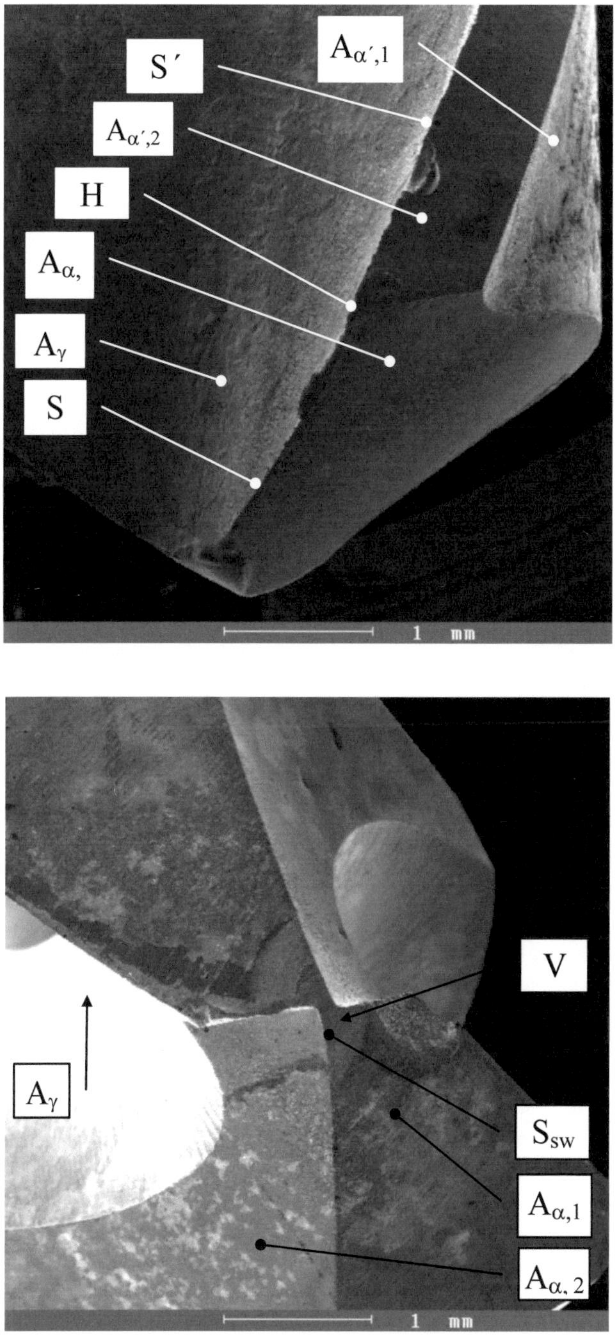

A_γ - face area, $A_{\alpha,1}$ - major flank area, $A_{\alpha,2}$ - major flank area - relieved,
$A_{\alpha',1}$ - minor flank area, $A_{\alpha',2}$ - minor flank area (margin), H-corner, V-peak,
S - major cutting edge, S´- minor cutting edge, S_{SW} - chisel cutting edge

Fig. 15. Main elements and sectors related to wear on the cutting part of the helical drill

Studies and research about machinability of materials often focus on assessment results of cutting tool life. Cutting tool wear is a parameter we can examine by means of an optical light microscope. The examination of the cutting zone (i.e. interaction between tool and workpiece) is analyzed using Scanning Electron Microscopy (SEM). Material were selected for the purposes of the experiment - ELC Cr16Ni7Ti steel - which were then compared to X2Cr18Ni8 steel. The chemical composition of machined materials is reported in Table 4. The microstructure of Cr16Ni7Ti steel is reported in Fig. 16.

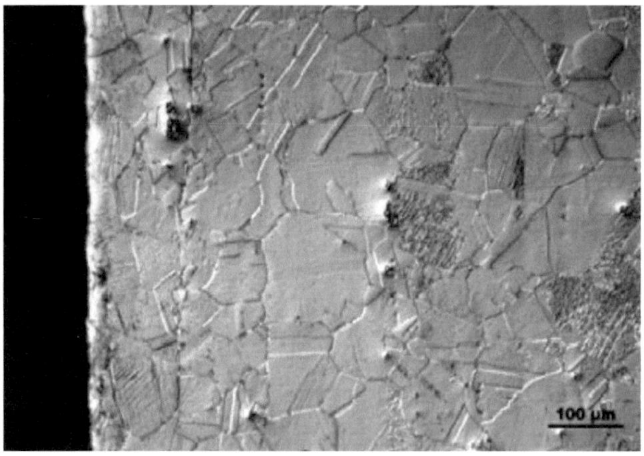

Fig. 16. Microstructure of X2Cr16Ni7Ti steel, etchant: Villela

C	Cr	Ni	Mn	Si	P	Ti	S
0.02	16.1	7.1	1.6	0.6	0.03	0.05	0.03

Table 4. Chemical composition of stainless steels in wt [%]

Cr18Ni8 steel microstructure contains larger quantities of complex carbides (as shown in Fig. 17.); these are generated along grain edges when compared to X2Cr16Ni7Ti steel. Due to the increased cutting speed of 60 m per min, drilling length was reduced to 1.6 m. Tool life was reduced from 15 minutes to 5 minutes. When drilling X2Cr18Ni8 steel at 30 m per min cutting speed, the following drilling length was achieved: 4.4 m (at feed f=0.06 mm per rev.). Due to the increased cutting speed of 60 m per min, drilling length was reduced to 1.4 m.

Tool life was reduced from 14 minutes to 3.6 minutes. When comparing the execution of holes for cutting tools into X2Cr16Ni7Ti steel, we observed that tool life was increased by 36% compared to X2Cr18Ni8 steel. The higher percentage of carbon does influence the increase of carbide formation, especially along grain edges (e.g. X2Cr16Ni7Ti steel); this fact has an impact on cutting tool lifes. By decreasing carbon content we eliminate carbide formation and therefore we increase cutting tool life.

Cutting tool damage follows a chain of events. After drilling holes the cutting edge was influenced by the formation of Built Up Edge-BUE, as shown in Fig.18. The built-up edge in the face and major flank area might be the result of adhesive wear close to the cutting edge. Holes were made in dry conditions in order to avoid any influence on experiment results with process media. Oxidation and diffusion (as wear mechanisms) were observed in deeper holes (3 to 5xD). One of the basic causes of such mechanisms is the quantity of heat generated in the cutting area when holes are made (here one of the main causes is represented by the low heat conductivity of steel). If we apply a process medium, the above mentioned mechanisms would be substantially influenced - and in some cutting conditions they would be almost non-existent.

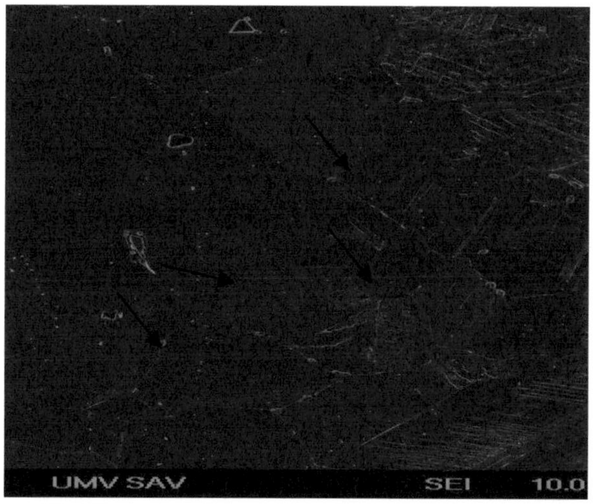

Fig. 17. Detail of X2Cr18Ni8 steel microstructure, generation of complex carbides along grain edges

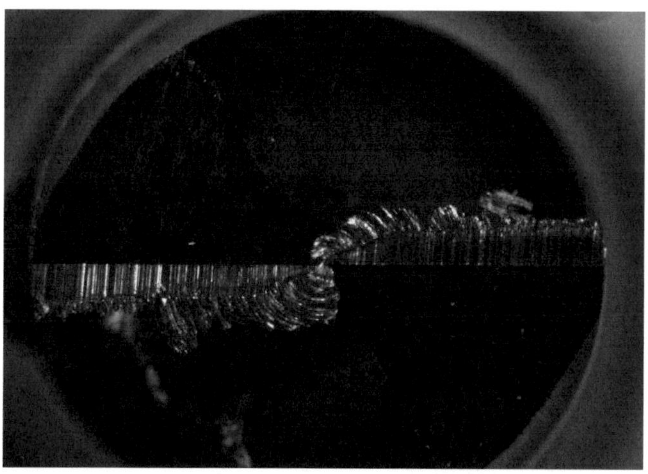

Fig. 18. Formation of built-up edges when drilling X2Cr16Ni7Ti steel

The tendency to generate built-up edges was more significant in the case of X2Cr18Ni8 steel than in the case of X2Cr16Ni7Ti steel. One example of BUE formation for X2Cr16Ni7Ti steel during drilling followed these cutting conditions: cutting speed v_c=45 m per min, feed f=0.06 mm per rev. and is shown in Fig.18. A cutting speed increase has an impact on BUE formation over the front area. By comparing the steels it was ascertained that in the case of X2Cr18Ni8 steel the corner was not worn (unlike X2Cr16Ni7Ti steel).

Fig.19 illustrates cutting zones. Qualitative and quantitative workpiece modifications do occur in the cutting zone. With regard to circumstances in the cutting zone, it is important to be aware of the input features of interacting objects (tool and workpiece) as well as of conditions influencing such interaction from the point of view of machining equipment. Output elements generated from the cutting: a workpiece with a machined surface and tool with possible cutting area damage.

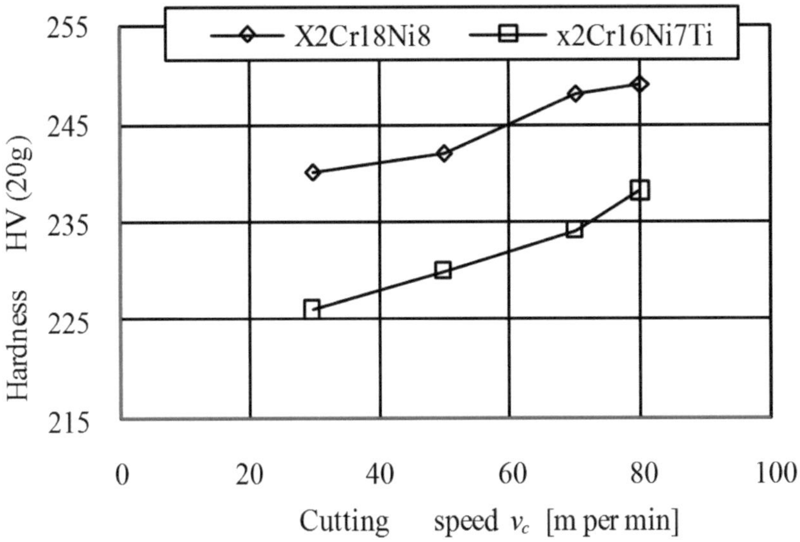

Fig. 19. Cutting zone - Drilling using a helical drill; interaction between the workpiece and tool cutting area

Tool wear occurred continuously for X2Cr18Ni8 steel with the use of a cutting tool in dry machining. In this respect, at increasing cutting speeds tool plastic deformation takes place with gradual laminar flaking of the surface on the cutting tool and with destruction of the coat (frittering) over the front area; tool wear is influenced by the formation of built-up edges and by coat flaking. Stainless steels are influenced by charging due to intensive mechanical reinforcement during machining. The examination of reinforced surfaces can be carried out by measuring the micro-hardness of the bottom part of the fragment; indeed, the bottom part of the fragment can be considered as the most deformed fragment zone. The results of micro-hardness examination are reported in Fig. 20 and Fig. 21, these results are as follows:

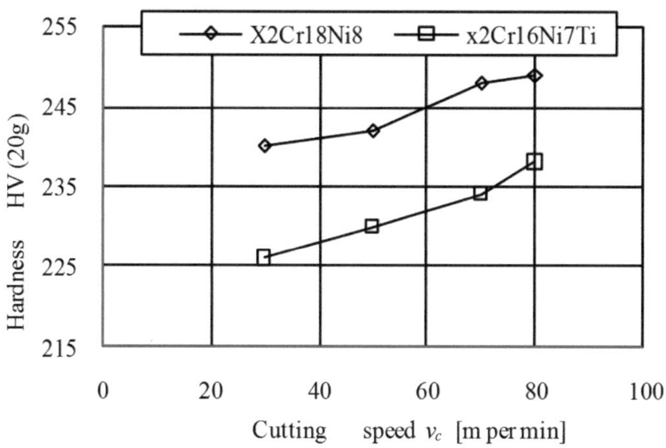

Fig. 20. Relation between machined workpiece micro-hardness and cutting speed when drilling holes into X2Cr18Ni8 and X2Cr16Ni7Ti steels – with feed 0.06 mm per rev.

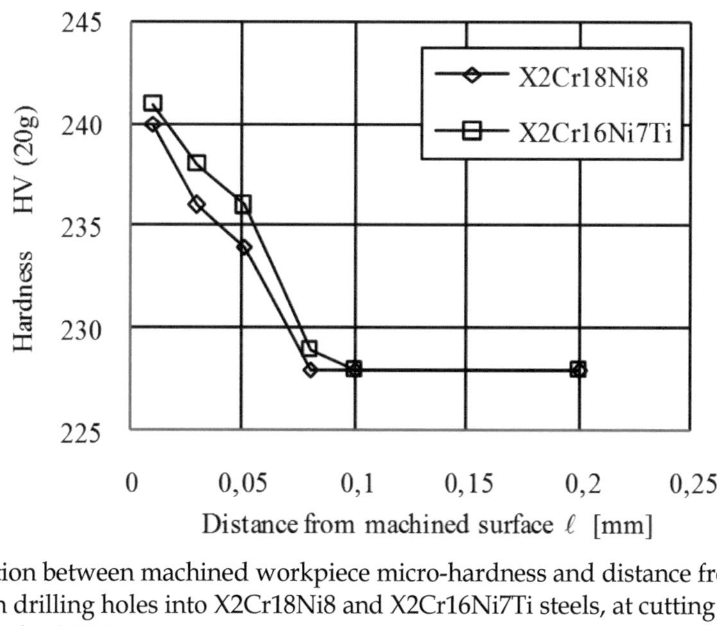

Fig. 21. Relation between machined workpiece micro-hardness and distance from machined surface when drilling holes into X2Cr18Ni8 and X2Cr16Ni7Ti steels, at cutting speed 45 m per min, with feed 0.06 mm per rev

- X2Cr18Ni8 steel: fragments are strongly deformed compared to X2Cr16Ni7Ti steel. Austenite fragment, bottom part, v_c=45 m per min, 305 HV (20 g) – if the bottom part of the fragment is measured: austenite 242 HV (20 g)
- X2Cr16Ni7Ti steel: v_c=45 m per min, austenite micro-hardness 228 HV (20 g) – if the bottom part of the fragment is measured: austenite 230 HV (20 g)
- Result experiment was also analyzed application CAX system. On figures is painted the cutting zone-Fig.22, and themperature influence on the cutting tool and on the workpiece- Fig.23.

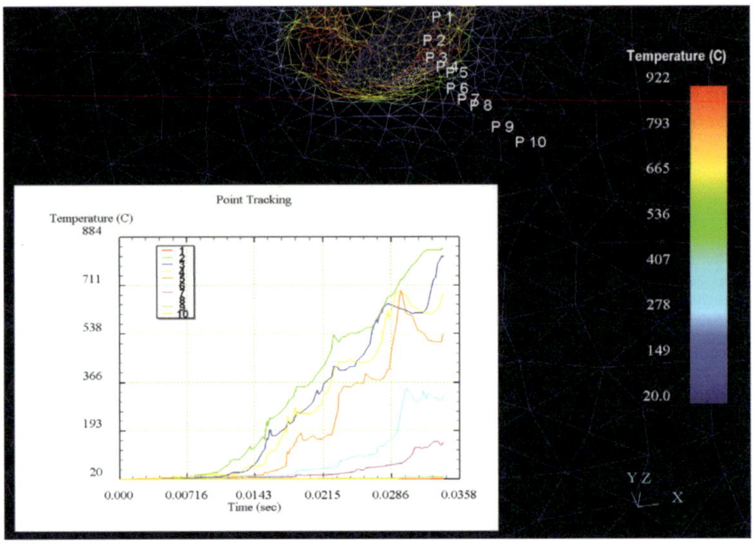

Fig. 22. The cutting zone-drilling of stainless steels X2Cr18Ni8

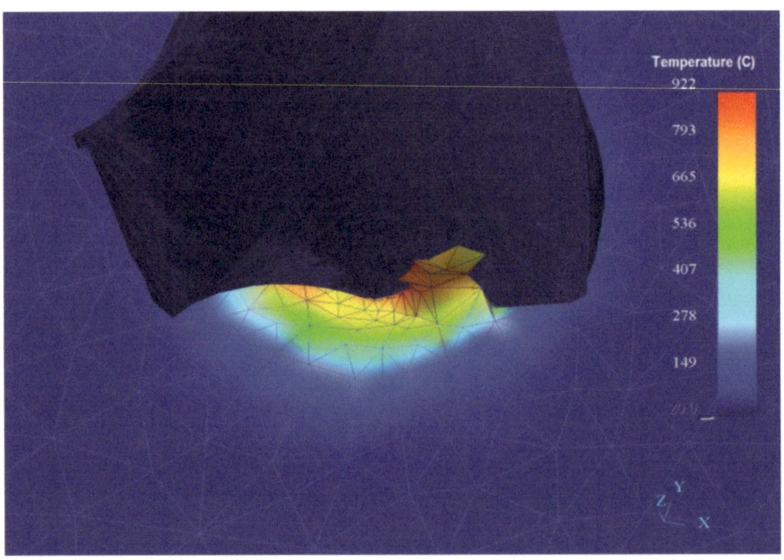

Fig. 23. Themperature influence - drilling of stainless steels X2Cr18Ni8, at cutting speed 45 m per min, with feed 0.06 mm per rev.

5. Conclusion

The main objective of the development chapter is to provide information on the new stainless steels for application in food production facilities. Other important part of the experimental results for the design of technological processes of production and information on these steels for the machinability - four basic criteria: the machining process kinematics, dynamics of the machining process, chip formation and shaping and surface quality. Specifically, the orientation to the presentation the results of experience machining stainless steels with very low carbon called Extra Low Carbon Stainless Steels. Results of experiments in the laboratory at the university and real-life engineering firms have been verified by applying the CA-system simulation software applications based on the finite element method. The paper described verification of CAD systems applied by analysis of drilling tools. Analysis of tool life is very important for proces effectivity. Every tool is damaged in the process of cutting. Wear mechanisms are activated in the cutting zone during the interaction of the elements of the cutting edge of the tool and the workpiece, and under the influence of temperature, and by the fact that friction depends on the interaction of the clean metal surface between the front plate of the cutting edge of the tool and the chip. According to DIN 50321 we recognize four fundamental mechanisms of tool wear: adhesive wear, abrasive wear, fatigue wear, tribochemical reaction wear. The mechanism of wear means the synergistic effect of factors that create a change in matter, a change in volume, i.e., a change in cutting edge dimension.

6. Acknowledgment

The authors would like to thank in words the grant agency for supporting research work and cofinancing the projects: KEGA #3/7166/2009 and VEGA #1/0048/2010.

7. References

Davim, J. P. (2008). *Machining - Fundamentals and Recent Advances*. Springer, ISBN: 978-1-84800-212-8, Dordtrecht, Netherlands

Jurko, J. & Zaborowski, T. (2009). *Drilling - the cutting process*. IBEN, ISBN 13-978-83-925108-2-6, Gorzów Wlkp., Poland

Džupon, M.; Gajdoš, M.; Jurko, J. ; Ferdinandy, M. & Jakubeczyová, D. (2011). Plastic Deformation Around Holes Drilled in Austenitic Steel 1.4301. *Chemické listy*, Vol.105, No. Special, (May 2011), pp. 606-608, ISSN 0009-2770

Jurko, J.; Panda, A. & Gajdoš,M. (2011). Study of changes under the machined surface and accompanying phenomena in the cutting zone during drilling of stainless steels with low carbon content. *Metalurgia*, Vol.50, No.2, pp. 113-117, ISSN 0543-5846

Jurko, J.; Panda, A. & Gajdoš,M. (2009). Accompanying phenomena in the cutting zone machinability during turning of stainless steels. *International Journal Machining and Machinability of Materials*, Vol.5, No.4, pp. 383-400, ISSN 1748-5711

Jurko, J.; Džupon, M.; Panda, A.; Gajdoš, M. & Pandová, I. (2011). Deformation of Material Under the Machined Surface in the Manufacture of Drilling Holes in Austenitic Stainless Steel. *Chemické listy*, Vol.105, No. Special, (May 2011), pp. 600-602, ISSN 0009-2770

Brinksmeier, E.W.; Reucher, A. & Griet-Solter, J.(2008). Influence of characteristic material properties on machinability under high speed cutting. *International Journal of Machining and Machinability of Materials*, Vol.4, No.4, pp. 419-428, ISSN 1748-5711

12

The Economics of Beer Processing

Gabriela Chmelikova and Mojmir Sabolovic
Mendel University
Czech Republic

1. Introduction

The brewing sector in the Czech Republic belongs to the most important agrarian business in the Czech Republic. Besides its long tradition (the first record of beer brewing in the Czech territory dates back to the year 993 and actually beer consumption per capita (158 litres per year) is the highest in Europe) it generates according to study of Ernst and Young (Leenen, 2010) nearly 7400 jobs directly in breweries and almost 12 300 jobs in the supplying sectors. Although this represents only minor part of all jobs in the Czech Republic, the industry represents an important factor in the local economic development, providing employment for relatively less skilled labour in the regions. Moreover in the hospitability sector approximately 32000 jobs can be attributed to the brewing sector while in retail around 2800 employees have jobs related to beer sales. These numbers also represent pretty benefits for the state budget from this sector. According to the Ernst and Young calculations (Leenen, 2010) the government revenues due to the production and sale of beer exceed actually to 676 million Euros, which create approximately 1,7 % of the state budget in 2010.

The number of industry breweries descends continuously from 72 industrial breweries in 1989 to 48 subjects in the Czech Republic in 2011. Contrariwise, the number of micro breweries concern 95 in the beginning of 2011 (Altova, 2011). This is the result of the progress from just one microbrewery to present number over the last 22 years. Despite the micro-brewing segment covers only approximately 0,5 % of total beer production in the Czech Republic, the growth of this segment is enormous. In 2006 the Czech Beer and Malt Association registered about 60 of them and it expects the number of these will exceed 100 in the end of 2011. Growth rate of this segment as well as the local character of this production is encouraging interest among researches and developing of economic analysis model for this segment is also a consequence of it.

2. Survey design

The survey involves the following structured sequence of steps. At the first stage brief overview of brewery industry is made. The method used is observation and description. At the second stage the research question is stated and null and alternative hypotheses are formulated. The method used is deduction. In addition the data sample and method of data collection is stated. At the third stage the theoretical framework is observed as the result of extensive theoretical literature review covering the state of the art of the business performance measurement system. Methods used are description, analysis and synthesis. At the fourth stage the particular steps in research methodology are designed. The method used is analysis and synthesis. At the fifth stage the research findings are explored. The descriptive statistic method is used. The sixth stage involves hypothesis testing and the answering the research question. The ANOVA method is used. At the seventh stage the theoretical model is designed. The synthesis and description method are used. In fine, the discussion and tasks for future research are articulated using deduction.

2.1 Research question articulation

The primary objective of entrepreneurship is the growth of stockholder value in general. Value based management disposes of tools for value enhancement. The main task is the quantity and selection of suitable variable as a proxy for value growth. The research question concerning identification of the most considerable factors of Economic Value Added and the value drivers of particular segment of breweries in the Czech Republic. The research problem is the formulation of theoretical multifactor model for explanation the particular factors based on research findings. For the response on stated research question we articulate null hypothesis H_0 and alternative hypothesis H_1 for existence difference explanation.

H_1: There is no significant difference of factors in Economic Value Added decomposition.

H_1: There is distinguishable impact of factors in Economic Value Added decomposition.

If the particular factors impact the business performance balanced, the subsequent theoretical model will cover the same set of variables for each factor. If the particular factors impact the business performance differently, the subsequent theoretical model for explanation requires appropriate set of variables for each factor.

2.2 Data collection

Data surveyed on target population are from secondary likewise primary resources. Method used for data collection and data processing is Stratified Random Sampling. It is assumed that this data is gathered in an unbiased manner. For some forms of analysis that use inferential statistical tests the data must be collected randomly, data observations should be independent of each other and the variables should be normally distributed. Secondary statistical and economic data are assembled from annual censuses of state agencies – Ministry of Agriculture of the Czech Republic, Ministry of Industry and Trade of the Czech Republic, Czech Statistical Office, Institute of Brewing and Malting. In term of legal form target population includes legal person as well self-employed persons. Flow indicators cover the whole structure of breweries according to the number of employees. A nationwide observation is carried out for enterprises with more than 50 employees. A selective survey is carried out for enterprises with 20 – 49 employees, and enterprises with less than 19 employees are calculated. Primary sampling frame comes from Creditinfo – Albertina database and Trade Register of the Czech Republic. Sample of analysed breweries is chosed according to market concentration analysis. Supplemental economic and market information were observed from particular WebPages of sample population of breweries. The target population of an analyzed subjects is geographically limited NUTS0, NUTS1 the Czech Republic. Analyzed data for Economic Value Added decomposition concerning period since 2000 till 2009.

3. Theoretical framework

Shift from the financial perspective to the non-financial one within the performance management invoked genesis of different performance measurement systems. According to Neely (Neely, 2002) a Perfomance Management System (PMS) is a balanced and dynamic system that is able to support the decision-making process by gathering, elaborating and analysing information. The concept of PMS was developed in response criticisms that traditional performance models are focused on financial measures, are historically oriented and do not cover all of the business areas. According to many scholars a well designed PMS should by using different kinds of measures represent whole organization. The balance approach offers by tying together various measures a holistic organizational view.

Interest on performance measurement management has started to increase in the 80s of the last century. Since then numerous of PMS models were developed and consequently theoretical (and very little empirical) research on PMSs has been carried out. The literature surveys tried to sort the particular models according to different criterions, such as attitude to firm's strategy, focus on stakeholders, balance, dynamic adaptability, process orientation, casual relationships or simplicity (Garageno et al., 2005). According Toni & Tonchia (Toni & Tonchia, 2001) the main models of PMSs can be referred to following typologies: hierarchical/vertical (cost and non-cost performance measures on different levels of aggregation), balanced scorecard/tableaux de board (several separate performances are considered independently), internal and external performances.

As our research focuses on performance management in small and medium-sized enterprises only those reviews concerning SME were taken into account. Garengo et al. (Garengo et al., 2005) focused their review on eight PMS models developed after the mid-1980s. The models considered were six of the most popular generic models and two PMS models designed specifically for SMEs. They focused on following models. Performance Measurement Matrix (Keegan et al., 1989): According to Garengo et al. (Garengo et al., 2005) and Neely et al. (Neely et al., 2000) this model uses the matrix combining the non-cost and cost perspective with external and internal perspective. The model is balanced and simple, for which it is sometimes criticized. Performance Pyramid System (Lynch & Cross; 1991) is designed as a pyramid with several levels linking the firm's strategy, business units and operations. Results and Determinants Framework (Fitzgerald et. Al, 1991): This model focuses on searching the relationship between the entrepreneur's results expressed in terms of competitiveness or financial performance and determinants of these results such as quality, innovations and flexibility. Balanced Scorecard (Kaplan & Norton, 1996): 4-box approach to performance measurement. In addition to financial measures, managers are encouraged to look at measures drawn from three other perspectives of the business: learning and growth, internal business processes and customer. The model is balanced and belongs to the most popular models both in the literature and in practice. Integrated Performance Measurement System (Bititci et al., 1997), who defined it as the information system by which the company manages its performance in line with its corporate and functional strategies and objectives, it is based on four levels. According to Hudson et al. (2001) this model fails to provide a structured process that specifies objectives and timescales for development and implementation. Performance Prism (Neely et al. 2000): According to Garengo et al. (Garengo et al., 2005) this model is three-dimensional, in correspondence with its name a prism graphically represents the architecture of the model.

Organizational Performance Measurement (Chennell et al., 2000), which was designed exclusively for SMEs. Is based on three principles (alignment, process thinking, and practicability) and is balanced. Integrated Performance Measurement for Small Firms (Laitinen, 2002). Within the model the internal dimension monitoring production process and the external dimension monitoring the competitive position are causally likened.

Hudson et al. (Hudson et al., 2001) evaluated ten PMSs. In contrast to Garengo et al. (Garengo et al., 2005) they included 4 different PM approaches. In addition to Garengo's selection following models were considered: Integrated Dynamic PMS (Ghalayini et al., 1997) which focuses on ensuring fast and accurate feedback. Integrated PM framework (Medori & Steeple, 2000) which is criticized for being complicated to understand and use. Integrated Measurement Model (Oliver & Palmer, 1998) defines the dimensions of performance and offers a mechanism for designing the measures. And finally Consistent PM Systems (Flapper et al., 1996) which is being criticized for weak balanced approach for critical dimensions of performance.

The common conclusions of the latest reviews show that there is a difference between models for big companies and models for SMEs. According to Garengo et al. (Garengo et al., 2005) most of the SMEs models are characterized by increasing strategy alignment, while continuing to focus on the most critical aspect for SMEs, i.e. operational aspects. Further all models are balanced, which is particularly important and which makes these models different form the traditional financially oriented ones. Finally clarity and simplicity characterize the most recent models.

3.1 The basis of performance system in Czech conditions

For centuries, economists have reasoned that for a firm to create wealth it must earn more than its cost of debt and equity capital - this principle is in the microeconomic terminology titled 'creating the economic profit'. A good financial performance measure should ask how well the firm has generated operating profits, given the amount of capital invested to produce these profits. In recent years the Stern Stewart & Company has operationalized this concept under the label Economic Value Added. EVA is defined as a spread between the return on capital invested and the cost of capital invested. It describes the ability of the firm to create the economic profit. Contrary to the traditional performance metrics, EVA manages to reflect real costs of the firm because it takes note of the equity costs as well as the other costs of the firm. The EVA metric is based on a simple and straightforward notion, as described in the following equation:

$$EVA = NOPAT - Capital \cdot WACC \quad (1)$$

Where *NOPAT is* Net Operating Profit After Taxes, *Capital* is Capital Employed to generate Operating Profit, and *WACC* is Weighted Average Cost of Capital.

The components of EVA are not directly obtainable from the financial statements, as EVA concept works with items referring entirely to operating activity. The EVA authors define operating activity as those operations that serve the basic entrepreneurial purpose. It is therefore necessary to convert the accounting data; under the Czech accounting rules, the "operating profit" and the corresponding capital include activities that are not directly aimed at fulfilling the basic entrepreneurial purpose - such as the investing of temporary free operating financial asset into the securities or creating constructions in progress (neither contributes to current operating activities). On the other hand, other activities necessary for meeting the basic entrepreneurial purpose of the firm are not covered under the operating profit and capital. The most important ones include financial and operative leasing, as well as capitalization and amortization of certain marketing costs, research and development costs, unrecorded goodwill, etc.

Similar to many accounting innovations, the concept of EVA promises better performance measurements, incentive schemes and equity valuation. The concept behind EVA is quite simple - maximize the spread between the return on capital used to generate profits and the costs of using that capital. Through its adoption, corporate executives hope that EVA will lead to increased efficiency in the allocation of all assets and hence increased shareholder wealth. In fact, Stern Stewart & Company has advocated that EVA can be used instead of earnings or cash from operations as a measure of performance. They claim that: "Eva is almost 50 % better than its closest accounting-based competitor in explaining changes in shareholder wealth" (Stewart, 1994), or "Forget EPS, ROE and ROI. Eva is what drives stock prices" (Stewart, 1995).

Though from the theoretical point of view EVA is seen as a superior performance metric, the results of some empirical studies do not support this claim. Numerous researchers have looked into the effectiveness of EVA using the independent empirical evidences (for instance: Biddle, Bowen, Wallace (Biddle, Bowen, Wallace; 1997); Turvey, Lake, Duren, Sparling (Turvey, Lake, Duren, Sparling; 2000); Feltham, Issac, Mbagwu, Vaidyanathan (Feltham, Issac, Mbagwu, Vaidyanathan; 2004); Bacidore, Boquist, Milbourn, Thakor (Bacidore, Boquist, Milbourn, Thakor; 1997); Berenstein (Berenstein, 1998); Kramer, Pushner (Kramer, Pushner, 1997) and did not indicate the superiority of EVA among other financial measures. Nevertheless, among both the Czech academic researches and practical financial analysts the usage of EVA is still limited because of the low empirical evidence of the behaviour of EVA within the Czech economy. A critical point of this research in the conditions of Czech economy is a lack of data about publicly trading companies, which at the same time, serve as an exogenous criterion for assessing the quality of the examined measure in the mentioned studies.

One of the most often claimed characteristics of EVA is its capability to inform owners about the creation of shareholder value, which could be in general described by the performance of capital market. In 2010 was carried out a study focusing on the relationship between ability of Czech firms to create economic value and performance of Czech capital market (Chmelíková, 2010). The research question was, whether performance metric EVA describes creation of shareholder value of the firms in the Czech Republic. The answer was found in the relationship between EVA and behaviour of capital market. As the development of these two categories proceeded in the same way it could be concluded, that EVA metric, with respect to its theoretical background, can be used as measure of shareholder wealth creation of the Czech firms. The behaviour of capital market was described by the stock exchange index PX. The official index of Prague stock exchange is currently the index PX, which is being the successor of the oldest Prague index PX 50. The index's values are published daily, which is in contrast to the information about creation of economic value added by firms in Czech Republic that are shown on year basis. This invokes the need to characterize the performance of capital market on the annual basis by using simple arithmetic average of daily index. Ministry of Industry and Trade of the Czech Republic monitors the creation of economic value added among the industry and construction firms in the Czech Republic. This analysis covers vast majority of all business in this sector (about 90%). Despite the number of business in this study is fluctuating in dependence on the number of currently operating firms, the trend of EVA development is well observable and enables the comparison with the development of capital market performance. The progress of these two categories indicated a general positive correspondence between the development of capital market performance and creation of economic value added among Czech firms. The regression results demonstrated high value of coefficient of determination R2, which gets to relatively high level of 0,83. This result is also supported by the research of the relationship between Economic Value Added, traditional performance measures (Return on Assets 'ROA' and Return on Equity 'ROE') and their ability to measure the creation of shareholder wealth of food-processing firms in the Czech Republic (Chmelíková, 2008). The intent of this research was fulfilled by providing a simple regression test of the hypothesis, that the EVA measure is more associated with improved shareholder wealth than traditional performance measures ROA and ROE. The results of regression analysis indicated in all cases a positive correspondence between EVA and financial performance metrics and show higher quality information content of EVA indicator in the relationship to the ability of shareholder wealth creation than traditional performance measures. This fact supports the tested hypothesis as well as the conclusions of corporate finance theory, that from the theoretical point of view EVA is seen as a superior performance metric. The results suggest that EVA should be considered when measuring performance of Czech-food processing firms and can become a basis of economic analysis in this sector.

When analyzing a firm current theory and praxis usually use three types of systems of measures: parallel systems, pyramidal systems and rating and bankruptcy indexes. Parallel systems concentrate measures into the groups according to the particular business areas. The advantage of this approach lies in the rich theoretical background and in the correspondence with functional structure of the firm. On the other hand the disadvantage is poor interconnection between particular groups of the system that leads to complicated interpretation of the results. Rating and bankruptcy indexes offer undemanding computative procedure unfortunately accompanied with rough information content of the results without identifying factors of the firm's efficiency. The advantage of pyramidal systems lies in the reflection of mutual interconnections between particular parts of the system with straightforward linking between the individual indicators and synthesis measure. On the other hand the pyramidal systems suffer from poor theoretical background and impose higher requirements on the analysts' qualification. The consequence is low popularity among financial analysts. Neumaierová (Neumaierová, 2008) claims, that current praxis prefers parallel evaluating systems. This is in contrast to the character of current situation, which is noted for high dynamical complexity due to the globalisation and rather than parallel systems of indicators requires the pyramidal ones. The keystone of pyramidal concepts is the involvement of interconnections between particular indicators, which makes these concepts the most compatible with the new environment. The basic principle of pyramidal system is decomposition of a top indicator with intention to identify the influence of its partial factors, when simultaneously the links between particular measures are represented by mathematical equations.

Enrichment of classical pyramidal system of any financial metric with the non-financial measures will offer a measurement system not dissimilar to the Balanced Scorecard. The Balanced Scorecard is a widely adopted performance management framework first described in the early 1990s through the work of Kaplan & Norton (Kaplan & Norton, 1992). Since then, the concept has become well known and its various forms widely adopted across the world. By combining financial and non-financial measures in a single report, the Balanced Scorecard aims to provide managers with richer and more relevant information about activities they are managing than is provided by financial measures alone. It is a performance management tool that enables a company to translate its strategy into a tangible set of performance measures. A Scorecard has to tell the story of a firm's strategy and the story is told by means of cause-and-effect model that links all the measures to the creating of shareholder value. The scorecard provides a view of a firm's overall performance by integrating financial measures with non-financial measures. This helps to manage the activities that stand beyond the control of financial measures in the framework of a holistic management system and overcomes the main disadvantage of pure financial analysis, which suffers form historic character of its information. The Balanced Scorecard contains a mix of leading and lagging indicators: Lag indicators represent the consequences of actions previously taken, while lead indicators are the measures that lead to the results achieved in the lagging indicators. Lagging indicators without performance drivers (usually described in non-financial terms) fail to inform managers of how to achieve the results. The authors of Balanced Scorecard Norton and Kaplan (Kaplan & Norton, 1992) claim that: "The balanced Scorecard retains traditional financial measures. But financial measures tell the story of past events, an adequate story for industrial age companies for which investments in long-term capabilities and customer relationships were not critical for success. These financial measures are inadequate, however, for guiding and evaluating the journey that information age companies must make to create future value through investment in customers, suppliers, employees, processes, technology, and innovation."

Balanced Scorecard is designed as a simple, 4-box approach to performance measurement. In addition to financial measures, managers are encouraged to look at measures drawn from three other perspectives of the business: Learning and Growth, Internal Business Processes and Customer. The power of the framework comes from a fact that it goes beyond an ad-hoc collection of financial and non-financial measures. Despite the apparent shortcomings of financial measures, a well-constructed Balanced Scorecard is not complete without them. Scorecard practitioners recognize this fact, and consider financial measures to represent the most important component of the Scorecard. Niven (Niven, 2006) claims, that "by using the Balanced Scorecard an organization has the opportunity to mitigate, if not eliminate entirely, many of the issues related to financial measures."

In building the scorecard, the process is just as important as the content. A scorecard devoid of process will be sterile and fail to mobilize both the executive team as well as the operational employees. To build a Balanced Scorecard for a specific company is a task for its whole executive team, since it is necessary to have specific information from all company's divisions. The choice of portfolio of non-financial measures depends on the character of a company. In order to be able to design a framework for economic analysis it is therefore necessary to specify at least the sector, or better a segment for future application. For this purposes the segment of microbreweries form the brewing sector of the Czech Republic was chosen.

4. Research method

Descriptive statistics are used for basic features of the data in the study. One-way ANOVA is used for hypothesis tests[1]. MS Excel is the tool for computation. Observed variables are computed for industry average and the sample of breweries. The results are compared and statistically tested.

4.1 Market concentration

Herfindahl Index (*HHI)* is used for concentration ratio analysis. The *HHI* is calculated by summing the squares of the individual firms' shares, see equation (2). The firms with larger market shares have proportionately greater weight in the results (Horizontal Merger Guidelines [HMG], 2010), thereafter (HMG, 2005). Breweries included in *HHI* constitute sample for Economic Value Added Decomposition.

$$HHI = \sum_{i=1}^{N} s_i^2 \qquad (2)$$

where *HHI* is Herfindahl Index, s_i is the market share of the firm *i* in the particular market, and *N* is the number of firms.

Markets are classified into three types (HMG, 2005):

- Highly competitive markets: $HHI < 0{,}10$,
- Unconcentrated markets: $0{,}10 < HHI < 0{,}15$,
- Moderately concentrated markets: $0{,}15 < HHI < 0{,}25$,
- Highly concentrated markets: $0{,}25 < HHI$.

4.2 INFA rating model

Beverage industry in general and brewery sector in particular are analysed by INFA Rating Model (Neumaierova & Neumaier, 2002, 2005, 2005) with particular emphasis on annual EVA decomposition (MPO, 2010). The model of EVA decomposition encompasses financial and risk controlling and analysis. INFA rating model is compiled from three stages of business performance measurement. The first stage considering creation of productive powers (*EBIT/Assets*) allows analyzing the product with no taxation impact. The second stage covers analysis of redistribution of *EBIT* among government (tax), creditors (interest), and shareholders (net profit). At the third stage involves financial stability analysis via useful life of assets and liabilities ratio. Algorithm of model is based on interdependencies among balance sheet, income statement and cash flow indicators.

INFA Rating Model is based on further simplistic assumptions (MPO, 2010).

---

[1] Fundamental statistics methods used in a standard way are not explained hereinafter.

- Financial interest is considered annually paid at the cost of debt,
- Market Value of debt is identified with the Book Value of interest-bearing debt,
- Independence of Weighted Average Cost of Capital on capital structure is assumed.
- Rate of EAT/EBT is used in the cost of capital instead of (1 - Tax) due to inclusion of the true impact of taxation.

4.3 Economic value added

Economic Value Added (EVA) modified by Neumaierova & Neumaier (Neumaierova & Neumaier, 2002, 2005, 2005) is primary in the form of shareholder claims articulation, see equation (3). The other explanations are not taken into account. According to methodology of Financial Analysis of Business the focus of EVA analysis is concerned on Value Spread (MPO, 2010). Value Spread *(ROE - r_e)* is difference of real return on equity and expected return on the corresponding risk r_e i.e. alternative cost of equity. If the Value Spread is positive the business reached positive EVA and thus shareholder value increases.

$$EVA = \left(ROE - r_e\right) \cdot E \tag{3}$$

where EVA is Economics Value Added, ROE is Return on Equits, r_e is Cost of Equity, and E is Equity.

4.3.1 Return on equity

The priority in economic value creation is a shareholder's perspective. The keen on intrinsic value growth is a cornerstone of entrepreneurial activity and business strategy (Damodaran, 2001). *ROE* is the result of INFA Rating Model financial controlling.

$$ROE = \frac{EAT}{EBT} \cdot \frac{\frac{EBIT}{A} - \left(In \cdot \left(\frac{CE}{A} - \frac{E}{A}\right)\right)}{\frac{E}{A}} \tag{4}$$

where ROE is Return to Equity, EAT is Earning After Taxes, EBT is Earning Before Taxes, $EBIT$ is Earning Before Interest and Taxes, A are total Assets, In are Interests, E is Equity, CE is Capital Employed (Equity, Debt, Obligations).

4.3.2 Cost of equity

Principle of cost of Equity r_e by course of INFA Rating Model contravenes mostly applied classical Modigliani - Miller theorem of capital structure (Modigliani & Miller, 1958; Brealey & Myers, 2008). The model of risk controlling comes from econometrics studies of rating agencies risk assessment. Mostly used Capital Assets Pricing Model is not suitable for emerging economics. As well, estimation of beta coefficient of non listed companies makes the model too subjective.

The Risk Premium represents the alternative Cost of Equity r_e *(5)*.. It is Return on Equity achievable from investment to alternative risk opportunity for investment.

$$r_e = r_f + r_{FINSTRU} + r_{FINSTAB} + r_B + r_{LS} \tag{5}$$

where r_f is Risk Free Rate, $r_{FINSTRU}$ is Financial Structure Risk Premium, $r_{FINSTAB}$ is Financial Stability Risk Premium, r_B Business Risk Premium, and r_{LS} Liquidity Risk Premium.

Risk Free Rate r_f is return on risk-free assets represented by annual yield on 10 years Czech government bond issued Czech National Bank.

Following risk premiums defined functions (6) in general shape. Because of lack of econometric studies suppose that from max certain level of indicators comprising the risk premium will be close to zero. Under these assumptions from min certain level the risk premium will converge to max value. The course of value of base indicator sets the interval of risk premium. Standard deviation measures the volatility of particular indicator in time series. Size of standard deviation indicates minimum value below which the risk premium cannot fall.

$$\begin{aligned} &X \le X_0 \Rightarrow r_x = \max \\ &X \ge X_1 \Rightarrow r_x = \min \\ &X \in \left(X_0, X_1\right) \Rightarrow r_x = a\left(X_1 - x\right)^b \end{aligned} \tag{6}$$

where X is the value of particular indicators constituting risk premiums, X_0 is the threshold value of an indicator by which achievement and lower values the risk premium converge to max, X_1 is the threshold value of an indicator by which achievement and higher values the risk premium converge to min, max is maximum risk premium, min is minimum risk premium, r_x is risk premium, a is constant force for equality $m = a(X_1 - X_0)^b$, a is constant indicating the course of function r_x, ($B = 1$ indicates linear function).

Liquidity Risk Premium r_{LS} characterises company size according to total Equity.

Business Risk Premium r_B is an indicator of creation of productive powers (*EBIT/Assets*) (7).

$$\begin{aligned} &Condition: \frac{\text{EBIT}}{\text{Assests}} \ge \frac{Interst}{Debt + Obligation} \cdot \frac{Equity + Debt + Obligation}{Assest} \\ &Let\ say: \frac{Interst}{Debt + Obligation} \cdot \frac{Equity + Debt + Obligation}{Assest} = X_1 \\ &If\ \frac{\text{EBIT}}{\text{Assests}} < \frac{Interst}{Debt + Obligation} \cdot \frac{Equity + Debt + Obligation}{Assest} \Rightarrow r_B = 10\% \\ &If\ \frac{\text{EBIT}}{\text{Assests}} > \frac{Interst}{Debt + Obligation} \cdot \frac{Equity + Debt + Obligation}{Assest} \Rightarrow r_B = \min\left(Sd\frac{EBIT}{Assest}\right) \\ &\text{If } 0 < \frac{\text{EBIT}}{\text{Assests}} < \frac{Interst}{Debt + Obligation} \cdot \frac{Equity + Debt + Obligation}{Assest} \Rightarrow r_B = \frac{\left(X_1 - \frac{EBIT}{Assets}\right)^2}{10 \cdot X_1^2} \end{aligned} \tag{7}$$

Financial Stability Risk Premium $r_{FINSTAB}$ is an indicator of financial stability by Liquidity Ratio:

- If $L \leq X_1$ then $r_{FINSTAB} = 10\ \%$,
- If $L \geq X_2$ then $r_{FINSTAB} = 0\ \%$
- If $X_1 < L < X_2$ then $r_{FINSTAB} = ((X_2 - L)^2/((X_2 - X_1))*0{,}1$
- Market Value of debt is identified with the Book Value of interest-bearing debt,

$$\text{If } \frac{\text{Current Assets}}{\text{Current Liabilities} + \text{Short - term} \cdots \text{Bank Loans}} \leq X_1 \Rightarrow r_{FINSTAB} = 10\%$$

$$\text{If } \frac{\text{Current Assets}}{\text{Current Liabilities} + \text{Short - term} \cdots \text{Bank Loans}} \geq X_1 \Rightarrow r_{FINSTAB} = 0\%$$

$$\text{If } 1 < \frac{\text{Current Assets}}{\text{Current Liabilities} + \text{Short - term} \cdots \text{Bank Loans}} \leq X_1 \Rightarrow r_{FINSTAB} = \frac{\left(X_1 - L\right)^2}{10 \cdot \left(X_1 - 1\right)^2}$$

Financial Structure Risk Premium $r_{FINSTRU}$ (7) is limited if $r_e = WACC$ than $r_{FINSTRU} = 0\ \%$. If $r_{FINSTRU} > 10\ \%$ then $r_{FINSTRU}$ is limited to $10\ \%$. The issue is in the case of extreme interest rate. Then interest rate shall be limited in the interval $0 \leq r_{FINSTRU} \leq 25\ \%$. Similarly tax burden is limited in the interval $0 \leq (EAT/EBT) \leq 100\ \%$. If the calculated value r_e is lower than $WACC$ then $r_e = WACC$.

$$r_{FINSTRU} = r_e - WACC \tag{8}$$

where $r_{FINSTRU}$ is Risk Premium for Financial Structure, r_e is Cost of Equity, and $WACC$ is Weighted Average Cost of Capital.

Analysis of sector's alternative cost of capital is calculated as weighted average of alternative cost of capital of particular subjects. As a concrete weights are supposed individual equities. Assumed economic profits are numbered and divided by sector's aggregated equity.

4.3.3 Economic value added decomposition

Economic Value Added in the INFA Rating Model is the crucial indicator of business performance. The peak indicator is influenced by particular factors for its determining, see Fig.1. The changes of the peak indicator are decomposed and the degrees of impacts are determined. Economic Value Added as a peak indicator can be decomposed by additive and multiplicative relationship (Neumaierova & Neumaier, 2002, 2005, 2005). According to empirical results the Logarithm Method and Index Method are used for calculation of changes in the degree of influence.

5. Results

In 2001 the Czech Republic ranked in the fifteenth position in the world beer production, see Fig. 2. It produced 18 mhl and it had 1,3 % of world and 3,8 %of European beer production. The biggest world beer producers were USA (231 mhl), China (215 mhl), Germany (109 mhl), Brazil (90 mhl) and Japan (71 mhl). According to Czech Association of Breweries and

Malt Plants the Czech breweries produced 17,881 mhl of beer, which means slight decrease on a year earlier (-0,25%). This change was caused by decline of 1,9 % in domestic consumption (by -309 thl to 16,026 mhl) and rise of 16,7 % in export (by 266 thl to 1,855 mhl). The average beer consumption per capita remained stable at the level of 160 ls. The decrease in the number of Czech breweries still continued. The number of 71 breweries in 1994 dropped to 54 in 2001. Shifting ownership from domestic to foreign one did influence neither the brand names nor the quality of the beer (Balsik, 2002).

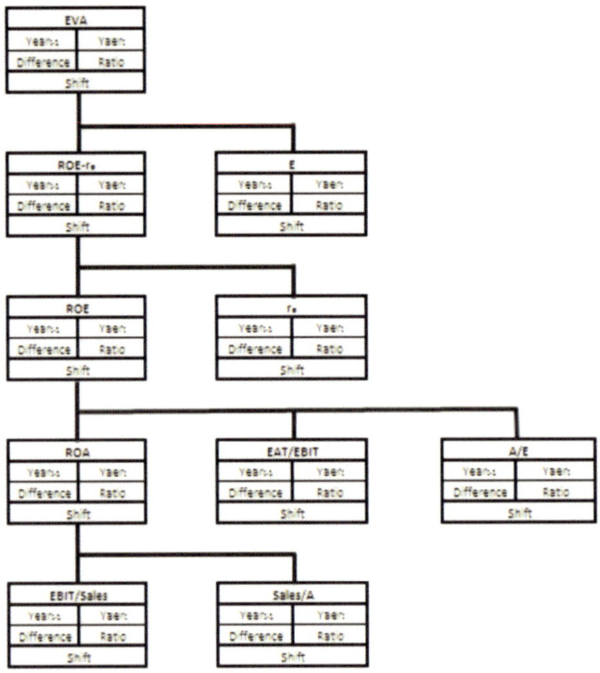

Fig. 1. Economic Value Added Decomposition

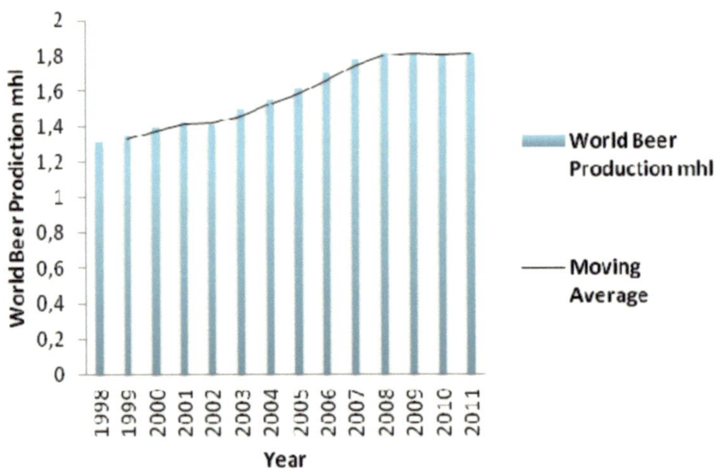

Fig. 2. World Beer Production

In 2002 he Czech Republic ranked in the fifteenth position in the world beer production. It produced 18 mhl and it had 1,3 % of world and 3,7 % of European beer production. The biggest world beer producers still remained USA (231,5 mhl), China (231,2 mhl), Germany (109 mhl), Brazil (85 mhl) and Japan (70,5 mhl). According to Czech Association of Breweries and Malt Plants the Czech breweries produced 18,178 mhl of beer, which means slight increase on a year earlier (101,7 %). This change (+ 297 thl) was caused by the rise of 1,1 % in domestic consumption (by 176 thl to 16,202 mhl) and rise of 6,5 % in export (by 120 thl to 1,975 mhl). The average beer consumption per capita remained stable at the level of 160 ls. The demand was composed largely from draft beer (67,6 %) which was cheaper and better complied with Czech lifestyle than lager beer with slightly growing market share (28,4 %). The consumption share of special beers declined, so did the share of non-alcoholic beer which was in 2002 0,6 % of total beer consumption in CR (Braznovsky, 2003). In 2003 the Czech Republic ranked in the fifteenth position in the world beer production. It produced 18 mhl and it had 1,2 % of world and 3,4 % of European beer production. The ranking in the world beer producers changed and China with 245 mhl was on the top. Then USA (235 mhl), Germany (105 mhl), Brazil (86 mhl) and newly Russia (79,8 mhl), which reached the highest growth of beer volume production (annual rate of 8%). According to Czech Association of Breweries and Malt Plants the Czech breweries produced 18,548 mhl of beer, which means slight increase on a year earlier (2,1 %). This change was caused by increase of 1,9 % in domestic consumption (by 216 thl to 16,418 mhl) and rise of 7,8 % in export (by 155 thl to 2,130 mhl). The number of industry breweries continued in the fall (Breweries Svitavy and Litoměřice were closed), the concentration of the market kept going which led to beer unification. On the other hand the development of number of micro breweries was growing and their supply of specials was aimed on local markets. The share of lager beer grew slightly and leveled off at 28,4 %, draft beer had 67,6 % and non-alcoholic 0,6 % (Altova & Braznovsky, 2004). In 2004 the Czech Republic ranked in the seventeenth position in the world beer production. It produced 18,1 mhl and it had 1,2 % of world and 3,4 % of European beer production. The biggest world beer producers were China (277,5 mhl), USA (238,0 mhl), Germany (104,5 mhl), Brazil (90 mhl) and Russia (83 million mhl). According to Czech Association of Breweries and Malt Plants the Czech breweries produced 18,753 mhl of beer, which means slight increase on a year earlier (1,1 %). This change was caused by decrease in domestic consumption (by 303 thl to 16,115 mhl) and year-on-year rise of 7,8 % in export (by 508 thl to 2,638 mhl). The number of industry breweries remained stable at the level of 53 breweries. The demand was composed largely from draft beer (61,3 %) which was cheaper and better complied with Czech lifestyle than lager beer with slightly growing market share (34,4 %) (Altova, 2010). In 2005 the Czech Republic ranked in the sixteenth position in the world beer production. It produced 19,0 mhl and it had 1,2 % of world and 3,5 % of European beer production. The biggest world beer producers were China (308,0 mhl), USA (232,7 mhl), Germany (105,8 mhl) and Russia (88,4 mhl). According to Czech Association of Breweries and Malt Plants the production of Czech breweries breached the boundary of 19 mhl and reached the volume of 19,069 mhl of beer, which means a year-to year increase by 1,7 %. Whereas the production for domestic market decreased by 145 thl to 15,970 mhl the share of export on total volume of beer produced in The Czech Republic rose from 14 % in 2004 to 16,3 % in 2005. The number of industry breweries decreased slightly to 47 plants owned by 38 companies. The number of microbreweries was still growing to the number of 30. The Czech beer market offered in 2005 more than 470 beer types. Despite the majority of them was produced by the Czech traditional way of beer production (Pilsner), there could be found also around 80 marks of special beers. Their specialty is based on a different way of fermentation, different yeasts and special flavor reached from herbals, fruit, honey etc. The number of industrial breweries (54) remained on the same level as in 2004. In comparison to previous year there is visible drop of 58,8 % in production of small breweries (in category up to 20 thousands hl) and growth of 1,1 % in the category 200 - 300 thl (Altova, 2010). In 2006 the Czech Republic ranked in the seventeenth position in the world beer production. It produced 19,2 mhl and it had 1,2 % of world and 3,5 % of European beer production. The biggest world beer producers were China (320 mhl), USA (223 mhl), Germany (105 mhl), and Russia (93 mhl). The Czech brewing segment continued in growing and it produced the highest volume of beer than ever before. According to Czech Association of Breweries and Malt Plants the Czech breweries produced 17,787 mhl of beer, what improved previous best record from 1992. The number of industrial breweries remained at the same amount, only the number of microbreweries grew and reached almost the level of 60. There has been a marked improvement in the production of non-alcoholic beer which was brewed by 19 breweries. The total volume of non-alcoholic beer reached 328 thousands hl what is 1,65 % of total beer production in the Czech Republic. These numbers indicates dramatic growth of 37 % in comparison to previous year (Altova, 2010). In 2007 the Czech Republic ranked in the seventeenth position in the world beer production. It produced 20 mhl and it had 1,1 % of world and 3,4 % of European beer production. The biggest world beer producers were China (370 mhl), USA (232,8 mhl), Russia (109,8 mhl) and Germany (106 mhl). The Czech brewing segment continued in growing and exceeded the previous year's record. According to Czech Association of Breweries and Malt Plants the Czech breweries produced 19,897 mhl of beer, what meant the growth by 100 thousands hl. on a year earlier, see Fig. 3. Regarding the concentration of the market it kept going which led to beer unification. 86 % of total volume of beer production was brewed in 7 biggest breweries. The total volume of non-alcoholic beer reached 497 thousands hl what is 2,5 % of total beer production in the Czech Republic. These numbers again indicates dramatic growth of 51,6 % in non-alcoholic beer production on a year earlier (Altova, 2010).

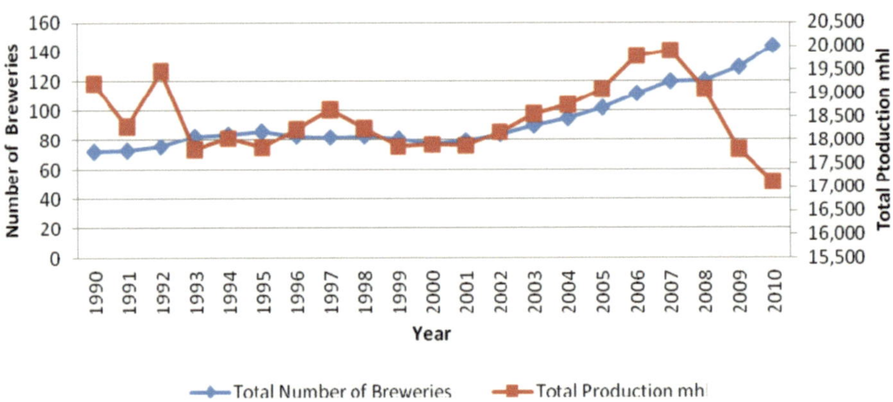

Fig. 3. Beer Production in the Czech Republic

In 2008 the Czech Republic ranked in the eighteenth position in the world beer production. It produced 20 mhl and it had 1,1 % of world and 3,3 % of European beer production. The biggest world beer producers were China (395 mhl), USA (236 mhl), Russia (119 mhl) and Germany (101 mhl). The growth of Czech brewing segment leveled off in 2008 at the volume of 19,806 mhl of beer. The volume of exported beer hectols exceeded in 2008 the bounder of 3 million. Regarding the concentration of the market it kept going, because 91,6 % of total beer production was produced in 10 breweries. In 2008 the flavored beers were on increase (by 38.8% to 6,8 thl) so did the specials with upper brewer's yeast which rose by 34,4 %. The non-alcoholic beers slowed down their expansion and grew by 16,4 % compared to last year's growth rate of 51,6 %. Czech breweries produced in 2008 578,9 thl (Altova, 2010). In 2009 was the total world beer production 1 802,7 mhl (99,3 % of the previous year's volume) and it recorded first downturn since 1999. In 2009 The Czech Republic ranked in the sixteenth position in the world beer production. It produced 20 mhl and it had 1,1 % of world and 3,5 % of European beer production. The biggest world beer producers were China (418,5 mhl), USA (234,1 mhl), Russia (110 mhl) and Brazil (107,3 mhl). Breweries associated with Czech Association of Breweries and Malt Plants produced in 2009 18,598 million hl of beer which means decrease by 5,9 % compared to previous year. This drop was caused mainly by the declining demand for draft beer of which production fell by 10,3 %. On the other hand the lager beer was on slight increase (5 %). The production of non-alcoholic beer declined for the first time during last decade by 1 % in comparison to 2008. The total decline in demand for beer can perhaps be explained by the fact that there was a drop in the number of tourists who visited Czech Republic in 2009 (Altova, 2010). World beer production decreased in 2009 for the first time since 1999. Moderate growth of 0,2% was recognised in 2010. The total world beer production was 1 811,4 mil. hl in 2010. In the terms of beer production the Czech Republic ranks the 17th from monitored countries by Hopsteiner. In the light of total yield Czech Republic covers 1,1 % of world beer production and 3,7 % of production in Europe. Czech beer production fell by 7,9 % following the decrease of tourism and consumption tax increase in 2010. The downturned was mostly caused by the industrial breweries' production gap. Light beer production slumped furthest by 13%. Traditional variety of the Czech brewing industry lay in the wide variety and uniqueness of the product range. Notwithstanding, production goes down but the number of brands increases. The diversification of production plays crucial role in customers' satisfaction. According to preliminary searching of Czech Association of Breweries and Malt Plants (August 2011) production turn into mild grow in 2011 (Altova, 2011).

5.1 Market structure of producers

A trend in brewery industry in the Czech Republic shows Fig. 4. Although the brewery industry's characteristics commemorate mature industry the number of microbreweries is constantly decreasing. A similar evolution was observed in USA (Carroll & Wade, 1991; Carroll & Swaminathan, 1992).

Fig. 5 comprises trend in concentration of production toward the segment of largest producers. The crucial role is played by breweries with annual production higher than 1000 khl. During the analysed period 2001 - 2008 the share of total production increased from 54% to 85%. Otherwise the largest decline experienced the breweries with production 500 - 1000 khl from 17% to 5%. The smallest drop experienced the breweries with production less than 120 khl 9% to 7%.

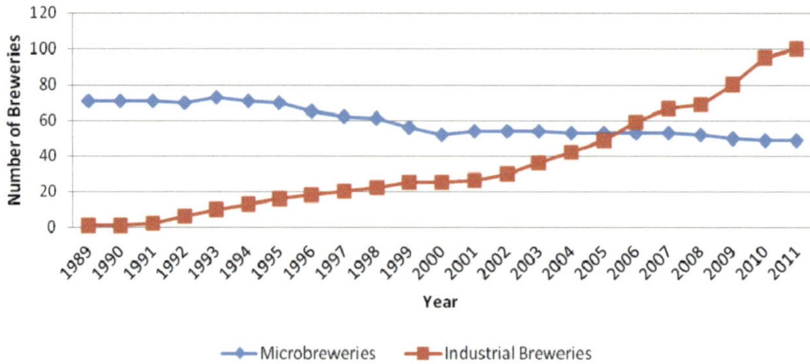

Fig. 4. Trends in the Number of Breweries in the Czech Republic

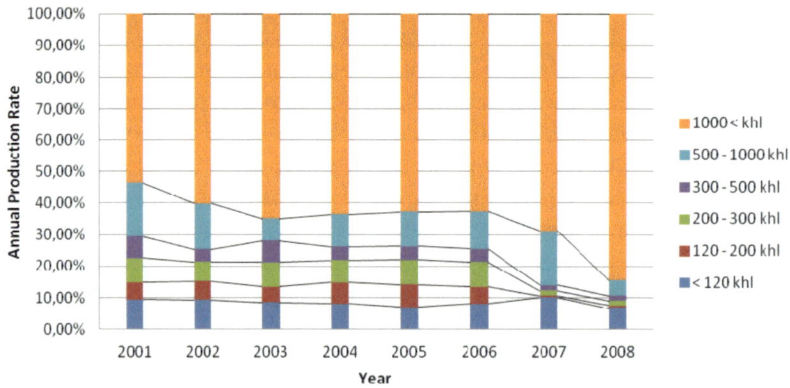

Fig. 5. Brewery Industry Concentration in the Czech Republic

Fig. 6 covers trend in market concentration measured by Herfindahl Index (HHI). Index is calculated on two levels. The first level covers the largest breweries with annual production higher than 500 khl and the second wider level covers the producers with annual production higher than 120 khl. The annual production of breweries included in the HHI captures decreasing concentration toward the largest producers.

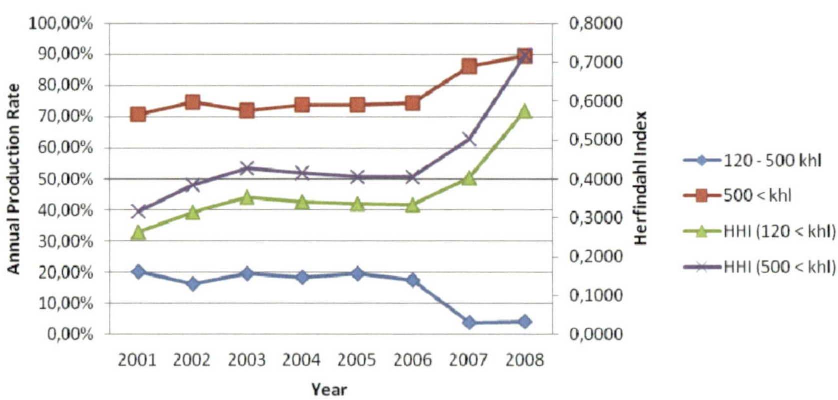

Fig. 6. Trend of HHI by the Size of Brewery in the Czech Republic

The distribution of annual production segmented by the size of the breweries covers Fig. 7. The trends of smallest and the largest producers are modelled by the appropriate functions. Trend of breweries with annual production 120 – 200 khl is modelled by the second-degree polynomial function. Trend of breweries with annual production more than 1000 khl is modelled by exponential function.

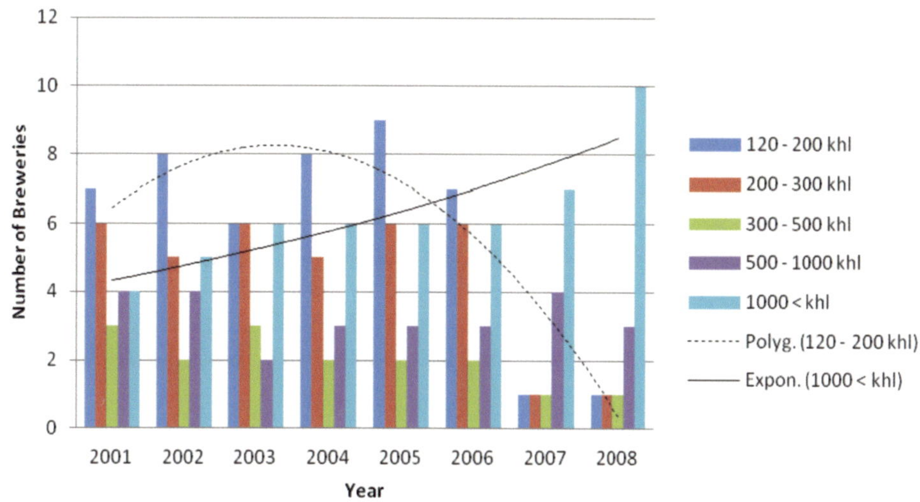

Fig. 7. Annual Production by the Size of Brewery in the Czech Republic

5.2 Economic value added decomposition

Economic Value Added is computed for beverage industry at the fist level and decomposition of changes of particular factors is carried out, see Table 1. The factors with the higher impact on peak variable differ. Factor with the highest frequency at the first level of decomposition is ROE-re.

Variables	2000	2001	2002	2003	2004	2005	2006	2007	2008	2009
EVA	232,09%	-45,30%	207,91%	723,09%	-18,65%	250,07%	-24,23%	-29,62%	-55,61%	-158,80%
E	-14,58%	13,88%	106,41%	-49,19%	8,58%	20,42%	4,89%	8,14%	10,39%	0,74%
ROE-re	229,46%	59,28%	204,51%	772,08%	-27,18%	229,65%	-24,39%	-37,76%	-66,00%	-159,54%
ROE	149,09%	-9,22%	71,56%	845,07%	-7,72%	126,18%	-20,05%	-4,32%	-58,18%	-31,53%
re	80,36%	68,50%	132,95%	-72,99%	-19,45%	103,47%	-4,34%	-33,45%	-7,82%	-128,01%
ROA	-26,84%	-5,89%	45,62%	558,10%	-6,60%	131,27%	-16,61%	-1,48%	-56,80%	-54,88%
EAT/EBIT	167,30%	-23,84%	70,99%	120,26%	22,61%	35,50%	4,17%	0,00%	6,87%	0,47%
A/E	8,63%	20,51%	-45,05%	166,70%	-23,74%	-40,59%	-7,62%	-2,84%	-8,25%	22,88%
EBIT/Sales	-26,76%	6,83%	60,08%	296,70%	-13,85%	119,79%	-22,39%	5,75%	-35,52%	-52,33%
Sales/A	-0,08%	-12,72%	-14,46%	261,40%	7,24%	11,48%	5,78%	-7,23%	-21,28%	-2,56%

Table 1. Industry Economic Value Added Decomposition

Descriptive statistics of particular factors of Economic Value Added of beverage sector shows Table 2.

Variable	EVA	E	ROE-re	ROE	re	ROA	EAT/EBIT	A/E	EBIT/Sales	Sales/A
Mean	1,0809561	0,1096818	1,1801132	1,060886	0,1192271	0,5658918	0,4043402	0,0906541	0,3383093	0,2275825
St. Error of the Mean	0,8150073	0,1230822	0,8433651	0,8497107	0,2611943	0,5831726	0,1930699	0,1896227	0,333387	0,2671901
Median	-0,2143662	0,0836167	0,1744268	-0,0601924	-0,0607929	-0,0624599	0,1473995	-0,0522999	-0,0404546	-0,0131765
Modus	#N/A	#N/A	#N/A	#N/A	#N/A	#N/A	#N/A	#N/A	#N/A	#N/A
Standard Deviation	2,5772794	0,38922	2,6669545	2,6870211	0,825969	1,8441536	0,6105405	0,5996397	1,0542624	0,8449294
Variance	6,6423689	0,1514922	7,1126461	7,2200823	0,6822247	3,4009026	0,3727597	0,3595677	1,1114691	0,7139056
Kurtosis	3,1332525	4,8397562	3,9405335	8,2231798	-0,7858908	7,6866278	0,7804933	6,3559884	4,3846358	9,5880005
Skewness	1,6582335	1,4945915	1,7989465	2,7934284	-0,1100759	2,7026855	1,2770505	2,3149773	2,049831	3,0725106
Difference Min-Max	8,8188763	1,5560068	9,3161726	9,0324608	2,6095707	6,1490056	1,9114327	2,117505	3,4902693	2,8268245
Min	-1,5879785	-0,4919113	-1,5953717	-0,581785	-1,2800608	-0,5679921	-0,2383978	-0,4504672	-0,5232674	-0,212813
Max	7,2308977	1,0640955	7,7208009	8,4506759	1,3295099	5,5810135	1,6730349	1,6670378	2,9670019	2,6140116
Sum	10,809561	1,0968181	11,801132	10,60886	1,1922714	5,6589176	4,0434018	0,9065409	3,3830925	2,275825
Number	10	10	10	10	10	10	10	10	10	10
Confidence Level	1,8436746	0,2784312	1,9078243	1,9221791	0,5908626	1,319228	0,4367544	0,4289564	0,7541739	0,6044261

Table 2. Industry Descriptive Statistics

Economic Value Added of sample of producers includes Fig. 8. The sample of producers concerns largest and most influential producers in the Czech Republic covered by Herfindahl index, see Fig. 6. Elements of the sample are denoted Brew.1 - Brew. 5. Factor with the highest frequency at the first level of decomposition for particular breweries are E, A/E, EBIT/Sales, E, and Sales/A.

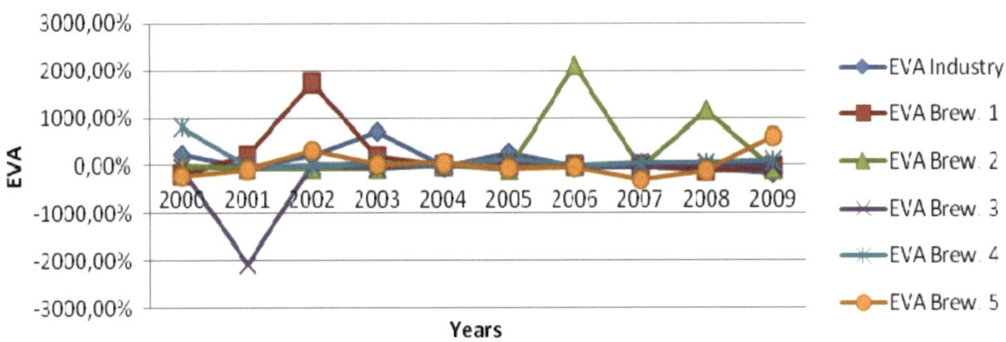

Fig. 8. Economic Value Added of Sample of Breweries

5.3 Hypothesis tests

Economic Value Added for particular breweries is decomposed and the results are used for verification of hypothesis and research question solution. The basic statistics of Economic Value Added decomposition for industry with comparison with particular breweries comprises Table 3.

Variables	Number	Sum	Mean	Variance
E Industry	10	1,096818096	0,10968181	0,151492231
E Brew. 1	10	11,71851773	1,171851773	10,54940141
E Brew. 2	10	12,50923797	1,250923797	9,107712839
E Brew. 3	10	-1,72798544	-0,172798544	0,481647761
E Brew. 4	10	6,916121354	0,691612135	1,564326416
E Brew. 5	10	9,921791203	0,99217912	5,894173
ROE-re Industry	10	11,8011317	1,18011317	7,112646076

Variables	Number	Sum	Mean	Variance
ROE-re Brew. 1	10	17,36378991	1,736378991	16,82857339
ROE-re Brew. 2	10	16,96957401	1,696957401	33,64941223
ROE-re Brew. 3	10	-18,6715505	-1,867155049	34,6138331
ROE-re Brew. 4	10	5,219831167	0,521983117	1,815547907
ROE-re Brew. 5	10	-8,02440309	-0,802440309	1,767543233
ROE Industry	10	10,60886029	1,060886029	7,220082313
ROE Brew. 1	10	15,19148314	1,519148314	14,21660606
ROE Brew. 2	10	1,926673091	0,192667309	0,756573518
ROE Brew. 3	10	-3,58545302	-0,358545302	1,214694686
ROE Brew. 4	10	5,464278608	0,546427861	1,804975771
ROE Brew. 5	10	-5,04353231	-0,504353231	1,693920646
re-re Industry	10	1,192271407	0,119227141	0,682224716
re Brew. 1	10	2,172306773	0,217230677	0,131062824
re Brew. 2	10	15,04290092	1,504290092	27,6222958
re Brew. 3	10	-15,0860975	-1,508609747	23,78197355
re Brew. 4	10	-0,24444744	-0,024444744	0,046228373
re Brew. 5	10	-2,98087078	-0,298087078	0,317321175
ROA-re Industry	10	5,658917576	0,565891758	3,400902569
ROA Brew. 1	10	13,36015824	1,336015824	12,72644729
ROA Brew. 2	10	-3,75063497	-0,375063497	1,436081246
ROA Brew. 3	10	-1,54108809	-0,154108809	0,371752623
ROA Brew. 4	10	6,605010194	0,660501019	2,259412793
ROA Brew. 5	10	0,841447855	0,084144786	2,194263131
EAT/EBIT Industry	10	4,043401764	0,404340176	0,372759683
EAT/EBIT Brew. 1	10	4,820393498	0,48203935	6,689907692
EAT/EBIT Brew. 2	10	1,571714428	0,157171443	0,119035818
EAT/EBI Brew. 3	10	-0,49597947	-0,049597947	0,049059267
EAT/EBIT Brew. 4	10	-0,20560792	-0,020560792	0,046590406
EAT/EBIT Brew. 5	10	-4,49029674	-0,449029674	0,709908584
A/E Industry	10	0,90654095	0,090654095	0,35956773
A/E Brew. 1	10	-2,9890686	-0,29890686	0,164112557
A/E Brew. 2	10	3,999529685	0,399952968	1,233232462
A/E Brew. 3	10	-1,25528772	-0,125528772	0,196760829
A/E Brew. 4	10	-0,93512366	-0,093512366	0,201868969
A/E Brew. 5	10	2,869958567	0,286995857	0,37470278
EBIT/Sales Industry	10	3,383092548	0,338309255	1,111469108
EBIT/Sales Brew. 1	10	14,80786278	1,480786278	14,70071229
EBIT/Sales Brew. 2	10	-0,93491412	-0,093491412	0,79979165
EBIT/Sales Brew. 3	10	0,495979469	0,049597947	0,049059267
EBIT/Sales Brew. 4	10	7,568201524	0,756820152	2,966541797
EBIT/Sales Brew. 5	10	-9,72405782	-0,972405782	4,821239837
Sales/A Industry	10	2,275825028	0,227582503	0,713905641
Sales/A Brew. 1	10	-1,44770454	-0,144770454	1,012756171
Sales/A Brew. 2	10	-6,00207537	-0,600207537	1,274417975
Sales/A Brew. 3	10	-2,40451342	-0,240451342	0,576558256
Sales/A Brew. 4	10	-0,96319133	-0,096319133	0,137428159
Sales/A Brew. 5	10	10,95205008	1,095205008	10,29613015

Table 3. Sample of Breweries Economic Value Added Statistics

Table 4. covers the results of ANOVA analysis. The variances of all set of factors of Economic Value Added decomposition are observed to recognise the factors with the highest influence for additional model constitution. Statistical significance P Value at the 5% for all variables falls to reject the null hypothesis.

E	Sum of Squares	Difference	Mean Square	F Value	P Value	F Stat
Model	17,17768339	5	3,4355367	0,7428521	0,5948124	2,3860699
Error	249,7387829	54	4,6247923			
Corrected Total	266,9164663	59				
ROE-re	**Sum of Squares**	**Difference**	**Mean Square**	**F Value**	**P Value**	**F Stat**
Model	106,7659603	5	21,353192	1,3375344	0,2625777	2,3860699
Error	862,0880034	54	15,964593			
Corrected Total	968,8539638	59				
ROE	**Sum of Squares**	**Difference**	**Mean Square**	**F Value**	**P Value**	**F Stat**
Model	31,46410009	5	6,29282	1,4032455	0,237851	2,3860699
Error	242,161677	54	4,4844755			
Corrected Total	273,625777	59				
re	**Sum of Squares**	**Difference**	**Mean Square**	**F Value**	**P Value**	**F Stat**
Model	46,89634399	5	9,3792688	1,070263	0,387112	2,3860699
Error	473,2299579	54	8,7635177			
Corrected Total	520,1263019	59				
ROA	**Sum of Squares**	**Difference**	**Mean Square**	**F Value**	**P Value**	**F Stat**
Model	19,6571875	5	3,9314375	1,0535876	0,3962382	2,3860699
Error	201,4997369	54	3,7314766			
Corrected Total	221,1569244	59				
EAT/EBIT	**Sum of Squares**	**Difference**	**Mean Square**	**F Value**	**P Value**	**F Stat**
Model	5,792401116	5	1,1584802	0,8702459	0,5072074	2,3860699
Error	71,88535304	54	1,3312102			
Corrected Total	77,67775416	59				
A/E	**Sum of Squares**	**Difference**	**Mean Square**	**F Value**	**P Value**	**F Stat**
Model	3,531577302	5	0,7063155	1,674894	0,1564189	2,3860699
Error	22,77220794	54	0,4217076			
Corrected Total	26,30378524	59				
EBIT/Sales	**Sum of Squares**	**Difference**	**Mean Square**	**F Value**	**P Value**	**F Stat**
Model	34,31330929	5	6,8626619	1,6841705	0,1541596	2,3860699
Error	220,0393255	54	4,0748023			
Corrected Total	254,3526348	59				
Sales/A	**Sum of Squares**	**Difference**	**Mean Square**	**F Value**	**P Value**	**F Stat**
Model	16,89886297	5	3,3797726	1,4473165	0,2224493	2,3860699
Error	126,1007671	54	2,3351994			
Corrected Total	142,9996301	59				

Table 4. ANOVA of Sample of Breweries

F Value is less than F stat for all variables. According to testing we fall to reject the null hypothesis H_0. We suppose data are not sufficiently persuasive for us to prefer the alternative hypothesis H_1 over the null hypothesis. Stated research question allows according to F Values at Table 4. statistical differences considered to be statistically significant enough for the design the model for explanation of particular factors. The statistical characteristics are so balanced and the draft of additional model concerns the whole field of value creation.

5.4 Economic value added with respect to soft skills value drivers

Despite the apparent shortcomings of financial measures, it is not possible to construct the economic analysis without them. These even represent the most important component of it. The results of above analysis identified performance measure Economic Value Added as basis for economic analysis of the firms form beverage sector in the Czech Republic. Most of analytical models are focused on the big businesses. We focus on sector of microbreweries hereinafter as the most dynamics segment in beverage industry in the Czech Republic, see Fig. 3. When designing the decomposition of EVA it is suitable to rewrite the equations (1) and (2) the following way:

$$EVA = \frac{NOPAT}{Sales} \cdot \frac{Sales}{Capital} - \frac{WACC}{Capital} \tag{9}$$

Where *NOPAT is* Net Operating Profit After Taxes, *Capital* is Capital Employed to Generate Operating Profit, and *WACC* is Weighted Average Cost of Capital.

This alternative expression of EVA measure determines three basic branches of its decomposition. The managers can use three ways how to drive the value of the firm:

- Through increasing the profit margin (NOPAT/SALES),
- Through increasing the turnover of total assets (SALES/CAPITAL),
- Through decreasing the riskiness of the firm (WACC/CAPITAL).

These free common financial measures represent in the framework of economic analysis the group of lag indicators that are the consequences of actions previously taken. But the managers are also interested in the question what are the lead indicators, what are the measures that lead to the results achieved in the lagging indicators? The framework of economic analysis should therefore contain a mix of leading and lagging indicators. Lagging indicators without performance drivers fail to inform of how to achieve the results. Conversely, leading indicators may signal improvements, but on their own they do not inform whether these improvements are improving the shareholder wealth.

Further EVA decomposition should therefore include also the lead measures, of which usage challenges leaving the financial perspective. Before incorporating the non-financial measures, one should first more specify the character of microbrewing segment in the Czech Republic. For the microbreweries according to Maier (Maier, 2009) are typical following characteristics: shipping does not exceed 5 000 hl/year, they have not a distribution network of its own, most of production is usually consumed in its own facility-restaurant, they do not export, the owner is usually a natural person or smaller legal entity, owner's relationship to the given sector is not only economic but also emotional. The quality of beer is believed to be the highest among other national brands and this fact is also connected with relatively higher selling price. This knowledge enables to design a framework of economic analysis suitable for business operating in this segment. According to stated research question and research problem we submit o multifactor model, see Fig. 9.

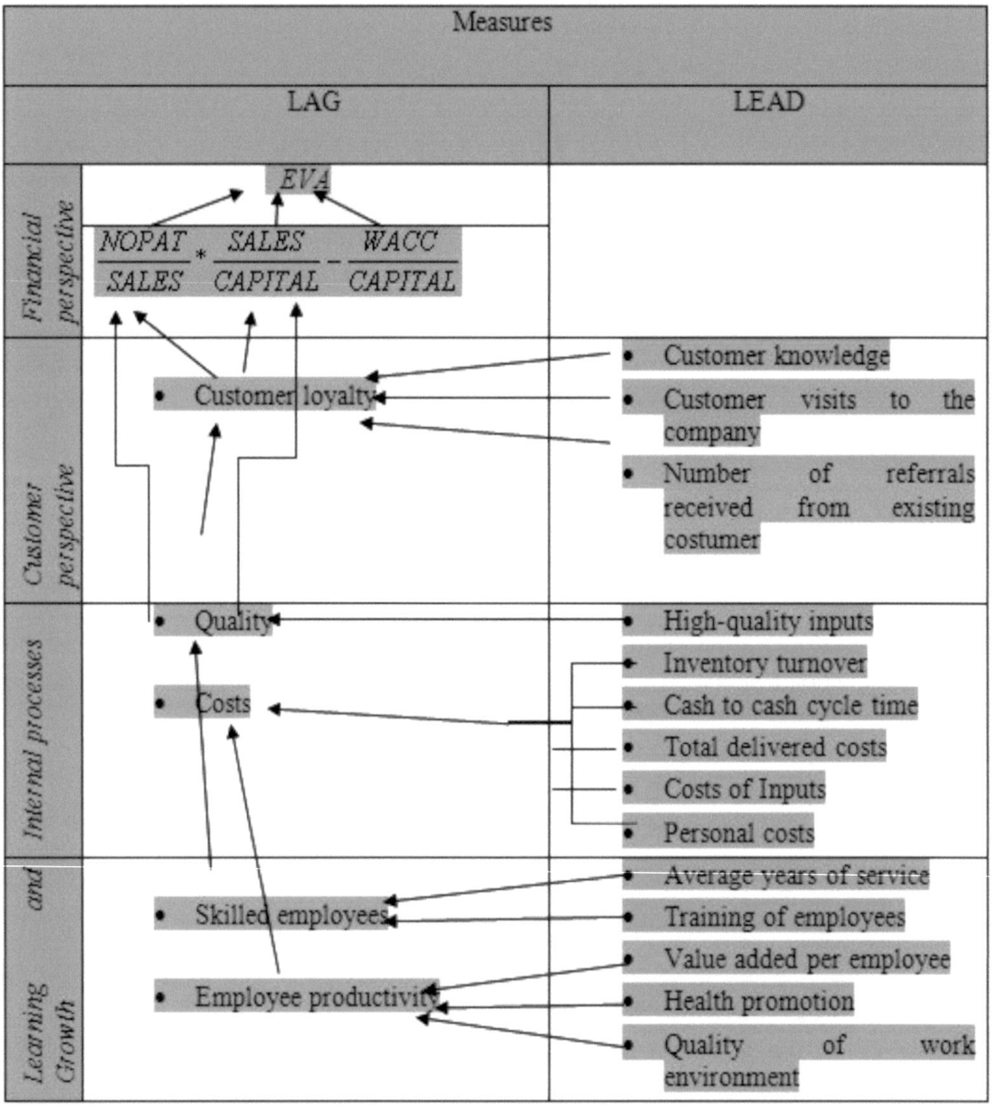

Fig. 9. Dynamics Soft Skills Model

According to both scholars and practitioners balanced top down approach is important for the successful managing of the firm, small firms are not excluded. Garengo et al. (Garengo et al., 2005) showed that even though the literature highlights the importance of using Performance Measurement System in small companies, very few firms carry out performance management. They see basically two main obstacles to introducing Performance Measurement in small firms - the lack of financial and human resources and the perception of Performance Measurement Systems as bureaucratic system that cause rigidity. As these obstacles were kept on mind when designing Performance Measurement System suitable for small breweries, the clarity and simplicity characterize this model. The framework of economic analysis built on the basis of Balanced Scorecard with the main performance indicator Economic Value Added represents from this view suitable tool for managing of Czech firms. No one pretends this is generally applicable for all microbreweries in the Czech Republic, nevertheless this procedure can be instrumental when building economic analysis framework in any microbrewery. Since this model was built solely on theoretical basis the next step in this research project is its empirical verification. This is also in accordance with the literature which claims that there is a significant gap between theory and practice. On one hand many PM models have been proposed but on the other hand very little empirical research has been carried out. In order better to understand the process performance measurement further empirical studies on this field are necessary. In spite of increasing interest on performance measurement systems during last 30 years, there is not visible any significant deviation from widely used financial measures in Czech business environment. These are generally criticized on account of several reasons: lag information content, bad fitting with information age competition and difficult communication to employees. Shift from the financial perspective to the non-financial one within the performance management invoked genesis of different performance measurement systems. The aim of this paper was therefore to establish the status of current knowledge in the area of performance measurement systems for small and medium enterprise. This theoretical phase of the research was based on the study of up-to-date reviews and it focused on the description of the most recent performance measurement systems. Further after considering Czech business specifics suitable base for performance measurement system was chosen and the framework of whole performance measurement system not dissimilar to Balanced Scorecard was designed. After considering the circumstances of the microbrewing segment in the Czech Republic this article resulted in designing an example system suitable for usage among Czech microbreweries. Since this model was built solely on theoretical basis the next step in this research project is its empirical verification. This is also in accordance with the literature which claims that there is a significant gap between theory and practice. On one hand many Performance Measurement models have been proposed but on the other hand very little empirical research has been carried out. In order better to understand the process performance measurement further empirical studies on this field are necessary.

6. Conclusion

The economics of beer processing covers the production part of the supply chain of beverage industry. The most considerable characteristics and industry movement observed in this chapter allows application, analysis and in addition development of modern economic theories and dynamic models. The primary objective of entrepreneurship is the growth of stockholder value. Value Based Management disposes of tools for value enhancement. The main task of the research was the quantity and selection of suitable variable as a proxy for value growth. The research question concerning identification of the most considerable factors of Economic Value Added and the value drivers of particular segment of breweries in the Czech Republic. The research problem was the formulation of theoretical multifactor model for explanation the particular factors based on research findings. For the response on stated research question we articulate null hypothesis H_0 and alternative hypothesis H_1. According to testing we fell to reject the null hypothesis H_0. Stated research question allows according Table 4. statistical differences considered to be statistically significant enough for the design the model for explanation of particular factors. The statistical characteristics are so balanced and the draft of additional model concerns the whole field of value creation. According to stated research question and research problem we submit o multifactor model based on Balanced Scorecard, Economic Value Added decomposition as a tool for Performance Management System including soft factors, see Fig. 9. The tasks for future research are induced from research findings. The challenge in the field of market structure is fulfilled by EVA decomposition for the whole target population with the emphasis on microbreweries and compares the brewery industry with comparable EU countries. The challenge in the field of researches on analytical tools is verification of the drafted model on empirical data and taxonomy of this model for the particular size of the breweries.

7. Acknowledgment

Hereby we would like to thank the Mendel University and the Ministry of Education, Youth and Sport for the support of the project No. 431100007 "The Agriculture and Food Industry Structure Formation and Trends of Behaviour of Economic Subjects in the Process of Integration of the Czech Republic into the European union".

8. References

Altova, M. & Braznovsky, I. (Eds.). (June 2004). Situační a výhledová zpráva chmel, pivo 2004 (Hops and Beer Situation Report 2004), Ministry of Agriculture of the Czech Republic, Prague, Czech Republic, ISBN 80-7084-326-8, 25.07.2011, Available from http://eagri.cz/public/web/file/2754/SZV_chmel_04.pdf

Altova, M. (Ed.). (June 2005). Situační a výhledová zpráva chmel, pivo 2005 (Hops and Beer Situation Report 2005), Ministry of Agriculture of the Czech Republic, Prague, Czech Republic, ISBN 80-7084-434-5, 25.08.2011, Available from http://eagri.cz/public/web/file/2750/chmel_2005.pdf

Altova, M. (Ed.). (June 2006). Situační a výhledová zpráva chmel, pivo 2006 (Hops and Beer Situation Report 2006), Ministry of Agriculture of the Czech Republic, Prague, Czech Republic, ISBN 80-7084-521-X, 14.08.2011, Available from http://eagri.cz/public/web/file/2758/chmel2006.pdf

Altova, M. (Ed.). (June 2007). Situační a výhledová zpráva chmel, pivo 2007 (Hops and Beer Situation Report 2007), Ministry of Agriculture of the Czech Republic, Prague, Czech Republic, ISBN 978-80-7084-601-8, 10.08.2011, Available from http://eagri.cz/public/web/file/2746/SVZ07_FINAL.pdf

Altova, M. (Ed.). (July 2008). Situační a výhledová zpráva chmel, pivo 2008 (Hops and Beer Situation Report 2008), Ministry of Agriculture of the Czech Republic, Prague, Czech Republic, ISBN 978-80-7084-696-4, 15.07.2011, Available from http://eagri.cz/public/web/file/2744/CHMEL_web2008.pdf

Altova, M. (Ed.). (July 2009). Situační a výhledová zpráva chmel, pivo 2009 (Hops and Beer Situation Report 2009), Ministry of Agriculture of the Czech Republic, Prague, Czech Republic, ISBN 978-80-7084-795-4, 15.08.2011, Available from http://eagri.cz/public/web/file/2752/CHMEL_7__2009.pdf

Altova, M. (Ed.). (July 2010). Situační a výhledová zpráva chmel, pivo 2010 (Hops and Beer Situation Report 2010), Ministry of Agriculture of the Czech Republic, Prague, Czech Republic, ISBN 978-80-7084-901-9, 19.07.2011, Available from http://eagri.cz/public/web/file/61081/CHMEL_7__2010.pdf

Altova, M. (Ed.). (August 2011). Situační a výhledová zpráva chmel, pivo 2011 (Hops and Beer Situation Report 2011), Ministry of Agriculture of the Czech Republic, Prague, Czech Republic, ISBN 978-80-7084-983-5, 15.09.2011, Available from http://eagri.cz/public/web/file/129275/CHMEL_8__2011.pdf

Bacidore, J. M., Boquist, J. A., Milbourn, T. T., Thakor, A. V. (1997). The search for the best financial performance measure, *Financial Analysis Journal,* Vol. 53, No. 3, pp. 11-20, ISSN: 0015-198X

Biddle, G., Bowen, R. M., Wallace, J. W. (1997). Does EVA beat earnings: Evidence on associations with stock returns and firm values, *Journal of Accounting and Economics,* Vol. 24, pp. 301-336, ISSN 0165-4101

Bititci, U.S., Carrie, A.S. and McDevitt, L. (1997). Integrated performance measurement systems: a development guide, International Journal of Operations and Production Management, Vol. 17, pp. 522-534, ISSN 0144-3577

Balsik, F. (Ed.). (May 2002). Situační a výhledová zpráva chmel, pivo 2002 (Hops and Beer Situation Report 2002), Ministry of Agriculture of the Czech Republic, Prague, Czech Republic, ISBN 80-7084-210-5, 05.08.2011, Available from http://eagri.cz/public/web/file/2756/svz_chmel_2002_05.pdf

Braznovsky, I. (Ed.). (May 2003). Situační a výhledová zpráva chmel, pivo 2003 (Hops and Beer Situation Report 2003), Ministry of Agriculture of the Czech Republic, Prague, Czech Republic, ISBN 80-7084-282-8, 05.08.2011, Available from http://eagri.cz/public/web/file/2748/CHMEL_05_03.pdf

Brealey, R. & Myers, S. C. (2008). *Principles of Corporate Finance* (9th edition), McGraw-Hill/Irwin, ISBN 9780073405100, Boston, United States of America

Carroll, G. R. & Swaminathan, A. (1992). The organizational ecology of strategic groups in the American Brewing industry from 1975-1990, Corporate and Industrial Change, No. 1(1992), pp. 65-97, ISSN 1464-3650

Carroll, G. R., & Wade, J. B. (1991). Density dependence in the evolution of the American brewing industry across different levels of analysis, *Social Science Research,* No. 20 (1991) , pp. 271-302, ISSN 0049-089X

Damodaran, A. (2001). Corporate Finance Theory and Practice (2nd edition), John Wiley & Sons, ISBN 0-471-28332, New York, United States of America

Feltham, G. D., et al. (2004). Perhaps EVA Does Beat Earnings - Revisiting Previous Evidence, Journal of Applied Corporate Finance, Vol. 16, No. 1, pp. 83–88, ISSN 1078-1196

Fitzgerald, L., Johnson, R., Brignall, S., Silvestro, R. And Voss, C. (1991). Performance measurement in Service Business, London: CIMA

Finanční analýza podnikové sféry za rok 2010 (Financial Analysis of Busines Sector 2010). (July 2011). Department of Economics Analysis, Ministry of Industry and Trade of the Czech Republic, 20.09.2011, Prague, Czech Republic, Available from http://download.mpo.cz/get/44436/49924/580371/priloha002.pdf

Flapper, S., Fortuin, L. And Stoop, P. (1996). Towards consistent performance management systems, *International Journal of Operations and Production Management*, Vol. 16, No. 7, pp. 27-37, ISSN 0144-3577

Garengo, P., Biazzo, S., Bititci, U. S. (2005). Performance Measurement Systems in SMEs: A Review for a Research Agenda, *International Journal of Management Reviews*, Vol.7, No. 1, pp. 25-47, ISSN 1468-2370

Ghalayini, A., Noble, J., and Crowe, T. (1997). An integrated dynamic performance measurement system for improving manufacturing competitiveness, *International Journal of Production Economics*, Vol. 48, pp. 207-225, 0925-5273

Horizontal Merger Guidelines. (August 2010). U.S. Department of Justice and the Federal Trade Commission, 15.08.2011, Available from http://www.justice.gov/atr/public/guidelines/hmg-2010.html

Hudson, M., Smart, P.A. and Bourne, M. (2001). Theory and practice in SME performance systems, International Journal of Operations and Production Management, Vol. 21, ISSN 1096-1116

Chennell, A., Dransfield, S., Field, J., Fisher, N., Saunders, I. and Shaw, D. (2000). Operational Performance Management: a system for organisational performance measurement, *Proceeding of the Performance Measurement Conference, Cambridge,* Great Britain, July 2000

Chmelíková, G., (2008). Economic Value Added versus traditional performance measures by food-processing firms in the Czech Republic, *International Food and Agribusiness Management Review*, 2008, Vol. 11, No. 4, ISSN 1096-7508

Chmelíková, G. (2010). Báze ekonomické analýzy českých podniků. *Acta Universitatis agriculturae et silviculturae Mendelianae Brunensis*, 2010. Vol. LVIII, No. 3, pp. 79--84. ISSN 1211-851

Kaplan, R.S., Norton, D.P., (1992). The Balanced Scorecard-Measures That Drive Performance, *Harvard Business Review*, Vol. 70, No. 1, ISSN 0017-8012

Kaplan, R.S., Norton, D.P., (1996). Translating Strategy into Action, HBS Press, USA, ISBN 0-87584-651-3

Keegan, D.P., Eiler, R.G. and Jones, C.R. (1989). Are your performance measure obsolete?, *Management Accounting*. Vol. 70, pp 45-50, ISSN 1092-8057

Kramer, J. K., Pushner, G., (1997). An empirical analysis of economic value added as a proxy for market value added, *Financial Practice and Education*, Spring/Summer, pp 41–49, ISSN 1082-0698

Laitinen, E.K. (2002). A dynamic performance measurement system: Evidence from small Finnish technology companies, Scandinavian Journal of Management, Vol. 18, pp 65-99, ISSN 0956-5221

Leenen, H et al. (March 2010). The Contribution made by Beer to rhe Europen Union Economy, Ernst & Young, Amserdam, Nederland, 25.07.2011, Available from http://www.brewersofeurope.org/docs/flipping_books/ey_2011/index.html#/3/zoomed

Lynch, R. and Cross, K. (1991). Measure Up! Yard-sticks for Continuous Improvement, Cambridge: Blackwell, ISBN 1557867186

Maier, T. (2009). Change in brand positioning in the Czech beer market in the period of 1998 - 2007, Proceedings of Conference on Economics of Beer and Brewing, Leuven, Belgium, May 27-30, 2009, 15.08.2011, Available from http://www.beeronomics.org/papers/5A%20Sustersic.pdf

Medori, D. and Steeple, D. (2000). A framework for auditing and enhancing performance measurement systems. *International Journal of Operationas and Production Management*, Vol. 15, No. 4: 520-533, ISSN 0144-3577

Modigliani, F. & Miller, M. (1958). The Cost of Capital, Corporation Finance and the Theory of Investment, *American Economic Review*, Vol. 48, No. 3, ISSN 0002-8282

Niven, P., (2006). *Balanced Scorecard Step-by-Step: Maximizing Performance and Maintaining Results*, Wiley, USA, ISBN 978-0-471-78049-6

Neely, A., Mills, J., Richards, H., Gregory, M., Bourne, J. And Kennerley, M. (2000). Performance measurement system design: developing and testing a process-based approach, *International Journal of Operations and Production Management*, Vol. 20, pp 1119–1145, ISSN: 0144-3577

Neumaierova, I. & Neumaier, I. (2002). *Výkonnost a tržní hodnoty firmy (Business Performance and Market Value)*, Grada, ISBN 80-247-0125-1, Prague, Czech Repulic

Neumaierova, I. & Neumaier, I. (2005). Benchmarkingový diagnostický systém finančních indikátorů INFA (INFA Financial Indicator Benchmarking System), Ministry of Industry and Trade of the Czech Republic, Prague, Czech Republic, 01.09.2011, Available from http://www.mpo.cz/cz/ministr-a-ministerstvo/ebita/

Neumaierova, I. & Neumaier, I. (2008). Proč se ujal index IN a nikoli pyramidový systém ukazatelů INFA (Why is Used Index IN and INFA System not ?), In : *Ekonomika a Management*, Vol. 2, No. 4, ISSN 1802-8934, 25.07.2011, Available from http://www.ekonomikaamanagement.cz/cz/clanek-proc-se-ujal-index-in-a-nikoli-pyramidovy-system-ukazatelu-infa.html

Oliver, L. and Palmer, E. (1998). An integrated model for infrastructural implementation of performance measurement. Performance Measurement-Theory and Practice (Conference proceedings) Cambridge University, Vol. 2, pp 695-702, ISSN 1321-5906

Stewart, S. (1994). EVA Roundtabel, *Journal of Applied Corporate Finance*, Vol. 7, pp. 46-70, ISSN 1745-6622

Stewart, S. (1995). Advertisment, *Harvard Business Review*, November-December: 20, ISSN 00178012

Tonchia, S., Toni, A. (2001). Performance Measurement Systems: Models, Characteristics and Measures. *International Journal of Operations and Production Management*. Vol. 21, No. 1-2, pp. 1-37, ISSN 0144-3577

Turvey, C. G., Lake, L., Duren, E., Sparling, D. (2000). The Relationship between Economic Value Added and the Stock Market Performance of Agribusiness Firms. *Agribusiness*, Vol. 16, No. 4, pp. 399-416, ISSN 1746-5664